AF852572

Student's Guide to FUNDAMENTALS OF CHEMISTRY

Fourth Edition

Student's Guide to

FUNDAMENTALS OF CHEMISTRY

Fourth Edition

Brescia, Arents, Meislich, Turk

Dr. Jo A. Beran
Texas A & I University

Academic Press
A Subsidiary of Harcourt Brace Jovanovich, Publishers
New York London Toronto Sydney San Francisco

Academic Press, Inc.
111 Fifth Avenue, New York, New York 10003

United Kingdom Edition published by
Academic Press, Inc. (London) Ltd.
24/28 Oval Road, London NW1

ISBN: 0-12-132397-8

Printed in the United States of America

CONTENTS

To the Student

You are about to embark upon one of the more rewarding disciplines in your college career. Basic chemical principles are used in nearly all the sciences, from agriculture to engineering and from biology to medicine. Although you may not plan to become a research chemist, the principles you will study in this course could be the very foundation of your expertise in your chosen profession.

This book is written to parallel *Fundamentals of Chemistry, Fourth Edition* by Brescia, Arents, Meislich, and Turk. The text introduces you to the basic chemical principles covered in most general chemistry courses. You will find that by excelling in this course, you will have a knowledge of chemistry that parallels any in the country.

This Study Guide is written to assist you in understanding the material in each section of every chapter in the accompanying text. Although the Study Guide is nearly self-contained, it is not designed to replace the text as a primary resource book. In the Study Guide each section is divided into an *Objective* and a *Focus*. Examples with worked-out solutions appear wherever appropriate. Representative study questions are located at the end of the chapter with answers.

To effectively use the Study Guide, you should take the following steps:

1. Read the *Objective(s)* in the Study Guide before reading the corresponding section in the text. This will identify the purpose and goal for that section in the text and prepare you for what is to come.
2. Read the material in the text and work the examples. Study the explanation for each principle.
3. Read the *Focus* and work through the examples in the Study Guide. The focus is written to help stress the important principle(s) covered in that section of the text. Rework the text examples, and then solve the corresponding study questions at the end of the Study Guide chapter. If you have difficulty with these questions, it indicates that you need to study the principles and work the Study Guide examples and questions again.
4. Now you should be ready to answer the problems at the end of the chapter in the text.

Where does the instructor enter the learning scheme? The chemistry instructor is another available source of information. The lecturer will explain the principles in the most effective manner for students at your school. However, you must learn the material. You should prepare for the lecture the same way you use this Study Guide: read the course syllabus and the corresponding sections of the chapter before class; check the portions of each chapter you do not understand; attend class regularly and listen especially carefully when those difficult areas are presented. Take notes in lecture but do *not* attempt to write everything the lecturer says; spend most of your time listening. Jot down notes when those difficult areas are being discussed. If those difficult areas are still unclear, see your instructor or the teaching assistant as soon as possible. Also review the Study Guide again (and again, if necessary). *Do not* slide over difficult areas. Chemistry builds, and what you learn today is tomorrow's building block.

In the Study Guide, the problems are solved using a basic problem-solving technique, that of dimensional analysis. In chemistry, every measurement has a unit. The solution to a problem follows a format such that units are multiplied, divided, added, and subtracted just like the numbers. When the units of the answer are known, the problem is solved by starting with the *given* quantity and units, and by applying the conversion factors and units that are necessary to obtain the *desired* quantity and units. Thus the units of the given and desired quantities serve as the guide in the selection of conversion factors. You will find the following format for a large number of problems in the Study Guide:

Given quantity and units: ______
Desired quantity and units: ______
Conversion factor(s): (These are the relationships that permit you to convert the given units into the desired units.)
Calculation:

$$\begin{aligned}\text{desired quantity and units} &= \text{given quantity and units} \times \text{conversion factor}\\ &\quad \times \text{conversion factor}\\ \text{units} &= \text{units}\end{aligned}$$

Above all, become comfortable with working problems. Proficiency in sports is similar to proficiency in chemistry; it requires hours of practice, mental discipline, and endurance.

The answers to the questions at the end of each Study Guide chapter may vary by one or two units in the last significant digit. This is due to the sequence in which numbers are rounded off when calculating the answer. Do not become alarmed because of this difference.

The approaches to chemical principles and problem solving that appear in this Study Guide have been developed through my experience in the instruction of over 1500 general chemistry students. Final suggestions were made by Drs. Jerry Mills of Texas Tech University, Wayne Boring of Stephen F. Austin State University, Emil Mucchetti of Texas A&I University, and by the authors of the text: Frank Brescia, John Arents, Herbert Meislich, and Amos Turk.

Most importantly, I express my heartfelt gratitude to Judi, Kyle and Greg for giving up familiy activities such as evenings out, the movies, weekends at the beach, and our summer vacation, thus allowing me time to complete this project.

Chapter 1

SOME FUNDAMENTAL TOOLS OF CHEMISTRY

SECTION 1.1 SCIENTIFIC METHOD FOR CHEMISTS

Objectives

- To explain how a chemist, engineer, or an auto mechanic seeks to find a logical explanation to a phenomenon.
- To define an element and to recognize the more common names and symbols.

Focus

Chemistry is a logical science. A logical thought process is used to account for the many observations made in the chemistry laboratory or in nature. From a qualitative point of view, a chemist is like an automobile mechanic. If an automobile engine is not operating smoothly, mechanics do not correct the problem by immediately changing spark plugs, points, condenser, oil, and so on. Instead they define the problem, perform a series of experiments (check the points, plugs, compression, and so on), and finally correct it. When mechanics work on other engines and their tested theory corrects a similar problem, they soon become "good" mechanics.

A chemist or, for that matter, anyone who uses a rational approach in explaining a phenomenon uses quantitative experimental methods to test a theory.

Historically, chemists have been able to explain and test many observations and theories by breaking down matter into two categories of pure substances —**elements** and chemical combinations of two or more elements called **compounds**. You should learn the names and symbols of the more common elements.

No statement is absolutely true to the scientist. Theories valid for years have been discarded because of further experimentation and interpretation. Throughout this course, remember that the principles you are learning are supported only by current experimental data. Your children or grandchildren, however, may consider today's understanding of chemistry as old-fashioned and incorrect as a result of their interpretation of new and more sophisticated experimentation. This does not mean that we should stop testing today. The test of time only makes good theories better.

SECTION 1.2 MEASUREMENT AND THE INTERNATIONAL SYSTEM OF UNITS

Objectives

- To name the common units of length, mass, time, and temperature in the International System of Units (SI System).
- To identify the prefixes used to express multiples or fractions of the common units.
- To calculate the density of a substance.
- To interconvert temperature readings from the Celsius, Kelvin, and Fahrenheit temperature scales.

Focus

The **SI System** is recognized as the set of standard weights and measures relative to which all scientific measurements are made throughout the world. For this reason, become familiar with the SI units of length, volume, mass, time, and temperature.

In addition, smaller or larger multiples of length, volume, and mass should be memorized, especially the prefixes from mega- through nano- listed in the text, Table 1.2. Remember that the prefix is equivalent to some multiple of a standard measurable unit. Also be familiar with expressing numbers in the exponential notation discussed in the text, Appendix B-1. To illustrate:

1. centi —(c) is equivalent to 10^{-2}. cm = 10^{-2} m produces two equivalent conversion factors:

$$\frac{\text{cm}}{10^{-2}\text{ m}} = \frac{10^{-2}\text{ m}}{\text{cm}} = 1$$

2. micro —(μ) is equivalent to 10^{-6}. μg = 10^{-6} g, resulting in two equivalent conversion factors:

$$\frac{\mu\text{g}}{10^{-6}\text{ g}} = \frac{10^{-6}\text{ g}}{\mu\text{g}} = 1$$

3. mega —(M) is equivalent to 10^{6}. MW (megawatt) = 10^{6} W and

$$\frac{\text{MW}}{10^{6}\text{ W}} = \frac{10^{6}\text{ W}}{\text{MW}} = 1$$

The use of these "equivalent" relationships are very useful in converting to new SI or English units.

Section 1.4 treats the mathematical method for these conversions. For now, study the meaning and the symbols of the SI prefixes and the following equivalent relationships between the English and the SI units:

$$1 \text{ liter} = 1.057 \text{ quarts}$$
$$2.54 \text{ cm} = 1 \text{ inch}$$
$$454 \text{ g} = 1 \text{ pound}$$

Also know the additional conversions in the text, Table 1.3.

Length (and Volume)

Chemists often use measurements of length when referring to the wavelength of light, to the dimensions of an atom, or to the distance between atoms in a molecule (bond length). Therefore, the **meter** (= 39.37 inches) is inconveniently large; instead chemists discuss length in units of nanometers (nm = 10^{-9} m), micrometers (μm = 10^{-6}), or ångstroms (an old unit where 1 ångstrom (Å) = 10^{-10} m).

The common SI unit for volume is the **liter** (L), equal to 1 dm^3. Chemists generally work with smaller volumes in the laboratory; the milliliter (mL = 10^{-3} L = cm^3) and the microliter (μL = 10^{-6} L) are common units of volume. Note that 1 mL = 1 cm^3.

Mass

The terms mass and weight are often interchanged by chemists; technically, however, the difference is important. Weight is a force, dependent upon gravitational pull,

$$\text{weight} = \text{mass} \times \text{acceleration due to gravity}$$

whereas mass is independent of gravitational pull. Balances are used in chemical laboratories to measure the *mass* of a sample by comparing it with standard masses. Spring-loaded scales on the other hand measure the weight of an object. An example of a scale is the ordinary bathroom scale.

The SI unit of mass is the **kilogram**. Because chemists use much smaller quantities in conducting research, the gram (g), milligram (mg), and occasionally the microgram (μg) are common units of mass.

The **density** of a sample, defined as its mass per unit volume, is generally expressed in units of g/mL, g/cm^3, or g/L, depending on its physical state (liquid, solid, or gas).

Example 1.1 A 27.8 g sample of magnesium metal displaces 16.0 cm^3 of water in a graduated cylinder. Calculate the density of magnesium metal.

Answer 1.74 g/cm^3

Solution Because the magnesium metal displaces 16.0 cm^3 of water, then its volume is also 16.0 cm^3. The density of the sample, defined as mass/volume, is

$$\text{density} = \frac{\text{mass}}{\text{volume}} = \frac{27.8\text{ g}}{16.0\text{ cm}^3} = 1.74\ \frac{\text{g}}{\text{cm}^3}$$

Time

Chemists frequently use time when discussing the rate of a chemical reaction. The reaction may occur within hours, seconds, milliseconds, and even microseconds.

Temperature

The **Celsius** and **Kelvin** temperature scales are commonly used in the chemistry laboratory. The Celsius scale, standardized to read 0° at the freezing point of water and 100° at its boiling point, is by far the most common. A one-degree interval on the Celsius scale is equal to a one-degree interval on the Kelvin scale, except that 0°C is equal to 273.15 K.

$$T(\text{K}) = t(°\text{C}) + 273.15$$

Absolute zero is defined as 0 K, or −273.15°C. The Fahrenheit scale, no longer used in the scientific fields, is still used by meteorologists.

A comparison of the two temperature scales shows how conversions can be made.

$$180 \text{ divisions} \left\{ \begin{matrix} 212°\text{F} & & 100°\text{C} \\ & & \\ 32°\text{F} & & 0°\text{C} \end{matrix} \right\} 100 \text{ divisions}$$

Example 1.2 Convert 65.0°F to the Celsius scale.

Answer 18.3°C

Solution The mercury level of the thermometer has passed through 33 (65-32) Fahrenheit divisions above the freezing point of water. Since there are 180 Fahrenheit divisions compared to only 100 Celsius divisions between the same two reference points on the two temperature scales, the mercury level passes through fewer Celsius divisions by a ratio of 100/180, which reduces to 5/9. Therefore the mercury level passes through 5/9 (33) or 18.3 Celsius divisions above the freezing point of water.

The formula for converting from °F to °C is:

$$°\text{C} = \frac{5°\text{C}}{9°\text{F}} \; (°\text{F} - 32°\text{F})$$

Example 1.3 Convert 65.0°F to Kelvin.

Answer 291.4K

Solution From the previous example, t = 18.3°C. Then, since T(K) = t(°C) + 273.15, T(K) = 18.3°C + 273.15 = 291.4K.

The equation for converting °F to °C, °C = 5°C/9°F (°F − 32°F), can also be used for converting from Celsius to Fahrenheit. Thus for the conversion only one formula needs to be known.

SECTION 1.3 SIGNIFICANT FIGURES IN MEASUREMENT

Objective

- To express measurements and data calculations using the correct number of significant figures.

Focus

If the mass of a copper penny is recorded as 3.74 grams, it is assumed that the first *two* digits are known accurately, but there is some uncertainty in the last recorded digit. This implies that the accuracy of the measurement is 3.74 ± 0.01 grams.

The number of significant figures in any recorded data for a physical measurement is equal to the number of digits known accurately plus one more that has some uncertainty. Therefore, 3.74 grams has three significant figures.

For very large or very small numbers in which zeros are used for placement of the decimal point, the numbers should be expressed in the proper scientific notation. For example, 0.0031 cm is expressed as 3.1×10^{-3} cm. The number 3.1 indicates two significant figures. Scientific notation is presented in Appendix B.1.

Measurement	Scientific Notation	Uncertainty	Significant Figures
4.223 g	4.223×10^0 g	4th digit	4
21.0 m	2.10×10^1 m	3rd digit	3
0.00067 mL	6.7×10^{-4} mL	2nd digit	2
600.0 mg	6.000×10^2 mg	4th digit	4
1001 Mg	1.001×10^3 Mg	4th digit	4

Learn rules 1, 2, and 3 in the text. These rules are extremely helpful in deciding how to transfer an answer from your calculator to an answer that indicates the precision of the data.

Example 1.4 Determine the total mass of the following mixture. Express the answer to the correct number of significant figures.

4.227 g gold
11.27 g silver
24.774 g water
163.2 g beaker

Answer 203.5 g

Solution The sum given by the calculator is 203.471 g. However, since the mass of the beaker is uncertain at the "tenths" digit (±0.1 g), then the total mass should be rounded off to ±0.1 g or to 203.5 g. (See rule 1.)

Example 1.5 Determine the density of an object that has a mass of 17.012 g and occupies a volume of 12.6 cm^3. Express your answer with the correct number of significant figures.

Answer 1.33 g/cm^3

Solution The calculator answer for the division

$$\frac{17.012 \text{ g (5 significant figures)}}{12.6 \text{ cm}^3 \text{ (3 significant figures)}}$$

is 1.3501587. Since the answer can have no more significant figures than the measurement which has the fewest number of significant figures, the density can be expressed to only three significant figures. Rounding off 1.3501587 to three significant figures produces an answer of 1.35 g/cm^3. See rule 2 for multiplication and division and rule 3 for rounding off.

Finally, when recording significant figures, remember that some conversions are expressed by exact numbers which do not impose any limit on the number of significant figures in the answer. For example, in the conversion factors

$$\frac{12\text{ in}}{1\text{ ft}} \quad \frac{100\text{ cm}}{1\text{ m}} \quad \frac{10^3\text{ g}}{1\text{ kg}} \quad \frac{3\text{ ft}}{1\text{ yd}}$$

all numbers are exact.

SECTION 1.4 CONVERSION FACTORS

Objectives

- To use the mathematical procedure of dimensional analysis in solving problems.
- To become more familiar with conversion factors in the SI and English System of measurement

Focus

Dimensional analysis is a very convenient method for solving problems in all the physical sciences for two reasons: (1) the units retain the physical meaning of the measured quantity in the calculation and (2) the units serve as a guide when doing algebraic calculations.

In nearly all chemistry problems there is a given or limiting quantity (and unit) from which the desired quantity (and unit) can be calculated. Starting with the given quantity and unit, appropriate conversion factor(s) are used so that like units can be canceled.

Example 1.6 Determine the width, in inches, of an airbase 6.0 miles wide.

Answer 3.8×10^5 inches

Solution The given quantity and unit is 6.0 miles, and the desired quantity and unit is the number of inches. In converting from miles, we know that 1 mile = 5280 feet and 1 foot = 12 inches. As conversion factors, these equivalent relationships become

$$\frac{1\text{ mi}}{5280\text{ ft}} \text{ or } \frac{5280\text{ ft}}{1\text{ mi}} \quad\text{and}\quad \frac{1\text{ ft}}{12\text{ in}} \text{ or } \frac{12\text{ in}}{1\text{ ft}}$$

The correct conversion factors are dependent on the logical approach to the solution. Let's set up our problem using the dimensional analysis approach:

$$\text{desired quantity} = \underset{\text{and unit}}{\text{given quantity}} \times \underset{\text{factor}}{\text{conversion}} \times \underset{\text{factor}}{\text{conversion}}$$

$$\text{length (inches)} = 6.0\,\cancel{\text{mi}} \times \frac{5280\,\cancel{\text{ft}}}{1\,\cancel{\text{mi}}} \times \frac{12\text{ in}}{\cancel{\text{ft}}}$$

↑ mi must be in the denominator to cancel with the mi in 6.0 mi in the numerator.

↑ ft must cancel with ft in the numerator.

When all like units are canceled, we have the equality inches = inches, and therefore our problem is set up correctly. The magnitude of the answer is, to within two significant figures (6.0 mi only has two significant figures), 3.8×10^5 inches.

Example 1.7 Calculate the mass, in micrograms (μg), of an object that weighs 60.1 centigrams (cg).

Answer $6.01 \times 10^5\ \mu\text{g}$

Solution Let's list our known information:

Given quantity and units: 60.1 cg

Desired quantity and units: mass (μg)

Conversion factors:

$$1\text{ cg} = 10^{-2}\text{ g};\ \left(\frac{\text{cg}}{10^{-2}\text{ g}} \quad \text{or} \quad \frac{10^{-2}\text{ g}}{\text{cg}}\right);$$

$$1\ \mu\text{g} = 10^{-6}\text{ g};\ \left(\frac{\mu\text{g}}{10^{-6}\text{ g}} \quad \text{or} \quad \frac{10^{-6}\text{ g}}{\mu\text{g}}\right)$$

Set up the problem, beginning with the "desired" and "given," and use the conversion factor appropriate for the cancellation of units.

desired = given × conversion factor × conversion factor

$$\text{mass }(\mu\text{g}) = 60.1\,\cancel{\text{cg}} \times \frac{10^{-2}\,\cancel{\text{g}}}{\cancel{\text{cg}}} \times \frac{\mu\text{g}}{10^{-6}\,\cancel{\text{g}}} = 6.01 \times 10^5\ \mu\text{g}$$

units cancel

unit desired = unit calculated.

Example 1.8 What is the volume in milliliters of a ½ gallon of milk?

Answer 1890 mL

Solution Given quantity and units: ½ gallon

Desired quantity and units: volume (mL)

Conversion factors:

$$1\text{ gal} = 4\text{ qt}\left(\frac{4\text{ qt}}{1\text{ gal}} \quad \text{or} \quad \frac{1\text{ gal}}{4\text{ qt}}\right)$$

$$1\text{ L} = 1.057\text{ qt}\left(\frac{1.057\text{ qt}}{1\text{ L}} \quad \text{or} \quad \frac{1\text{ L}}{1.057\text{ qt}}\right)$$

$$1\text{ mL} = 10^{-3}\text{ L}\left(\frac{1\text{ mL}}{10^{-3}\text{ L}} \quad \text{or} \quad \frac{10^{-3}\text{ L}}{1\text{ mL}}\right)$$

These are the conversion factors necessary to convert from gallons to milliliters. We are now ready to set up the problem:

$$\text{volume (mL)} = \tfrac{1}{2}\,\cancel{\text{gal}} \times \frac{4\,\cancel{\text{qt}}}{1\,\cancel{\text{gal}}} \times \frac{\cancel{\text{L}}}{1.057\,\cancel{\text{qt}}} \times \frac{\text{mL}}{10^{-3}\,\cancel{\text{L}}}$$

(desired) = (given) × (conversion factor to cancel gal) × (conversion factor to cancel qt) × (conversion factor to cancel l)

A review of the units on each side of the equation shows that volume (mL) = mL. The answer on the calculator is 1892.1475. Assuming three significant figures for ½ gal, four significant figures in the 1 L/1.057 qt, and an infinite number of significant figures in the definitions 4 qt/gal and 1 mL/10^{-3} L, the answer must be expressed using no more than three significant figures or 1890 mL.

SECTION 1.5 PRECISION AND ACCURACY

Objective

- To distinguish between the precision and accuracy of laboratory data.

Focus

The **precision** of a measurement refers to the closeness with which the measurement can be reproduced. For example, suppose that three students, all using the same thermometer, determine the boiling point of water to be 98.3°, 98.2°, and 98.3°C at one atmosphere of pressure. The precision in this series of measurements is good, but the accuracy of the measurements is poor because the accepted value for the boiling point of water is 100.0°C. A possible reason for the poor accuracy was that the thermometer was improperly calibrated at the factory.

These measurements teach a lesson. A measuring instrument (a balance, a meter stick, a graduated cylinder, and so on) can be improperly calibrated and still yield data with good precision. However, the accuracy cannot be definitely known unless the device is previously calibrated against a known standard.

When required to make several weighings on a sample in the laboratory, you are more likely to have good precision by using the same balance. Changing balances, or using the same balance that has been moved or re-leveled, introduces error into the precision of your weighing. In addition, different balances may be calibrated differently. Keep these sources of error at a minimum. The accuracy of a measurement, however, is improved with proper calibration of the balance. Leveling and zeroing the balance are only two steps in its calibration.

SECTION 1.6 PURE SUBSTANCES

Objectives

- To differentiate between mixtures and pure substances.
- To differentiate between an element and a compound.

Focus

A **pure substance** has its own fixed percentage composition by mass and a unique set of chemical and physical properties. The two types of substances are elements and compounds. Elements are composed of atoms. Compounds are composed of chemical combinations of different atoms.

In the laboratory, chemists frequently work with pure substances. Various separation methods are used to remove impurities from the pure substance. These impurities are also pure substances although their concentrations may vary considerably.

QUESTIONS

Section 1.1

1.1 Without referring to the text, complete the following table. If necessary, guess and then check the text to see if the guess is correct.

Element	Symbol
Aluminum	
	S
Chlorine	
	Fe
Carbon	
	Mg
Silicon	
	Na

1.2 What is the origin of the name mercury?

Section 1.2

1.3 Complete the following table without looking in the text, Table 1.2.

Multiple or fraction	Prefix	Symbol
10^3		
	centi	
		M
	micro	
10^{-9}		

1.4 Express the weight of 5 lb in grams and kilograms.

1.5 Rearrange $°C = \frac{5°C}{9°F}(°F - 32°F)$ to solve for °F.

1.6 The density of gasoline is about 0.70 g/mL. Calculate the volume occupied by 20.0 g gasoline.

1.7 What is the SI unit of volume? Is this volume larger or smaller than a quart? By what factor?

Section 1.3

1.8 Express each of the following measurements in scientific notation:

(a) 0.00430 nanograms (ng)

(b) 601,000 amperes (three significant figures)

(c) 3100 ångstroms (two significant figures)

(d) 1434.33×10^{-6} g

1.9 Indicate the number of significant figures in each of the following physical measurements:

(a) 0.7130 g ________

(b) 43.681 mg ________

(c) 2.54 cm ________

(d) 0.006 mm ________

(e) 3.67×10^{-4} cm ________

(f) 301 mL ________

(g) 40.00 mL ________

(h) 0.987 L ________

1.10 Round off each of the following physical measurements to three significant figures:

(a) 2.643×10^{-2} L ________

(b) 417,698 g ________

(c) 2.0090 ounces ________

(d) 20,000 sec ________

1.11 Assuming that each quantity represents a physical measurement, express the answer in each of the following problems to the correct number of significant figures:

(a) $3.87\,(4.0 \times 10^{-2} + 17.71)$

(b) $(4.37 \times 10^{-4} \times 0.22 \times 10^{-2}) + (6.2 \times 10^{-5} \times 2.003 \times 10^{-2})$

(c) $276.4 \times \frac{293}{263} \times \frac{784}{754 - 14.7}$

(d) $3.22(6.02 \times 10^{23})$

(e) $37.3 \times \frac{1}{2(55.847) + 3(16.0)}$

Section 1.4

1.12 Determine the volume in μL of 2.0 mL of gasoline.

1.13 Determine the volume in liters of 28.7 dm^3 of oil.

1.14 What is the mass in kg of a 4.33 mg nugget of gold?

1.15 The volume of a barrel of oil is about 160 L. Express this volume in kL.

1.16 The wavelength of violet light is approximately 4000 Å. What is its length expressed in nm?

1.17 Convert the velocity 8.8 m/sec to km/hr.

1.18 The mass of a pencil is 17.5 g. Express this mass in mg.

1.19 A fashionable house has a floor area of 450 m^2. What is the floor area in dm^2?

1.20 A 250 mL beaker is equivalent to how many pints? (2 pints = 1 quart.)

Section 1.5

1.21 How can you assure that precise measurements are also accurate?

1.22 Comment on the mistaken use of a damaged balance in determining the mass of a silver coin with respect to precision and accuracy.

1.23 A meter stick, subdivided to 1 mm units, is used for determining the tread wear for one month on an automobile tire. Comment on the precision and accuracy of the measurement.

ANSWERS

1.1

Aluminum	Al
Sulfur	S
Chlorine	Cl
Iron	Fe
Carbon	C
Magnesium	Mg
Silicon	Si
Sodium	Na

1.2 See text Table 1.1.

1.3

10^3	kilo	k
10^{-2}	centi	c
10^6	mega	M
10^{-6}	micro	μ
10^{-9}	nano	n

1.4 2270 g
2.27 kg

1.5 $°F = \frac{9°F}{5°C}\ °C + 32°F$

1.6 29 mL

1.7 liter; larger than 1 qt; 1.057

1.8 (a) 4.30×10^{-3} ng
(b) 6.01×10^5 amperes

(c) 3.1×10^3 Å
(d) 1.43433×10^{-3} g

1.9 (a) 4 (e) 3
(b) 5 (f) 3
(c) 3 (g) 4
(d) 1 (h) 3

1.10 (a) 2.64×10^{-2} L (c) 2.01 ounces
(b) 418,000 g (d) 2.00×10^4 sec

1.11 (a) 6.9×10^1
(b) 2.2×10^{-6}
(c) 327
(d) 1.94×10^{24}
(e) 0.234

1.12 2.0×10^3 μL

1.13 28.7 L

1.14 4.33×10^{-6} kg

1.15 0.160 kL

1.16 400 nm

1.17 32 km/hr

1.18 1.75×10^4 mg

1.19 4.50×10^4 dm^2

1.20 0.53 pint

1.21 By calibrating the instrument (or measuring device) with a known value.

1.22 Repeated weighings of the silver coin produce a reliable precision, but because of the damaged balance, the accuracy may be poor.

1.23 An instrument that measures with greater accuracy is necessary. Both the accuracy and precision of the measurement are questionable.

Chapter 2

ATOMS, MOLECULES AND IONS

SECTION 2.1 INTRODUCTION

Objective

- To present a historical development of the atom concept.

Focus

Philosopher-scientists conceived the concept of the atom long before any real data were available. Because of the lack of experimental evidence, the atomic theory of matter did not make an impact on those early scientists. The most popular philosophers of the time influenced many people. Aristotle and Plato did not teach the atomic concept of matter, as did their predecessors Leucippos and Democritos; rather they stressed the continuous nature of matter by which one substance could be transformed into another. The possibility of transmuting one substance (such as iron) into another (such as gold) affected scientific thought throughout the Middle Ages. By 1800, with the development of the printing press, the increased emphasis on quantitative measurements, and the growing amount of travel and communication between countries, scientists began to record and interpret their data and to exchange their ideas.

SECTION 2.2 LAW OF CONSERVATION OF MATTER

Objective

- To give examples illustrating the conservation of mass in a chemical reaction.

Focus

The total mass contained in a chemical system remains unchanged during a chemical reaction.

Example 2.1 A mass of 914 g octane is burned using 3207 g oxygen to produce carbon dioxide and water vapor. If a mass of 2822 g CO_2 is produced, what mass of water vapor (in grams) must also be produced?

Answer 1299 g H_2O

Solution Because mass is conserved, the mass sum of the reactants must equal the mass sum of the products or

$$m_{\text{octane}} + m_{\text{oxygen}} = m_{\text{carbon dioxide}} + m_{\text{water}}$$

$$m_{\text{water}} = (914\text{ g} + 3207\text{ g}) - 2822\text{ g} = 1299\text{ g}$$

SECTION 2.3 THE LAW OF DEFINITE COMPOSITION

Objective

- To calculate the mass percent composition of a compound.

Focus

The mass composition of a given compound is constant no matter what its source or method of preparation. For example, the percent composition by mass for several common substances is:

Octane: 84.2% carbon, 15.8% hydrogen
Water: 88.9% oxygen, 11.1% hydrogen
Ammonia: 82.4% nitrogen, 17.6% hydrogen
Table salt (sodium chloride): 39.3% sodium, 60.7% chlorine

Example 2.2 Analysis shows that a 3.80 g sample of a pure compound of sulfur and fluorine contains 1.37 g S. Calculate the mass percentage of sulfur and fluorine in the compound.

Answer 36.1% sulfur, 63.9% fluorine

Solution The fraction of the mass of the compound that is sulfur is 1.37 g S/3.80 g cmpd; its mass percentage is therefore

$$\frac{1.37\text{ g S}}{3.80\text{ g compd}} \times 100\% = 36.1\%\text{ S}$$

The mass of fluorine in the 3.80 g sample of compound is 2.43 g (3.80 g − 1.37 g). Multiplied by 100, its mass percentage is

$$\frac{2.43\text{ g F}}{3.80\text{ g compd}} \times 100\% = 63.9\%\text{ F}$$

or 100% compd − 36.1% S = 63.9% F.

SECTION 2.4 THE ATOMIC THEORY

Objectives

- To state the various postulates of Dalton's atomic theory.
- To use the postulates of Dalton's atomic theory to explain the conservation of mass and the law of definite composition.

Focus

John Dalton's atomic theory of matter accounts for many observations recorded in the late 1700s. With only a few minor refinements, the theory is still supported by the current understanding of the nature of the atom. Know the five basic postulates of the theory and be able to account for (1) the conservation of matter and (2) the law of definite composition.

Example 2.3 The relative mass of the phosphorus atom is 31.0 and that of the sulfur atom is 32.1. Calculate the mass ratio of phosphorus to sulfur in the molecule expressed by the formula P_4S_3.

Answer $\frac{\text{mass P}}{\text{mass S}} = \frac{124}{96.3} = \frac{1.29}{1}$

Solution There are four atoms of phosphorus for every three atoms of sulfur; the mass ratio is

$$\frac{\text{mass P}}{\text{mass S}} = \frac{31.0 \times 4}{32.1 \times 3} = \frac{124}{96.3} = \frac{1.29}{1}$$

Example 2.4 What is the phosphorus to sulfur mass ratio for 7.00×10^{20} molecules of P_4S_3?

Answer $\frac{1.29}{1}$

Solution For 7.00×10^{20} molecules of P_4S_3 the total relative mass of phosphorus must be

$$7.00 \times 10^{20} \cancel{\text{molecules } P_4S_3} \times \frac{4 \cancel{\text{atoms P}}}{\cancel{\text{molecule } P_4S_3}} \times \frac{31.0}{\cancel{\text{atom P}}} = 8.68 \times 10^{22}$$

and the total relative mass of sulfur must be

$$7.00 \times 10^{20} \cancel{\text{molecules } P_4S_3} \times \frac{3 \cancel{\text{atoms S}}}{\cancel{\text{molecule } P_4S_3}} \times \frac{32.1}{\cancel{\text{atom S}}} = 6.74 \times 10^{22}$$

$$\frac{\text{relative mass P}}{\text{relative mass S}} = \frac{8.68 \times 10^{22}}{6.74 \times 10^{22}} = \frac{1.29}{1}$$

Therefore the mass ratio of the elements in a given compound is fixed (law of definite composition) regardless of the number of molecules in the sample because (1) each molecule is made up of whole atoms, (2) the ratio of the various kinds of atoms in the molecule is constant, and (3) each atom has a fixed mass.

SECTION 2.5 ATOMS AND ELECTRONS

Objectives

- To define the properties of an electron; its mass and charge.
- To characterize the cathode ray; the condition for current flow, its deflection in electric and magnetic fields, and its straight-line movement.
- To explain the formation of positive and negative ions.
- To distinguish between cathode rays and positive rays.

Focus

The postulates of Dalton's atomic theory do not permit any prediction of the chemical properties of an element (or atom) or the electrical nature of matter. Atoms are not billiard balls of uniform mass, but instead are composed of negatively charged **electrons** which surround a positively charged **nucleus**.

The experiments of Faraday, Crookes, and others indicated that electrons are subatomic particles, common to the structure of all atoms. An electron has a mass of 9.1×10^{-28} g and a negative electrical charge of 1.6×10^{-19} coulombs. The mass of an electron is insignificant compared to the mass of the atom because a positively charged atom resulting from the loss of an electron has nearly the same mass as the neutral atom. The properties of an electron are independent of the atom from which it is separated. It is therefore considered a fundamental particle of all matter.

The electric current in an electric discharge tube is the movement of electrons in a straight line from the cathode (negatively charged electrode) to the anode (positively charged electrode). An electric and magnetic field deflects these cathode rays in a direction that shows they are negatively charged.

A positively charged ion, which results from the loss of an electron from the neutral atom, is much heavier than the electron. Unlike an electron the mass of a positive ion is dependent on the nature of the atom from which it is obtained. Positive ions obtained from helium, argon, sodium, or mercury atoms have different masses. The positive ion, which has a charge of $+1.6 \times 10^{-19}$ coulombs, because of the loss of an electron, is attracted toward the cathode (negatively charged electrode).

Example 2.5 In an electric discharge tube, the cathode rays are dependent on the type of cathode used. True or false?

Answer False

Solution The rays (electron beam) emitted from the cathode are independent of the type of cathode used. The rays are always deflected by electric and magnetic fields in the same direction and are independent of the nature of the gas in the discharge tube.

Example 2.6 In an electric discharge tube, the mass of positively charged particles, like electrons, is independent of the nature of the gas in the discharge tube. True or false?

Answer False

Solution When electrons are stripped from a neutral gaseous atom in the discharge tube, a positive ion that has the same approximate mass as the neutral gaseous atom is produced. But because atoms of different elements have different masses, the positive ions also have different masses.

In the presence of an electric and/or magnetic field, the hydrogen ion, H^+, shows the largest deflection of all positive ions in the discharge tube. This means that for the same charge, e, the H^+ must have the smallest mass of all positive ions. It was concluded that H^+, like the electron, is a subatomic particle called a **proton**.

Example 2.7 If the e/m ratio for a proton (H^+) is 9.58×10^4 C/g, then (a) what is the mass of a proton? The charge for the electron is -1.60×10^{-19} C. (b) what is the mass ratio of the proton to the electron? The mass of the electron is 9.11×10^{-28} g.

Answer (a) 1.67×10^{-24} (b) 1830/1

Solution (a) The charge on a proton is the same (but opposite in sign) as the charge on an electron, or $+1.60 \times 10^{-19}$ C.

$$\text{mass (proton)} = 1.60 \times 10^{-19}\,\cancel{C} \times \frac{g}{9.58 \times 10^4\,\cancel{C}}$$
$$= 1.67 \times 10^{-24}\,g$$

(Note the use of dimensional analysis in problem solving.)

(b)
$$\frac{\text{mass (proton)}}{\text{mass (electron)}} = \frac{1.67 \times 10^{-24}\,g}{9.11 \times 10^{-28}\,g} = \frac{1830}{1}$$

The mass of the proton is approximately 1830 times heavier than an electron!

SECTION 2.6 THE RUTHERFORD THEORY OF THE ATOM

Objectives

- To state the experimental interpretations used in postulating the presence of a positively charged nucleus in an atom.
- To identify the composition of the nucleus.
- To define the terms: mass number, atomic number, isotopes, nucleons.

Focus

The research of the Rutherford group was a major breakthrough in modern science. To appreciate their results, become familiar with the thought process (the scientific method) that led Rutherford to postulate the presence of a positively charged nucleus in an atom:

1. The Thomson model of the atom: a puddinglike mass of positive charge containing enough individual electrons to make the atom electrically neutral.
2. The spread of the positive charge over the volume of an atom in which electrons are embedded would make the Thomson atom appear virtually neutral to an approaching positively charged particle.
3. An alpha particle, He^{2+}, has a mass about four times that of the proton and a charge twice that of the proton.
4. Clearly, then, if an alpha particle, traveling at a velocity of about 1.5×10^7 m/sec (about 10,000 mi/sec), is shot at a Thomson atom, the "little" protons should have virtually no real effect on the pathway of the alpha particle.
5. Experimentation, however, shows that high-speed He^{2+} particles are deflected through large angles; in fact, the pathway of a few He^{2+} particles is reversed.
6. In order for the pathway of the He^{2+} particles to be so drastically altered, Rutherford reasoned that the atom must have a center of high *positive charge density*. A positive charge spread over the volume of an atom just does not have the electric force to repel a high-speed He^{2+} particle. Only when the positive charge of an atom is concentrated into an extremely small volume (compared to the volume of an atom) does the electric force become sufficiently large to repel He^{2+} particles.
7. Since the proton is about 2000 times heavier than the electron, it is logical to assume that most of the mass of the atom is confined to a very small volume.
8. Rutherford named this concentrated volume of positive charge and mass the **nucleus of the atom**.
9. The Rutherford atom is thus a Thomson atom in which the positive charge is reduced to almost zero volume, leaving the electrons outside of the positive charge. The Rutherford atom is mostly empty space.
10. Rutherford concluded that the electrons must be located in this space.

Rutherford determined that the number of protons in an atom accounts for about one-half the mass of an atom. He postulated that the presence of a neutral particle having about the same mass as a proton accounts for the remainder of the mass. The neutral particle, the **neutron**, was subsequently discovered in 1932, some 20 years after his prediction.

According to modern atomic theory, the proton, neutron, and electron are the fundamental building blocks of all matter.

The following terms aid in the discussion of the atom:

1. Number of protons in the nucleus = **atomic number**.
2. For a neutral atom, number of electrons = number of protons = **atomic number**.
3. Number of protons plus number of neutrons in the nucleus = **mass number**.
4. Number of protons and neutrons collectively = **nucleons**.

Isotopes are atoms of the same element (thus having the same atomic number) which have a different number of neutrons (thus having a different mass number).

Example 2.8 Determine the number of protons, neutrons, electrons, and nucleons in lead-206.

Answer 82 protons; 124 neutrons; 82 electrons; 206 nucleons

Solution The isotope, lead-206, is represented by the symbol

$$\begin{matrix}\text{mass number} \rightarrow \\ \text{atomic number} \rightarrow\end{matrix} {}^{206}_{82}\text{Pb}$$

The atomic number, the whole number above the symbol in the periodic table, refers to the number of protons in the nucleus and the number of electrons in the space of a neutral atom; for all atoms of lead there are

82 protons and 82 electrons

The mass number is equal to the number of protons plus neutrons in the atom. Therefore

$$\begin{aligned}\text{number of neutrons} &= \text{mass number} - \text{number of protons} \\ &= 206 - 82 \\ &= 124\end{aligned}$$

Since protons and neutrons are collectively called **nucleons**, the number of nucleons must be equal to the mass number: 206 nucleons.

SECTION 2.7 FORMULAS

Objective

- To write chemical symbols and formulas.

Focus

You should study the names and symbols for the more common elements that you will continually encounter in this course.

Example 2.9 Complete the following table:

Name	Symbol	Name	Symbol
	Li		Pb
Sodium		Tin	
	K		Cr

(continued)

Name	Symbol	Name	Symbol
Magnesium		Manganese	
	Ba		Ca
Sulfur			Fe
Carbon		Fluorine	
Oxygen			Co
	N	Nickel	
Chlorine			Br
	I	Copper	
	P		Zn
Silicon			Ag
	As	Gold	
Mercury			

Answer See table inside the back cover.

Compounds are formed as a result of a chemical combination between two or more atoms. Ionic compounds result from the electrostatic attraction between a positively charged ion, called a cation, and a negatively charged ion, called an anion.

Compounds are neutral substances. Therefore in writing the formula for an ionic compound, cations and anions are combined such that the net charge of the compound is zero.

Example 2.10 Using the text, Table 2.2, write the formula for (a) copper(II) (cupric) chloride; (b) aluminum hydroxide, (c) ammonium carbonate; and (d) iron(III) (ferric) sulfate.

Answer (a) $CuCl_2$; (b) $Al(OH)_3$; (c) $(NH_4)_2CO_3$; (d) $Fe_2(SO_4)_3$.

Solution (a) The copper(II) (cupric) ion is Cu^{2+} and the chloride ion is Cl^- (Table 2.2). To form a neutral combination, one Cu^{2+} ion balances two Cl^- ions. The formula of the compound must be $CuCl_2$. (The absence of a subscript beside a symbol in the compound implies the presence of one atom (or ion) in the compound; for example, $Cu_1Cl_2 = CuCl_2$.)

(b) The aluminum ion is Al^{3+}. The hydroxide ion is OH^-. One Al^{3+} ion balances three OH^- ions to form the neutral compound with the formula $Al(OH)_3$.

(c) The ammonium ion is NH_4^+. The carbonate ion is CO_3^{2-}. Two NH_4^+ ions are required to balance one CO_3^{2-} ion. Therefore the formula is $(NH_4)_2CO_3$.

(d) The iron(III) (ferric) ion is Fe^{3+} and the sulfate ion is SO_4^{2-}. To produce a formula representing a neutral compound, two Fe^{3+} ions (a total positive charge of 6^+) and three SO_4^{2-} ions (a total negative charge of 6^-) are needed, giving the formula $Fe_2(SO_4)_3$.

SECTION 2.8 EQUATIONS

Objective

- To explain the meaning of a balanced equation.

Focus

A balanced equation represents the rearrangement of the atoms in one set of compounds on the left (reactants) to form a second set of compounds on the right (products). A coefficient in the balanced equation represents 2×, 3×, 4×, and so on, the formula that follows it.

Example 2.10 Nitrogen, N_2, reacts with hydrogen, H_2, to produce ammonia, NH_3. Write a balanced equation for the reaction.

Answer $N_2 + 3H_2 \rightarrow 2NH_3$

Solution The format for writing an equation is

$$A + B \rightarrow C$$

where the "+" means *reacts with* and "→" means *to produce*. Therefore,

$$N_2 + H_2 \rightarrow NH_3$$

We now use coefficients in front of the formulas in order to have the same number of atoms appearing on both sides of the equation. Balance the N atoms with a coefficient of 2 in front of the formula for ammonia.

$$N_2 + H_2 \rightarrow 2NH_3$$

Two N atoms appear on each side of the equation. Balance the H atoms by placing a coefficient of 3 in front of H_2 or

$$N_2 + 3H_2 \rightarrow 2NH_3$$

Six H atoms appear on each side of the equation. The equation is now balanced. The balanced equation tells us that 1 molecule of N_2 reacts with 3 molecules of H_2 to produce 2 molecules of NH_3.

More detail for balancing chemical equations appears in Chapter 4.

QUESTIONS

Section 2.2

2.1 Aluminum metal reacts with oxygen in air to produce aluminum oxide, Al_2O_3. If 42.7 g Al produces 87.0 g Al_2O_3, what mass of oxygen reacts?

Section 2.3

2.2 From the data in Question 2.1, calculate the % Al and % O by mass in Al_2O_3.

2.3 The percentage of iron in Fe_3O_4 is 72.36%. Calculate the mass of iron and oxygen in 2.00×10^2 g Fe_3O_4.

Section 2.5

2.4 Which has the larger *e/m* ratio, H^+ or e^-? Explain.

2.5 What is the difference between an electron and a negative ion?

2.6 What is an alpha particle?

Section 2.6

2.7 What would Rutherford have concluded if high-energy electrons had been used instead of α-particles in his experiment?

2.8 Complete the following table:

Symbol	Protons	Electrons	Neutrons	Nucleons	Net Charge
	24		28		0
$^{127}_{53}I^-$					1^-
		47		109	0
$^{23}_{11}Na$					0
	58		78		0
$^{88}_{38}Sr^{2+}$					2^+

Section 2.7

2.9 Determine the number of atoms of each element in the following compounds: $(NH_4)_3PO_4$; C_2H_5OH; C_6H_5COOH; CH_3CHNH_2COOH.

2.10 With the aid of the text, Table 2.2, complete the following table:

Name	Formula
Silver oxide	
Mercury(II) chloride	
Barium sulfite	
Magnesium iodide	
Calcium carbonate	
Ammonium nitrate	
Iron(III) (ferric) hydroxide	
Potassium sulfate	
Copper(II) (cupric) phosphate	

Section 2.8

2.11 Write a balanced equation for each of the following. Use Table 2.2 wherever necessary.
(a) Aluminum metal, Al, reacts with oxygen gas, O_2, to produce aluminum oxide, Al_2O_3.
(b) Calcium chloride, $CaCl_2$, reacts with silver nitrate, $AgNO_3$, to produce calcium nitrate, $Ca(NO_3)_2$, and silver chloride, AgCl.
(c) Zinc metal, Zn, reacts with hydrochloric acid, HCl, to produce zinc chloride, $ZnCl_2$, and hydrogen gas, H_2.

ANSWERS

2.1 38.0 g O_2

2.2 52.9% Al; 47.1% O

2.3 145 g Fe; 55 g O

2.4 e^-: Since e^- has a smaller mass than H^+, the e/m ratio for the e^- is larger.

2.5 An e^- is a basic subatomic particle of all matter. A negative ion is an atom with a negative charge.

2.6 An alpha particle is a doubly ionized He atom, He^{2+}.

2.7 The electrons would have been attracted to the nucleus with some slight deflections of their trajectory, but none would have bounced back. The nucleus would not have been proposed.

2.8

Symbol	Protons	Electrons	Neutrons	Nucleons	Net Charge
$^{52}_{24}Cr$	24	24	28	52	0
$^{127}_{53}I^-$	53	54	74	127	1−
$^{109}_{47}Ag$	47	47	62	109	0
$^{23}_{11}Na$	11	11	12	23	0
$^{136}_{58}Ce$	58	58	78	136	0
$^{88}_{38}Sr^{2+}$	38	36	40	78	2+

2.9 $(NH_4)_3PO_4$: 3N, 12H, 1P, 4O
C_2H_5OH: 2C, 6H, 1O
C_6H_5COOH: 7C, 6H, 2O
CH_3CHNH_2COOH: 3C, 7H, 2O, 1N

2.10

Ag_2O	NH_4NO_3
$HgCl_2$	$Fe(OH)_3$
$BaSO_3$	K_2SO_4
MgI_2	$Cu_3(PO_4)_2$
$CaCO_3$	

2.11 (a) $4Al + 3O_2 \rightarrow 2Al_2O_3$
(b) $CaCl_2 + 2AgNO_3 \rightarrow Ca(NO_3)_2 + 2AgCl$
(c) $Zn + 2HCl \rightarrow ZnCl_2 + H_2$

Chapter 3

ATOMIC AND MOLECULAR WEIGHTS

SECTION 3.1 INTRODUCTION

Objective

- To discuss the historical determination of atomic weights for the elements.

Focus

The atomic concept of matter became more widely accepted with its increasing success in determining molecular and atomic weights and increasing use of the analytical balance. Because the mass of individual atoms could not be determined, after years of experimentation a system of relative masses of atoms evolved. The set of atomic weights was established from the interpretation of many related, quantitative, and tedious experiments.

SECTION 3.2 THE MOLE

Objective

- To define a mole in terms of (1) the number of particles and (2) the mass of a substance.

Focus

A **mole** (abbreviated mol) is 6.02×10^{23} particles of any substance (it is not an abbreviation for molecule). The particles may be atoms, molecules, ions, electrons, and even apples, oranges, or people. Just as 12 particles is referred to as a dozen, a mole is often referred to as a "chemist's dozen." The number, 6.02×10^{23}, is called Avogadro's number, N_A.

Just as the mass of one dozen eggs does not equal the mass of one dozen apples, although there are 12 of each, the mass of 1 *mole* of carbon atoms does not equal the mass of 1 *mole* of uranium atoms, although there are 6.02 $\times 10^{23}$ atoms of each.

Experimentally, the mass of 1 mole (6.02×10^{23} atoms) of uranium is approximately 20 times the mass of 1 mole (6.02×10^{23} atoms) of carbon. This also means that one atom of uranium is about 20 times the mass of one atom of carbon. Therefore, on a relative scale, if a carbon atom is assigned a mass of 12, then the mass of a uranium atom would be approximately 240. The relative masses of other atoms may be measured by determining the mass of 1 mole.

Suppose when using a balance that has no mass scale, you determined that the mass of one dozen cantaloupes is twice the mass of one dozen grapefruit. We rationalize that the mass of one cantaloupe is then twice the mass of one grapefruit. If the cantaloupe is assigned a relative weight of 50, then the relative weight of the grapefruit is 25. Now that a reference has been established, the relative weights of other fruits and vegetables can be established (relative to cantaloupe = 50), even though the balance can only weigh "dozen" quantities. A similar relative weight scale is used for atoms and moles of atoms.

Chemists do not work with single atoms or molecules in the laboratory, but instead they work with mole quantities.

Example 3.1 Calculate the number of H_2O molecules in 3.7 moles of H_2O.

Answer 2.2×10^{24} molecules

Solution Use the method of dimensional analysis described in Chapter 1.
Given quantity and units: 3.7 mol H_2O
Desired quantity and units: number of H_2O molecules
Conversion factor: 1 mol $H_2O = 6.02 \times 10^{23}$ H_2O molecules

$$\left(\frac{1 \text{ mol } H_2O}{6.02 \times 10^{23} \text{ } H_2O \text{ molecules}} \quad \text{or} \quad \frac{6.02 \times 10^{23} \text{ } H_2O \text{ molecules}}{1 \text{ mol } H_2O} \right)$$

If 1 mole of H_2O contains Avogadro's number, N_A, (6.02×10^{23}) of molecules, then 3 moles of H_2O would contain $3 \times N_A$ molecules H_2O. Remember to select the conversion factor that gives the correct units after all like units have been canceled.

$$\text{number of } H_2O \text{ molecules} = 3.7 \cancel{\text{mol } H_2O} \times \frac{6.02 \times 10^{23} \text{ } H_2O \text{ molecules}}{1 \cancel{\text{mol } H_2O}}$$

$$= 2.2 \times 10^{24} \text{ } H_2O \text{ molecules}$$

SECTION 3.3 ATOMIC WEIGHTS AND MASS SPECTROSCOPY

Objectives

- To describe the use of a mass spectrometer in determining atomic weights.
- To calculate the atomic weight of an element, given the atomic weight and the percent abundance of each isotope.

Focus

The **mass spectrometer** is an instrument that determines the isotopic composition and the relative masses of the isotopes of an element. The relative atomic weight of an element is calculated from the relative abundance of the various isotopes. An element's atomic weight is set relative to that of the carbon-12 isotope, ^{12}C. By definition, 1 mole of ^{12}C weighs exactly 12 grams.

Example 3.2 The mass spectrum of magnesium shows three isotopes. Their atomic masses relative to ^{12}C and their peak heights are as follows:

	Relative Atomic Weight	Relative Peak Height
^{24}Mg	23.98504	0.7870
^{25}Mg	24.98584	0.1013
^{26}Mg	25.98259	0.1117

Determine the atomic weight for magnesium.

Answer 24.31

Solution Each isotope, with its respective mass, is a fraction of the total atomic weight of magnesium. The contribution of the ^{24}Mg isotope is

$$23.98504 \times 0.7870 = 18.88$$

The contribution of the ^{25}Mg isotope is

$$24.98584 \times 0.1013 = 2.531$$

The contribution of the ^{26}Mg isotope is

$$25.98259 \times 0.1117 = 2.902$$

The sum of the three isotopes of magnesium is its atomic weight:

$$18.88 + 2.531 + 2.902 = 24.31$$

This example shows that the atomic weight of magnesium is 24.31 relative to ^{12}C and that 1 mole of Mg atoms has a mass of 24.31 grams. The molar mass of Mg is 24.32 g/mol.

Therefore,

$$1 \text{ mole } ^{12}C = \text{exactly } 12 \text{ g} \quad \text{(definition)}$$
$$1 \text{ mole Mg} = 24.31 \text{ g Mg}$$

In addition,

$$1 \text{ mole P} = 30.97 \text{ g P}$$
$$1 \text{ mole S} = 32.06 \text{ g S}$$
$$1 \text{ mole C} = 12.01 \text{ g C}$$
$$1 \text{ mole Au} = 197.0 \text{ g Au}$$

SECTION 3.4 MOLECULAR WEIGHTS OF GASES

Objectives

- To calculate the molecular weight of a compound from the sum of the atomic weights of its elements.
- To measure the molecular weight of a gas, using data from a mass spectrometer.

Focus

The mass spectrometer is also used to determine the molecular weights of compounds relative to ^{12}C.

In interpreting the mass spectrum for a gaseous compound, M, if only one major M^+ peak appears, the mass number of the peak is the molecular weight of the gas molecule. If several major M^+ peaks appear, caused by the isotopic distributions of the atoms in the molecule, then the molecular weight of the gas is calculated from the weighted sum of the mass numbers of the various peaks. (Review Section 3.3)

The molecular weight of a compound is the sum of the atomic weights of all the elements in the compound. The atomic weight for each element appears on the periodic table.

Example 3.3 Determine the molecular weight of (a) CCl_4, (b) CH_2F_2, (c) P_4O_{10}, (d) SF_4.

Answer (a) 153.823; (b) 52.023; (c) 283.886; (d) 108.05

Solutions (a) From the periodic table: C = 12.011, Cl = 35.453. The mass of four chlorine atoms is 4(35.453) = 141.812. The sum of the atomic weights is 12.011 + 141.812 = 153.823.

(b) 1C = 12.011, 2H = 2(1.008) = 2.016, 2F = 2(18.998) = 37.996. CH_2F_2 = 12.011 + 2.016 + 37.996 = 52.023.

(c) 4P = 4(30.974) = 123.896, 10O = 10(15.999) = 159.990. P_4O_{10} = 12.896 + 159.990 = 283.886.

(d) S = 32.06, 4F = 4(18.998) = 75.992. SF_4 = 32.06 + 75.992 = 108.05.

The molecular weights of compounds like atomic weights for elements are relative to ^{12}C. The molar mass of each compound has the same number of molecules (6.02×10^{23}) as are in exactly 12 grams or 1 mol ^{12}C. Therefore 1 mole of each compound in Example 3.3 has a mass, in grams, equal to its molecular weight:

$$6.02 \times 10^{23} \text{ molecules } CCl_4 = 1 \text{ mol } CCl_4 = 153.823 \text{ g } CCl_4$$

$$6.02 \times 10^{23} \text{ molecules } CH_2F_2 = 1 \text{ mol } CH_2F_2 = 52.023 \text{ g } CH_2F_2$$

$$6.02 \times 10^{23} \text{ molecules } P_4O_{10} = 1 \text{ mol } P_4O_{10} = 283.886 \text{ g } P_4O_{10}$$

$$6.02 \times 10^{23} \text{ molecules } SF_4 = 1 \text{ mol } SF_4 = 108.05 \text{ g } SF_4$$

$$6.02 \times 10^{23} \text{ atoms C} = 1 \text{ mol C} = 12.011 \text{ g C}$$

$$6.02 \times 10^{23} \text{ atoms } ^{12}C = 1 \text{ mol } ^{12}C = 12 \text{ g C (exactly)}$$

SECTION 3.5 MOLE RELATIONSHIP: SAME QUANTITY, DIFFERENT UNITS

Objective

- To convert the number of atoms or molecules of a substance to moles and to mass or vice versa.

Focus

In previous sections of this chapter, we illustrated the relationships of the mole to the number of molecules (or atoms) of a substance and the mole to the mass of a substance. These unit conversions are a very important part of chemistry. You should continue to familiarize yourselves with these conversions.

Example 3.4 Calculate the number of grams in 6.82 moles of antimony, Sb.

Answer 831 g Sb

Solution From the periodic table (or table of atomic weights) antimony has an atomic weight of 121.8 or 1 mol Sb = 121.8 g Sb.

Given quantity and units: 6.82 mol Sb
Desired quantity and units: g Sb
Conversion factor: 1 mol Sb = 121.8 g Sb:

$$\left(\frac{121.8\text{ g Sb}}{1\text{ mol Sb}} \quad \text{or} \quad \frac{1\text{ mol Sb}}{121.8\text{ g Sb}}\right)$$

Since 1 mol Sb has a mass of 121.8 g, then 6.82 mol Sb has

$$\text{g Sb} = 6.82\ \cancel{\text{mol Sb}} \times \frac{121.8\text{ g Sb}}{1\ \cancel{\text{mol Sb}}}$$

$$= 831\text{ g Sb}$$

Example 3.5 Calculate the number of molecules of CH_3HgCl in 2.60×10^{-9} g CH_3HgCl.

Answer 6.23×10^{12} molecules CH_3HgCl

Solution

Given quantity and units: 2.60×10^{-9} g CH_3HgCl
Desired quantity and units: molecules CH_3HgCl
Conversion factors: Determine the mass of 1 mole CH_3HgCl which, in grams, is equal to the molecular weight of CH_3HgCl:

$$\begin{array}{rcl} 1\text{C} & = & 12.01\text{ g} \\ 3\text{H} = 3(1.01) & = & 3.03\text{ g} \\ 1\text{Hg} & = & 200.59\text{ g} \\ 1\text{Cl} & = & 35.45\text{ g} \\ \hline CH_3HgCl & = & 251.08\text{ g} \end{array}$$

Therefore 1 mol CH_3HgCl = 251.08 g CH_3HgCl:

$$\left(\frac{1\text{ mol }CH_3HgCl}{251.08\text{ g }CH_3HgCl} \quad \text{or} \quad \frac{251.08\text{ g }CH_3HgCl}{1\text{ mol }CH_3HgCl}\right)$$

The number of molecules in 1 mol CH_3HgCl is

$$1 \text{ mol } CH_3HgCl = 6.02 \times 10^{23} \text{ molecules } CH_3HgCl$$

$$\left(\frac{6.02 \times 10^{23} \text{ molecules } CH_3HgCl}{1 \text{ mol } CH_3HgCl} \quad \text{or} \quad \frac{1 \text{ mol } CH_3HgCl}{6.02 \times 10^{23} \text{ molecules } CH_3HgCl}\right)$$

Using the method of dimensional analysis, we have

$$\text{molecules } CH_3HgCl = 2.60 \times 10^{-9} \cancel{\text{g } CH_3HgCl} \times \underset{\text{(converts grams to moles)}}{\frac{1 \text{ mol } CH_3HgCl}{251.08 \cancel{\text{g } CH_3HgCl}}}$$

$$\times \underset{\text{(converts mole to molecules)}}{\frac{6.02 \times 10^{23} \text{ molecules } CH_3HgCl}{1 \text{ mol } CH_3HgCl}} = 6.23 \times 10^{12} \text{ molecules } CH_3HgCl$$

Conversions to the mole are made in two ways: (1) by using the atomic weight or molar mass when given the mass of the substance or (2) by using Avogadro's number when given the number of its atoms (or molecules).

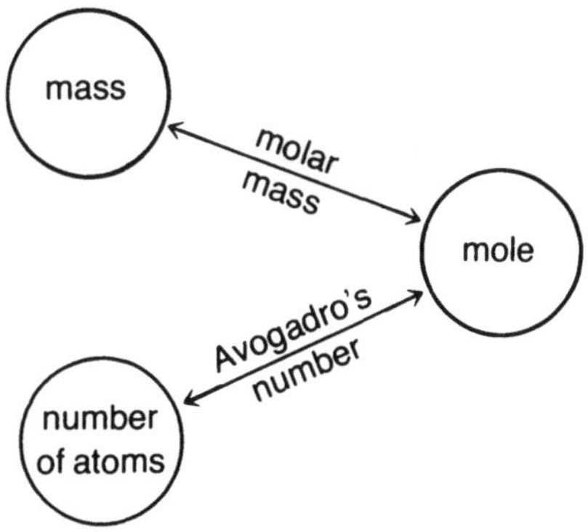

QUESTIONS

Section 3.2

3.1 Determine the number of moles NH_3 in 6.77×10^{25} molecules NH_3.

3.2 One mole Cu is 2.35 times heavier than one mole Al. If 1 mole Al has a mass of 27.0 g, what is the mass of 1 mole Cu? What is the mass of ½ mole Cu? How many atoms are in ½ mole Cu?

Section 3.3

3.3 The mass number of an isotope is the ________ ________ nearest to its atomic weight.

3.4 The element europium, atomic number 63, consists of two isotopes. Calculate its atomic weight.

Mass Number	Atomic Weight	% Abundance
151	150.9196	47.82
153	152.9209	52.18

3.5 Write the mass number for the three isotopes of silicon, with atomic weights of 27.9769, 28.9765, and 29.9738.

Section 3.4

3.6 Determine the mass of 1 mole of

(a) $AuCl_3$

(b) $Fe(ClO_4)_2$

(c) NaCl (table salt)

(d) H_2SO_4 (sulfuric acid)

(e) $NaHCO_3$ (baking soda)

(f) NaOCl (bleaching agent)

(g) $HC_2H_3O_2$ (acetic acid in vinegar)

Section 3.5

3.7 Calculate the mass, in grams, of each of the following:

(a) 1.00 mol Cl_2

(b) 3.77 mol Ag

(c) 12.1 mol H_2S

(d) 3.71×10^{-4} mol CS_2

(e) 0.742 mol $KMnO_4$

3.8 Calculate the number of molecules and/or atoms in

(a) 0.266 mol Cu

(b) 21.1 mol PCl_3

(c) 4.77×10^{-8} mol NO_2

(d) 9.84 mol $C_{12}H_{22}O_{11}$ (sugar)

(e) 14.6 mol N_2

3.9 Calculate the mass in grams of each of the following:

(a) 4.77×10^{20} atoms Au

(b) 3.33×10^{26} molecules SO_2

(c) 2.00 atoms Mg

(d) 1.99×10^{23} molecules Cl_2

(e) 4.77×10^{25} electrons (the relative atomic weight of an electron is 0.000549)

3.10 Calculate the number of moles in

(a) 4.23 g $C_9H_8O_4$ (aspirin)

(b) 1.00 lb $C_{12}H_{22}O_{11}$ (sugar)
(c) 9.66×10^{-12} g Si
(d) 9.66×10^{-12} g Pb

3.11 Calculate the number of moles in
(a) 180 g $Al_2(SO_4)_3$
(b) 13.3 lb $Ca(NO_3)_2$
(c) 2.40 kg $NiCl_2$
(d) 1.33 g Fe_2O_3
(e) 2.47 μg K_2CrO_4
(f) 6.67×10^{22} electrons

ANSWERS

3.1 112 mol NH_3

3.2 63.5 g Cu; 31.7 g Cu; 3.01×10^{23} atoms Cu

3.3 whole number

3.4 151.96

3.5 28; 29; 30

3.6 (a) 303.33
(b) 254.75
(c) 58.44
(d) 98.08
(e) 84.00
(f) 74.44
(g) 60.05

3.7 (a) 70.9 g
(b) 407 g
(c) 412 g
(d) 2.82×10^{-2} g
(e) 117 g

3.8 (a) 1.60×10^{23} atoms Cu
(b) 1.27×10^{25} molecules PCl_3
(c) 2.87×10^{16} molecules NO_2
(d) 5.93×10^{24} molecules $C_{12}H_{22}O_{11}$
(e) 8.79×10^{24} molecules N_2

3.9 (a) 0.156 g Au
(b) 3.55×10^{4} g SO_2
(c) 8.07×10^{-23} g Mg
(d) 23.5 g Cl_2
(e) 4.35×10^{-2} g electron

3.10 (a) 0.0235 mol $C_9H_8O_4$

(b) 1.33 mol $C_{12}H_{22}O_{11}$

(c) 3.44×10^{-13} mol Si

(d) 4.66×10^{-14} mol Pb

3.11 (a) 0.526 mol $Al_2(SO_4)_3$

(b) 36.8 mol $Ca(NO_3)_2$

(c) 18.5 mol $NiCl_2$

(d) 8.33×10^{-3} mol Fe_2O_3

(e) 1.27×10^{-8} mol K_2CrO_4

(f) 0.111 mol electron

Chapter 4

CHEMICAL FORMULAS, EQUATIONS, AND NOMENCLATURE

SECTION 4.1 CHEMICAL FORMULAS

Objectives

- To calculate the molecular weight of a compound from its formula.
- To calculate the empirical formula of a compound, given its mass or percent composition.
- To write the molecular formula of a compound, given its empirical formula and molecular weight.

Focus

The **molecular weight** of a molecule is determined by adding the atomic weights of its atoms. The molecular formula represents the number and ratio of atoms in the compound. As a review of Section 3.4, let us look at another example for determining molecular weight.

Example 4.1 What is the molecular weight of sodium cyclamate (artificial sweetener)? Its molecular formula is $NaC_6H_{12}NSO_3$. What is the mass of 1 mole of $NaC_6H_{12}NSO_3$?

Answer 201.22; 201.22 grams

Solution

In one molecule of $NaC_6H_{12}NSO_3$ there are:

1 atom	Na	=	1(22.990)	=	22.990
6 atoms	C	=	6(12.011)	=	72.066
12 atoms	H	=	12(1.008)	=	12.096
1 atom	N	=	1(14.007)	=	14.007
1 atom	S	=	1(32.06)	=	32.06
3 atoms	O	=	3(15.999)	=	47.997
					201.22

Also in 1 mol $NaC_6H_{12}NSO_3$ there are:

1 mol	Na	=	22.990 g
6 mol	C	=	72.066 g
12 mol	H	=	12.096 g
1 mol	N	=	14.007 g
1 mol	S	=	32.06 g
3 mol	O	=	47.997 g
			201.22 g

One mole sodium cyclamate has a mass of 201.22 g.

The **empirical formula** (simplest formula) is the *simplest* whole number mole ratio of elements in a compound, whereas the molecular formula is the *actual* mole ratio of elements in the compound. In addition, for some compounds the molecular formula is the same as the empirical formula. Some examples include:

Empirical Formula	Molecular Formula
H_2O	H_2O
HO	H_2O_2 (hydrogen peroxide)
CH_2	C_2H_4 (ethylene)
CH_2	C_3H_6 (propylene)
NH_2	N_2H_4 (hydrazine)
NH_3	NH_3 (ammonia)

When chemists prepare new compounds in the laboratory, they determine the mass composition (or percentage composition) of each element in the compound. Like the chemist, we can determine the empirical formula of the compound by converting the mass of each element to moles and expressing the mole ratio in whole numbers.

Example 4.2 A 21.700 g sample of a sulfur-fluorine compound contains 7.812 g S. Determine the empirical (simplest) formula for the compound.

Answer SF_3

Solution The empirical formula is the simplest whole number *mole* ratio of elements in a compound. First determine the number of moles of sulfur and fluorine in the 21.700 g sample.

$$\text{mol S} = 7.812\,\cancel{\text{g S}} \times \frac{1\text{ mol S}}{32.06\,\cancel{\text{g S}}} = 0.2437\text{ mol S}$$

To calculate the mol F, we must find the mass of fluorine in the sample. If the mass of the sample is 21.700 g, of which 7.812 g is S, then (21.700 − 7.812) g or 13.888 g F must be present.

$$\text{mol F} = 13.888\,\cancel{\text{g F}} \times \frac{1\text{ mol F}}{19.00\,\cancel{\text{g F}}} = 0.7309\text{ mol F}$$

The mole ratio in the 21.700 g sample is

$$S_{0.2437}:F_{0.7309}$$

To express the ratio in whole numbers, indicating that compounds are made up of whole atoms, we divide through by the smallest number or

$$S_{\frac{0.2437}{0.2437}}:F_{\frac{0.7309}{0.2437}} = S_1F_{2.999}\text{ or }SF_3$$

If the molecular weight of the compound can be found from another experiment, we can ascertain its molecular (or actual) formula. The molecular weight is either equal to or is a whole number multiple of the empirical formula weight.

Example 4.3 The molecular weight of the sulfur-fluorine compound in Example 4.2 is approximately 178. What is the molecular formula of the compound? What is the actual molecular weight of the compound?

Answer S_2F_6; 178.12

Solution The empirical formula weight of SF_3 is

$$\begin{array}{lll} 1 \text{ atom S} & = 1(32.06) = & 32.06 \\ 3 \text{ atom F} & = 3(19.00) = & \underline{57.00} \\ & & 89.06 \end{array}$$

Since the molecular weight is 178 or approximately twice (178/89.06 ≈ 2), the empirical formula weight, the molecular formula must be twice the empirical formula or

$$(SF_3)_2 \quad \text{or} \quad S_2F_6$$

The actual molecular weight of the compound is

$$\begin{array}{lll} 2 \text{ atoms S} & = 2(32.06) = & 64.12 \\ 6 \text{ atoms F} & = 6(19.00) = & \underline{114.00} \\ & & 178.12 \end{array}$$

One mole of S_2F_6 has a mass of 178.12 grams.

Laboratory analyses are frequently expressed in terms of the mass percentage of each element in the compound; in these cases we can assume a 100 gram sample in the calculation.

Example 4.4 The approximate mass percentage of a compound is 12.12% C, 16.16% O, and 71.71% Cl. Determine the empirical formula of the compound.

Answer $COCl_2$

Solution In a 100 gram sample, there are 12.12 g C, 16.16 g O, and 71.71 g Cl. First, calculate the number of moles of each element in the compound:

$$\text{mol C} = 12.12 \cancel{\text{g C}} \times \frac{1 \text{ mol C}}{12.011 \cancel{\text{g C}}} = 1.009 \text{ mol C},$$

$$\text{mol O} = 16.16 \cancel{\text{g O}} \times \frac{1 \text{ mol O}}{15.999 \cancel{\text{g O}}} = 1.010 \text{ mol O}$$

$$\text{mol Cl} = 71.71 \cancel{\text{g Cl}} \times \frac{1 \text{ mol Cl}}{35.453 \cancel{\text{g Cl}}} = 2.023 \text{ mol Cl}$$

$$C_{1.009} \quad O_{1.010} \quad Cl_{2.023}$$

Divide by the smallest number or

$$C_{\frac{1.009}{1.009}} \quad O_{\frac{1.010}{1.009}} \quad Cl_{\frac{2.023}{1.009}} = C_1O_{1.001}Cl_{2.005} \quad \text{or} \quad COCl_2$$

SECTION 4.2 MASS COMPOSITION AND CONCENTRATION

Objectives

- To calculate the mass composition of a substance.
- To define the solvent and solute in a solution.
- To calculate the molarity, M, of a solute in solution and use molarity for related calculations.

Focus

We can calculate the **mass composition** of an element in a compound by determining the mass of the element in 1 mole of the compound.

$$\frac{\text{mass of element in 1 mole compound}}{\text{mass of 1 mole compound}} \times 100 = \text{mass percentage of element}$$

Once we know the mass composition, we can calculate the yield of an element from a given quantity of a compound.

Example 4.5 Calculate the mass percentage of tungsten in wolframite, $FeWO_4$.

Answer $\frac{183.9 \text{ g W}}{303.8 \text{ g } FeWO_4}$ or 60.53% W

Solution First, determine the molecular weight of $FeWO_4$:

Fe:	55.85
W:	183.9
4O:	64.0
	303.8

Of the 303.8, 183.9 is tungsten. Therefore the mass fraction of W is

$$\frac{183.9 \text{ g W}}{303.8 \text{ g } FeWO_4}$$

or, expressed as a percentage,

$$\frac{183.9}{303.8} \times 100 = 60.53\% \text{ W}$$

Example 4.6 How many lb W can be refined from 6.00 lb $FeWO_4$? The mass percentage of W is 60.53%.

Answer 3.63 lb W

Solution The mass composition of tungsten in $FeWO_4$ is 183.9/303.8. Since this is a mass or weight ratio of W to $FeWO_4$, the units of this ratio may be any unit of mass:

$$\frac{183.9 \text{ g W}}{303.8 \text{ g } FeWO_4} \qquad \frac{183.9 \text{ lb W}}{303.8 \text{ lb } FeWO_4} \qquad \frac{183.9 \text{ kg W}}{303.8 \text{ kg } FeWO_4}$$

or, as a mass percentage the mass units are also arbitrary:

$$\frac{60.53 \text{ g W}}{100 \text{ g } FeWO_4} \qquad \frac{60.53 \text{ lb W}}{100 \text{ lb } FeWO_4} \qquad \frac{60.53 \text{ kg W}}{100 \text{ kg } FeWO_4}$$

(Use the method of dimensional analysis for problem solving.)

Given quantity and units: 6.00 lb $FeWO_4$
Desired quantity and units: lb W

Conversion factor: $\dfrac{183.9 \text{ lb W}}{303.8 \text{ lb } FeWO_4}$

$$\text{lb W} = 6.00 \text{ lb } \cancel{FeWO_4} \times \frac{183.9 \text{ lb W}}{303.8 \text{ lb } \cancel{FeWO_4}} = 3.63 \text{ lb W}$$

A **solute** added to a **solvent** produces a homogeneous mixture called a *solution*. The concentration of solute dissolved in a solution is often expressed as moles of solute in 1 liter of solution. This is known as the **molarity** of the solution.

$$\text{molarity, M} = \frac{\text{mol solute}}{\text{L solution}}$$

Example 4.7 A mass of 3.77 g NaCl is dissolved and diluted with water to the "mark" in a 250 mL volumetric flask†

(a) What is the molarity of the solution?
(b) How many grams of NaCl are in 50.0 mL of the solution?

Answer (a) 0.258 M NaCl; (b) 0.755 g NaCl

Solution

(a) Given quantity and units: $\dfrac{3.77 \text{ g NaCl}}{250 \text{ mL solution}}$

Desired quantity and units: $M = \dfrac{\text{mol NaCl}}{\text{L solution}}$

Conversion factors: one mole of NaCl has a mass of 58.5 grams (23.0 + 35.5) or

$$1 \text{ mol NaCl} = 58.5 \text{ g NaCl}$$

$$10^{-3} \text{ L} = 1 \text{ mL}$$

$$M\left(\frac{\text{mol NaCl}}{\text{L solution}}\right) = \frac{3.77 \cancel{\text{ g NaCl}}}{250 \cancel{\text{ mL}}} \times \underset{\text{(converts grams to moles)}}{\frac{1 \text{ mol NaCl}}{58.5 \cancel{\text{ g NaCl}}}} \times \underset{\text{(converts mL to L)}}{\frac{1 \cancel{\text{ mL}}}{10^{-3} \text{ L}}}$$

$$= \frac{0.258 \text{ mol NaCl}}{\text{L}} = 0.258 \text{ M NaCl}$$

† In expressing the correct number of significant figures for a measured quantity in the study guide, we will assume all zeros to be significant. For example, 250 mL has three significant figures, 1000°C has four significant figures, and 100 mg has three significant figures.

(b) Given quantity and units: 50.0 mL solution
Desired quantity and units: g NaCl
Conversion factors: 0.258 mol NaCl = 1 L = 1000 mL solution

1 mol NaCl = 58.5 g NaCl

$$\text{mass, g NaCl} = 50.0\,\cancel{\text{mL}} \times \underset{\substack{\text{(conversion to mol}\\ \text{NaCl in 50.0 mL)}}}{\frac{0.258\,\cancel{\text{mol NaCl}}}{1000\,\cancel{\text{mL}}}} \times \underset{\substack{\text{(conversion to}\\ \text{g NaCl)}}}{\frac{58.5\text{ g NaCl}}{\cancel{\text{mol NaCl}}}} = 0.755\text{ g NaCl}$$

Example 4.8 Calculate the mass, in grams, of $KMnO_4$ required to prepare 600 mL of 0.811 M $KMnO_4$.

Answer 76.9 g $KMnO_4$

Solution

Given quantity and units: 600 mL solution
Desired quantity and units: g $KMnO_4$
Conversion factors:

$$0.811\text{ M KMnO}_4 \quad \text{means} \quad \frac{0.811\text{ mol KMnO}_4}{1\text{ L solution}}$$

or

$$0.811\text{ mol KMnO}_4 = 1\text{ L or }1000\text{ mL solution,}$$

$$1\text{ mol KMnO}_4 = 158\text{ g KMnO}_4$$

(*Do not* attempt to use the symbol M in your calculation! Convert the definition of M to mol solute/L solution.)

$$\text{mass, g KMnO}_4 = 600\,\cancel{\text{mL}} \times \underset{\substack{\text{(conversion to mol}\\ \text{KMnO}_4\text{ in 600 mL)}}}{\frac{0.811\,\cancel{\text{mol KMnO}_4}}{1000\,\cancel{\text{mL}}}} \times \underset{\substack{\text{(conversion to}\\ \text{g KMnO}_4)}}{\frac{158\text{ g KMnO}_4}{\cancel{\text{mol KMnO}_4}}} = 76.9\text{ g KMnO}_4$$

To prepare the solution, 76.9 g $KMnO_4$ is diluted to 600 mL with distilled water. Adding 600 mL of water to 76.9 g $KMnO_4$ does not mean that the total volume of the solution is 600 mL.

SECTION 4.3 WRITING FORMULAS

Objectives

- To define valence.
- To write the formulas for compounds using the concept of valence and ionic charge.

Focus

The **valence** of an atom is a number that represents its capacity to combine with another atom. H is assigned a valence of 1. Therefore in NH_3, N has a valence of 3 and in NF_3, F has a valence of 1.

Example 4.9 Determine the valence of iron and sulfur from the formulas H_2O; NaH; Na_2S; Fe_2S_3.

Answer S = 2; Fe = 3

Solution H has a valence of 1. Therefore in NaH, Na also has a valence of 1. In Na_2S, two Na atoms combine with one S atom, so S must have a valence of 2. Given that S has a valence of 2, the sum of the valences of three S atoms in Fe_2S_3 must be 6. Therefore the sum of the valences of the two Fe atoms must be 6, or each Fe atom has a valence of 3.

As illustrated in Example 4.9, each element and polyatomic group of elements such as the sulfate group can be assigned a valence. Metals and nonmetals that have a charge are called ions. The positive or negative charge on the ion is the same as its valence. Generally a positive sign is associated with a metal and a negative sign with most nonmetal and polyatomic groups.

To make formula writing easier, memorize Table 4.1. The formula of a compound must be written as electrically neutral.

Example 4.10 Write the formula for the compound formed between (a) K^+ and SO_4^{2-}; (b) NH_4^+ and S^{2-}; (c) Na^+ and PO_4^{3-}.

Answer (a) K_2SO_4; (b) $(NH_4)_2S$; (c) Na_3PO_4.

Solution (a) Since K^+ has a charge of 1+ and SO_4^{2-} has a charge of 2−, then two K^+ ions must combine with one SO_4^{2-} ion. The formula therefore must be K_2SO_4. The subscript "2" after K represents the number of K^+'s in the formula.

(b) Two NH_4^+ ions must combine with one S^{2-} ion. The formula is $(NH_4)_2S$, in which the "$( )_2$" indicates that two NH_4^+'s are necessary. $(NH_4)_2S$ is preferred to $NH_{42}S$ because the latter formula can be mistakenly read as having 1 N, 42 H, and 1 S per molecule. The parentheses are needed for clarity in writing the formula.

(c) Three Na^+ balance the 3− charge on one PO_4^{3-}. The formula is Na_3PO_4.

SECTION 4.4 NOMENCLATURE OF BINARY COMPOUNDS

Objectives

- To write the formula of a compound, given its name.
- To write the name of a compound, given its formula.

Focus

The formulas for compounds can be easily written from their names.

Example 4.11 Using the names of the ions listed in Table 4.1, write the formula of the following compounds:

(a) iron(III) or ferric chloride
(b) magnesium nitrate
(c) tin(IV) or stannic fluoride

(d) barium hydride
(e) aluminum phosphate
(f) potassium chromate
(g) mercuric oxide
(h) sodium hydrogen phosphate

Answer (a) $FeCl_3$; (b) $Mg(NO_3)_2$; (c) SnF_4; (d) BaH_2; (e) $AlPO_4$; (f) K_2CrO_4; (g) HgO; (h) Na_2HPO_4.

Solution

(a) iron(III) or ferric, Fe^{3+}; chloride, Cl^- (one Fe^{3+} combines with three Cl^-); $FeCl_3$
(b) magnesium, Mg^{2+}; nitrate, NO_3^- (one Mg^{2+} combines with two NO_3^-); $Mg(NO_3)_2$
(c) tin(IV) or stannic, Sn^{4+}; fluoride, F^- (one Sn^{4+} combines with four F^-); SnF_4
(d) barium, Ba^{2+}; hydride, H^- (one Ba^{2+} combines with two H^-); BaH_2
(e) aluminum, Al^{3+}; phosphate PO_4^{3-} (one Al^{3+} combines with one PO_4^{3-}); $AlPO_4$
(f) potassium, K^+; chromate, CrO_4^{2-} (two K^+ combine with one CrO_4^{2-}); K_2CrO_4
(g) mercuric, Hg^{2+}; oxide, O^{2-} (one Hg^{2+} combines with one O^{2-}); HgO
(h) sodium, Na^+; hydrogen phosphate, HPO_4^{2-} (two Na^+ combine with one HPO_4^{2-}); Na_2HPO_4

SECTION 4.5 CHEMICAL EQUATIONS

Objective

- To balance chemical equations.

Focus

In a chemical reaction, mass is conserved. An equation that represents a chemical reaction must have the same number and kind of atoms appearing on both sides of the equation in order for it to be balanced with respect to mass. Two rules to remember in balancing a chemical equation are the following:

1. Balance only by using coefficients before the formulas of the compound.
2. *Do not* change subscripts in formulas; this represents a change in the composition of the reactants and/or products of the reaction.

Example 4.12 Balance the equation for the oxidation of aluminum metal to aluminum oxide.

$$Al(c) + O_2(g) \rightarrow Al_2O_3(c)$$

Answer $4Al(c) + 3O_2(g) \rightarrow 2Al_2O_3(c)$

Solution In balancing the equation, first focus attention on the most complicated-looking formula. Al_2O_3 is the most complex formula. Notice that the Al_2O_3 formula has 3 O atoms, whereas there must be an even number of oxygen atoms in the form of O_2 on the left side of the equation. Thus, to balance the oxygen atoms, we multiply the Al_2O_3 by 2 and the O_2 by 3 to obtain

$$Al(c) + 3O_2(g) \rightarrow 2Al_2O_3(c) \qquad \text{(not balanced)}$$

Now there are 4 Al atoms on the right side and only one on the left side. Placing the coefficient 4 in front of Al(*c*) will balance the Al atoms.

$$4Al(c) + 3O_2(g) \rightarrow 2Al_2O_3(c)$$

Notice that only coefficients balance the equation; subscripts are never changed to achieve a mass balance in the equation.

The use of fractional coefficients is discouraged in balancing equations although it is not incorrect to do so.

Example 4.13 Balance the equation for the neutralization of $Al(OH)_3$ with H_2SO_4.

$$Al(OH)_3 + H_2SO_4 \rightarrow H_2O + Al_2(SO_4)_3$$

Answer $2Al(OH)_3 + 3H_2SO_4 \rightarrow 6H_2O + Al_2(SO_4)_3$

Solution The most complex-looking formula is $Al_2(SO_4)_3$. Placing a coefficient of 2 before $Al(OH)_3$ balances the Al atoms:

$$2Al(OH)_3 + H_2SO_4 \rightarrow H_2O + Al_2(SO_4)_3 \qquad \text{(not balanced).}$$

Note that the SO_4 group appears unchanged in the $Al_2(SO_4)_3$ and the H_2SO_4 formulas; balance the SO_4 groups by using a coefficient of 3 for H_2SO_4:

$$2Al(OH)_3 + 3H_2SO_4 \rightarrow H_2O + Al_2(SO_4)_3 \qquad \text{(not balanced)}$$

Only the H atom in H_2SO_4 and the O atoms in $Al(OH)_3$ remain to be balanced with H_2O on the right-hand side. If we rewrite H_2O as HOH, we see that the 6OH groups in the $2Al(OH)_3$ and 6H atoms in $3H_2SO_4$ combine to form 6HOH or $6H_2O$:

$$2Al(OH)_3 + 3H_2SO_4 \rightarrow 6H_2O + Al_2(SO_4)_3$$

Notice again that the subscripts remain unchanged.

Groups of atoms like the SO_4 in Example 4.13 often remain intact in a chemical reaction. When this occurs, balance the group by the same procedure as you would any atoms. Become familiar with the various groups of atoms listed in Table 4.1.

SECTION 4.6 QUANTITATIVE INFORMATION FROM CHEMICAL EQUATIONS

Objective

- To use a chemical equation for calculating quantitative relationships between reactants and products.

Focus

The coefficient and formula of a compound in a balanced equation have several meanings. For example, suppose that in a balanced equation for the oxidation of ammonia gas, ammonia has a coefficient of 4 or $4NH_3(g)$ and oxygen gas has a coefficient of 5 or $5O_2(g)$. The "$4NH_3(g)$" and the "$5O_2(g)$" mean each of the following:

$$4NH_3(g) + 5O_2(g) \rightarrow$$

$$4 \text{ mol } NH_3 + 5 \text{ mol } O_2 \rightarrow$$

$$4 \text{ mol } NH_3 \times \frac{6.02 \times 10^{23} \text{ molecules } NH_3}{\text{mol } NH_3} + 5 \text{ mol } O_2 \times \frac{6.02 \times 10^{23} \text{ molecules } O_2}{\text{mol } O_2} \rightarrow$$

$$4 \text{ mol } NH_3 \times \frac{17.0 \text{ g } NH_3}{\text{mol } NH_3} + 5 \text{ mol } O_2 \times \frac{32.0 \text{ g } O_2}{\text{mol } O_2} \rightarrow$$

All of these quantities are equivalent. This means that the amounts of NH_3 reacting with O_2 can be expressed using grams, molecules, or moles. For examples,

$$4 \text{ mol } NH_3 \text{ reacts with } 5 \text{ mol } O_2 \times \frac{6.02 \times 10^{23} \text{ molecules } O_2}{\text{mol } O_2}$$

$$4 \text{ mol } NH_3 \times \frac{17.0 \text{ g } NH_3}{\text{mol } NH_3} \text{ reacts with } 5 \text{ mol } O_2 \times \frac{6.02 \times 10^{23} \text{ molecules } O_2}{\text{mol } O_2}$$

$$4 \text{ mol } NH_3 \times \frac{6.02 \times 10^{23} \text{ molecules } NH_3}{\text{mol } NH_3} \text{ reacts with } 5 \text{ mol } O_2$$

Let us take this concept and extend it to the complete balanced equation.

Example 4.14 How many moles of O_2 react with 33.1 moles NH_3?

$$4NH_3(g) + 5O_2(g) \rightarrow 4NO(g) + 6H_2O(g)$$

Answer 41.4 moles O_2

Solution As in earlier sections, we should formalize our calculations by using dimensional analysis.

Given quantity and units: 33.1 mol NH_3
Desired quantity and units: mol O_2
Conversion factor: Derive the mol NH_3/mol O_2 relationship from the balanced equation.
4 mol NH_3 requires 5 mol O_2; the conversion factor is:

$$\frac{5 \text{ mol } O_2}{4 \text{ mol } NH_3}$$

$$\text{mol } O_2 = 33.1 \cancel{\text{mol } NH_3} \times \frac{5 \text{ mol } O_2}{4 \cancel{\text{mol } NH_3}}$$

$$= 41.4 \text{ mol } O_2$$

Example 4.15 Calculate the mass, in grams, of $H_2O(g)$ produced from the oxidation of 2.88 g $NH_3(g)$.

$$4NH_3(g) + 5O_2(g) \rightarrow 4NO(g) + 6H_2O(g)$$

Answer 4.57 g H_2O

Solution

Given quantity and units: 2.88 g NH_3
Desired quantity and units: g H_2O
Conversion factor: Obtain the g NH_3/g H_2O relationship from the balanced equation and express in corresponding units:

$$4 \text{ mol } NH_3 \text{ yields } 6 \text{ mol } H_2O$$

$$4 \cancel{\text{mol } NH_3} \times \frac{17.0 \text{ g } NH_3}{\cancel{\text{mol } NH_3}} \text{ yields } 6 \cancel{\text{mol } H_2O} \times \frac{18.0 \text{ g } H_2O}{\cancel{\text{mol } H_2O}}$$

or

$$\frac{6 \times 18.0 \text{ g } H_2O}{4 \times 17.0 \text{ g } NH_3}$$

$$\text{mass } H_2O = 2.88 \cancel{\text{g } NH_3} \times \frac{6 \times 18.0 \text{ g } H_2O}{4 \times 17.0 \cancel{\text{g } NH_3}} = 4.57 \text{ g } H_2O$$

Example 4.16 A mass of 245 g H_2O_2 reacts with excess N_2H_4 to produce HNO_3 and H_2O according to the equation.

$$7H_2O_2(g) + N_2H_4(g) \rightarrow 2HNO_3(g) + 8H_2O(g)$$

(a) How many moles of $HNO_3(g)$ are produced?
(b) How many molecules of $N_2H_4(g)$ react?

Answer (a) 2.06 mol $HNO_3(g)$
(b) 6.20×10^{23} molecules $N_2H_4(g)$

Solution

(a) Given quantity and units: 245 g H_2O_2
Desired quantity and units: mol HNO_3
Conversion factor: Obtain the H_2O_2 to HNO_3 mole relationship from the balanced equation and express it in corresponding units:

$$7 \text{ mol } H_2O_2 \text{ yields } 2 \text{ mol } HNO_3$$

$$7 \text{ mol } H_2O \times \frac{34.0 \text{ g } H_2O_2}{\text{mol } H_2O} \text{ yields } 2 \text{ mol } HNO_3$$

or

$$\frac{2 \text{ mol } HNO_3}{7 \times 34.0 \text{ g } H_2O_2}$$

$$\text{mol } HNO_3 = 245 \cancel{\text{g } H_2O_2} \times \frac{2 \text{ mol } HNO_3}{7 \times 34.0 \cancel{\text{g } H_2O_2}}$$

$$= 2.06 \text{ mol } HNO_3$$

(b) Given quantity and units: 245 g H_2O_2
Desired quantity and units: molecules of N_2H_4
Conversion factor:

$$7 \text{ mol } H_2O_2 \text{ reacts with } 1 \text{ mol } N_2H_4$$

$$7 \cancel{\text{mol } H_2O_2} \times \frac{34.0 \text{ g } H_2O_2}{\cancel{\text{mol } H_2O_2}} \text{ reacts with } 1 \cancel{\text{mol } N_2H_4} \times \frac{6.02 \times 10^{23} \text{ molecules } N_2H_4}{\cancel{\text{mol } N_2H_4}}$$

or

$$\frac{6.02 \times 10^{23} \text{ molecules } N_2H_4}{7 \times 34.0 \text{ g } H_2O_2}$$

$$\text{molecules } N_2H_4 = 245 \cancel{\text{g } H_2O_2} \times \frac{6.02 \times 10^{23} \text{ molecules } N_2H_4}{7 \times 34.0 \cancel{\text{g } H_2O_2}}$$

$$= 6.20 \times 10^{23} \text{ molecules } N_2H_4$$

Therefore, we can calculate the mass or the number of molecules or atoms of reactants needed or products produced using the diagram shown below. This is an addition to the diagram appearing in the *Study Guide*, section 3.5.

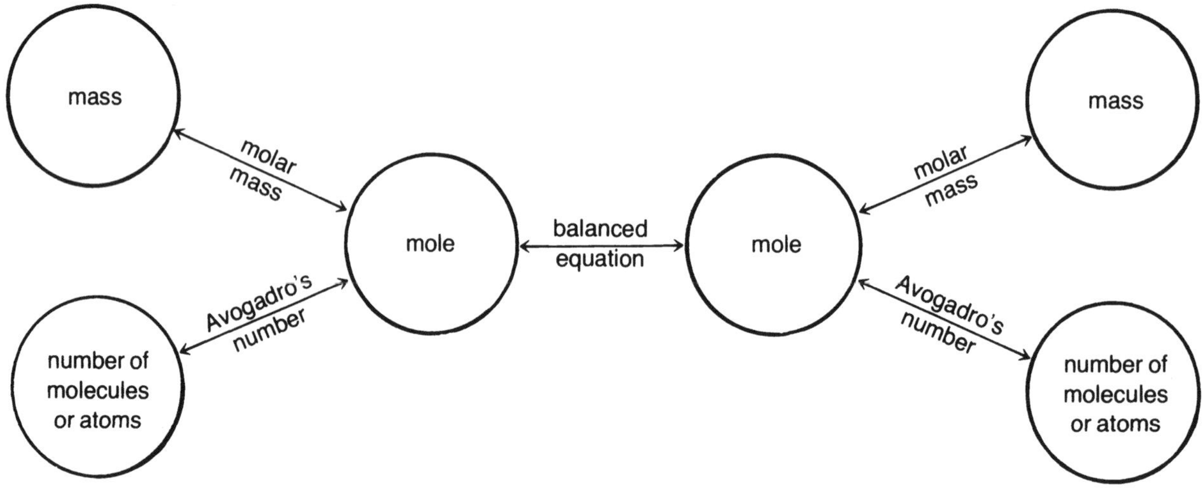

SECTION 4.7 EXCESS REACTANT

Objectives

- To calculate the maximum yield of a product, given limited amounts of reactants.
- To find which reactant is in excess.
- To calculate the quantity of excess reactant.

Focus

Most chemical reactions are conducted without regard to the balanced equation. The reactant that is consumed entirely (called the limiting reagent) fixes the maximum amount of product formed. Once a reactant is depleted, the reaction stops and no additional product can form. The other reactant which cannot be entirely consumed is termed the **excess reagent**.

Example 4.17 A total of 538 g H_2O_2 and 74.0 g N_2H_4 are placed in a reaction vessel.

$$7H_2O_2(g) + N_2H_4(g) \rightarrow 2HNO_3(g) + 8H_2O(g)$$

(a) Calculate the maximum amount of HNO_3, in grams, that can be produced.
(b) Identify the excess reactant and calculate the grams of it that remain unreacted.

Answer (a) 285 g HNO_3
(b) N_2H_4; 1.7 g N_2H_4

Solution

Given quantity and units: 538 g H_2O_2 and 74.0 g N_2H_4
Desired quantity and units: g HNO_3

(a) To calculate the mass of HNO_3 produced in the reaction, we should first determine the amounts of HNO_3 each reactant produces.

(i) The mass of HNO_3 produced from 538 g H_2O_2:
Conversion factors: From the balanced equation

$$7 \text{ mol } H_2O_2 \text{ yields } 2 \text{ mol } HNO_3$$

$$7 \text{ mol } H_2O_2 \times \frac{34.0 \text{ g } H_2O_2}{\text{mol } H_2O_2} \text{ yields } 2 \text{ mol } HNO_3 \times \frac{63.0 \text{ g } HNO_3}{\text{mol } HNO_3}$$

$$\text{mass } HNO_3 = 538 \text{ g } H_2O_2 \times \frac{2 \times 63.0 \text{ g } HNO_3}{7 \times 34.0 \text{ g } H_2O_2}$$

$$= 285 \text{ g } HNO_3$$

(ii) The mass of HNO_3 produced from 74.0 g N_2H_4:

$$1 \text{ mol } N_2H_4 \text{ yields } 2 \text{ mol } HNO_3$$

$$1 \text{ mol } N_2H_4 \times \frac{32.0 \text{ g } N_2H_4}{\text{mol } N_2H_4} \text{ yields } 2 \text{ mol } HNO_3 \times \frac{63.0 \text{ g } HNO_3}{\text{mol } HNO_3}$$

$$\text{mass } HNO_3 = 74.0 \text{ g } N_2H_4 \times \frac{2 \times 63.0 \text{ g } HNO_3}{1 \times 32.0 \text{ g } N_2H_4}$$

$$= 291 \text{ g } HNO_3$$

Only 285 g HNO_3 can be produced from 538 g H_2O_2; 291 grams HNO_3 cannot be produced from 538 g H_2O_2. Therefore when the 538 g H_2O_2 (limiting reagent) is consumed, the reaction is complete. Some N_2H_4 remains in excess.

(b) To determine the excess N_2H_4 subtract the quantity of N_2H_4 that reacts with 538 g H_2O_2 from the 74.0 g in the system. Therefore:

Given quantity and units: 538 g H_2O_2
Desired quantity and units: g N_2H_4 consumed
Conversion factor:

$$7 \text{ mol } H_2O_2 \text{ reacts with } 1 \text{ mol } N_2H_4$$

$$7 \text{ mol } H_2O_2 \times \frac{34.0 \text{ g } H_2O_2}{\text{mol } H_2O_2} \text{ reacts with } 1 \text{ mol } N_2H_4 \times \frac{32.0 \text{ g } N_2H_4}{\text{mol } N_2H_4}$$

$$\text{mass } N_2H_4 = 538 \text{ g } H_2O_2 \times \frac{1 \times 32.0 \text{ g } N_2H_4}{7 \times 34.0 \text{ g } H_2O_2}$$

$$= 72.3 \text{ g } N_2H_4$$

Excess g N_2H_4 = 74.0 g − 72.3 g = 1.7 g N_2H_4

SECTION 4.8 PERCENT YIELD

Objective

- To calculate the percent yield of a chemical reaction.

Focus

The **theoretical yield** of a product is the maximum amount of the product that can be isolated from the given quantities of reactants.

The yields that were calculated from a chemical equation in previous sections were all assumed to be a maximum or theoretical yield of 100%. Most chemical reactions, however, do not produce a 100% yield; nonideal experimental conditions cause the production of by-products.

Example 4.18 According to the equation, copper metal displaces silver from a silver nitrate solution:

$$Cu(c) + 2\,AgNO_3(aq) \rightarrow 2Ag(s) + Cu(NO_3)_2$$

A mass of 28.88 g Ag metal is produced from a solution containing 6.774 g Cu. Calculate the percent yield for the extraction of Ag metal from the solution.

Answer 90.78%

Solution

Given quantity and units: 6.774 g Cu
Desired quantity and units: g Ag
Conversion factor:

1 mol Cu yields 2 mol Ag

$$1\,\cancel{\text{mol Cu}} \times \frac{63.55\text{ g Cu}}{\cancel{\text{mol Cu}}} \text{ yields } 2\,\cancel{\text{mol Ag}} \times \frac{107.9\text{ g Ag}}{\cancel{\text{mol Ag}}}$$

or

$$\frac{2 \times 107.9\text{ g Ag}}{1 \times 63.55\text{ g Cu}}$$

$$\text{mass Ag} = 6.774\,\cancel{\text{g Cu}} \times \frac{2 \times 107.9\text{ g Ag}}{1 \times 63.55\,\cancel{\text{g Cu}}}$$
$$= 23.00\text{ g Ag}$$

The 23.00 g Ag is the maximum (or theoretical) yield. If only 20.88 g Ag are displaced from the solution, the percent yield is

$$\%\,\textbf{yield} = \frac{\textbf{actual yield}}{\textbf{theoretical yield}} \times 100\% = \frac{20.88\,\cancel{\text{g Ag}}}{23.00\,\cancel{\text{g Ag}}} \times 100\%$$
$$= 90.78\%$$

Example 4.19 The Haber process for the commercial production of ammonia requires the passage of $N_2(g)$ and $H_2(g)$ over a catalyst at elevated temperatures and pressures.

$$N_2(g) + 3H_2(g) \xrightarrow[\text{temp, pres}]{\text{catalyst}} 2HN_3(g)$$

A mass of 165 g $NH_3(g)$ is produced from a reaction mixture of 60.0 g H_2 and 200 g N_2. Calculate the percent yield of NH_3.

Answer 67.9%

Solution

Given quantity and units: 60.0 g H_2 and 200 g N_2

Desired quantity and units: g NH_3

Since we have a specific mass of each reactant, we must make two calculations to see which reactant limits the amount of NH_3 formed.

(i) The g NH_3 from 200 g N_2:

Conversion factors:

$$1 \text{ mol } N_2 \text{ yields } 2 \text{ mol } NH_3$$

$$1\,\cancel{\text{mol } N_2} \times \frac{28.0 \text{ g } N_2}{\cancel{\text{mol } N_2}} \text{ yields } 2\,\cancel{\text{mol } NH_3} \times \frac{17.0 \text{ g } NH_3}{\cancel{\text{mol } NH_3}}$$

or

$$\frac{2 \times 17.0 \text{ g } NH_3}{1 \times 28.0 \text{ g } N_2}$$

$$\text{mass } NH_3 = 200\,\cancel{\text{g } N_2} \times \frac{2 \times 17.0 \text{ g } NH_3}{1 \times 28.0\,\cancel{\text{g } N_2}} = 243 \text{ g } NH_3$$

(ii) The g NH_3 from 60.0 g H_2:

$$3 \text{ mol } H_2 \text{ yields } 2 \text{ mol } NH_3$$

$$3\,\cancel{\text{mol } H_2} \times \frac{2.02 \text{ g } H_2}{\cancel{\text{mol } H_2}} \text{ yields } 2\,\cancel{\text{mol } NH_3} \times \frac{17.0 \text{ g } NH_3}{1\,\cancel{\text{mol } NH_3}}$$

or

$$\frac{2 \times 17.0 \text{ g } NH_3}{3 \times 2.02 \text{ g } H_2}$$

$$\text{mass } NH_3 = 60.0\,\cancel{\text{g } H_2} \times \frac{2 \times 17.0 \text{ g } NH_3}{3 \times 2.02\,\cancel{\text{g } H_2}} = 337 \text{ g } NH_3$$

200 g N_2 produces a maximum of 243 g NH_3, the theoretical yield of NH_3. Since only 165 g NH_3 is actually produced, the percent yield is

$$\% \text{ yield} = \frac{\text{actual yield}}{\text{theoretical yield}} \times 100\%$$

$$= \frac{165\,\cancel{\text{g } NH_3}}{243\,\cancel{\text{g } NH_3}} \times 100\% = 67.9\%$$

QUESTIONS

Section 4.1

4.1 What is the empirical formula for each?

(a) Au_2Cl_6

(b) $C_6H_9Cl_3$

(c) Na_3AlF_6

(d) C_8H_{18}
(e) $C_{20}H_{14}O_4$ (phenolphthalein)

4.2 Determine the empirical formula of cryolite, a substance used in the manufacture of aluminum; its mass composition is 32.79% Na, 13.02% Al and 54.19% F.

4.3 (a) Determine the empirical formula of paraldehyde which has a mass composition of 62.04% C, 10.41% H and 27.55% O.
(b) The molecular weight of paraldehyde is 116; write its molecular formula.
(c) What is its accurate molecular weight?

Section 4.2

4.4 (a) What is the mass percentage of C, H, and O in $C_{20}H_{14}O_4$ (phenolphthalein)?
(b) How many g C are in 200 mg of phenolphthalein?

4.5 A mass of 0.248 g Na_3AlF_6 is used to prepare 2.00 L solution. What is the molarity of the solution?

4.6 Calculate the mass, in grams, of $AgNO_3$ necessary to prepare 180 mL of a 0.350 M $AgNO_3$ solution.

4.7 (a) Calculate the volume, in mL, of 0.0600 M HCl required to precipitate 7.32 mg Ag^+ from a $AgNO_3$ solution. $Ag^+ + Cl^- \rightarrow AgCl(c)$.
(b) What mass of AgCl precipitated?

Section 4.3

4.8 Write the formula for the compound formed between:
(a) Hg^{2+} and S^{2-}
(b) Sn^{4+} and NO_2^-
(c) Fe^{2+} and CO_3^{2-}
(d) Ag^+ and SO_4^{2-}
(e) Co^{2+} and $Cr_2O_7^{2-}$
(f) Cr^{3+} and O^{2-}

Section 4.4

4.9 Write the formulas for each of the following compounds:
(a) chromium(III) acetate
(b) magnesium chromate
(c) cobalt(II) iodide
(d) copper(I) cyanide
(e) iron(II) or ferrous hydroxide
(f) mercury(II) or mercuric chloride

4.10 Write the name of the compounds that have the formulas:
(a) $KMnO_4$
(b) $NaHSO_4$
(c) $Al_2(CO_3)_3$
(d) PbI_2
(e) $Cu_3(PO_4)_2$
(f) $CaSO_3$
(g) SnF_4
(h) SnF_2

Section 4.5

4.11 Balance the following equations:
(a) $Ca(CN)_2 + H_2O \rightarrow Ca(OH)_2 + HCN$

(b) $P_4O_6 + I_2 \rightarrow P_2I_4 + P_4O_{10}$
(c) $B_2O_3 + C \rightarrow B_4C + CO$
(d) $P + Cl_2 \rightarrow PCl_3$
(e) $CaCl_2 + AgNO_3 \rightarrow AgCl + Ca(NO_3)_2$

Section 4.6

4.12 (a) Calculate the mass, in grams, of Ca_3N_2 produced from the reaction of 37.0 g Ca with excess NH_3.

$$6Ca + 2NH_3 \rightarrow 3\,CaH_2 + Ca_3N_2$$

(b) If only 95% of the Ca reacts, what is the yield of CaH_2?

4.13 How many molecules of PCl_3 are produced from the reaction of 7.66 μg P with excess Cl_2?

$$2P + 3Cl_2 \rightarrow 2PCl_3$$

4.14 What volume of 0.630 M HNO_3 dissolves 60% of the 28.0 g Cu in a copper ore?

$$3Cu(c) + 8HNO_3(aq) \rightarrow 3Cu(NO_3)_2(aq) + 2NO(g) + 4H_2O$$

4.15 What mass, in grams, of $FeCl_2$ reacts completely with 4.22 g Na_3PO_4?

$$FeCl_2 + Na_3PO_4 \rightarrow Fe_3(PO_4)_2 + NaCl \qquad \text{(unbalanced)}$$

4.16 According to the following equations, how many grams of $FeCl_3$ are needed for the production of 100 g $SiCl_4$?

$$4FeCl_3 + 3O_2 \rightarrow 2Fe_2O_3 + 6Cl_2$$

$$Si + 2Cl_2 \rightarrow SiCl_4$$

(*Hint*: How many moles of Cl_2 are needed? Where is the Cl_2 produced?)

Section 4.7

4.17 (a) How many grams of Bi_2S_3 form when 12.3 g H_2S and 126 g $Bi(NO_3)_3$ are mixed in an aqueous solution?

$$2Bi(NO_3)_3(aq) + 3H_2S(g) \rightarrow Bi_2S_3(c) + 6HNO_3(aq)$$

(b) What is the quantity of excess reactant?

4.18 Masses of 7.45 g FeO and 0.111 mol Al were melted in a high temperature oven. Calculate the maximum amount of iron produced.

$$Al + FeO \rightarrow Fe + Al_2O_3 \qquad \text{(unbalanced)}$$

4.19 A total of 2.50 g $K_2Cr_2O_7$ and 5.00 g $Fe(NO_3)_2$ is dissolved in a HNO_3 solution.

(a) Calculate the mass, in grams, of $Cr(NO_3)_2$ produced.

$$14HNO_3 + K_2Cr_2O_7 + 6Fe(NO_3)_2 \rightarrow 2Cr(NO_3)_2 + 6Fe(NO_3)_2 + 7H_2O + 2KNO_3$$

(b) What is the quantity of excess reactant?

Section 4.8

4.20 In the combustion of 8.20 g C_2H_6,

$$2C_2H_6 + 7O_2 \rightarrow 4CO_2 + 6H_2O$$

only 6.33 g CO_2 are produced. Calculate the percent yield of CO_2 in the reaction.

4.21 Only 50.2 g HCl is produced from the reaction of 63.7 mol CH_4 with excess Cl_2.

What is the percent yield?

$$3Cl_2 + CH_4 \rightarrow CHCl_3 + 3HCl$$

ANSWERS

4.1 (a) $AuCl_3$
(b) C_2H_3Cl
(c) Na_3AlF_6
(d) C_4H_9
(e) $C_{10}H_7O_2$

4.2 Na_3AlF_6

4.3 (a) C_3H_6O
(b) $C_6H_{12}O_2$
(c) 116.18

4.4 (a) 75.45% C; 4.44% H; 20.1% O
(b) 151 mg C

4.5 5.90×10^{-4} M Na_3AlF_6

4.6 10.7 g $AgNO_3$

4.7 (a) 1.13 mL 0.0600 M HCl
(b) 9.73 mg AgCl

4.8 (a) HgS
(b) $Sn(NO_2)_4$
(c) $FeCO_3$
(d) Ag_2SO_4
(e) $CoCr_2O_7$
(f) Cr_2O_3

4.9 (a) $Cr(C_2H_3O_2)_3$
(b) $MgCrO_4$
(c) CoI_2
(d) CuCN
(e) $Fe(OH)_3$
(f) $HgCl_2$

4.10 (a) potassium permanganate
(b) sodium hydrogen sulfate or sodium bisulfate
(c) aluminum carbonate
(d) lead(II) iodide
(e) copper(II) phosphate or cupric phosphate
(f) calcium sulfite
(g) tin(IV) fluoride or stannic fluoride
(h) tin(II) fluoride or stannous fluoride

4.11 (a) $Ca(CN)_2 + 2H_2O \rightarrow Ca(OH)_2 + 2HCN$
(b) $5P_4O_6 + 8I_2 \rightarrow 4P_2I_4 + 3P_4O_{10}$
(c) $2B_2O_3 + 7C \rightarrow B_4C + 6CO$
(d) $2P + 3Cl_2 \rightarrow 2PCl_3$
(e) $CaCl_2 + 2AgNO_3 \rightarrow 2AgCl + Ca(NO_3)_2$

4.12 (a) 22.8 g Ca_3N_2
(b) 18.4 g CaH_2

4.13 1.49×10^{17} molecules PCl_3

4.14 1.12 L 0.630 M HNO_3

4.15 4.90 g $FeCl_2$

4.16 127 g $FeCl_3$

4.17 (a) 61.9 g Bi_2S_3
(b) 30.8 g $Bi(NO_3)_3$

4.18 5.81 g Fe

4.19 (a) 1.63 g $Cr(NO_3)_2$
(b) 1.14 g $K_2Cr_2O_7$

4.20 26.4%

4.21 0.720%

Chapter 5

GASES

SECTION 5.1 THE PROPERTIES OF GASES: THE BAROMETER

Objectives

- To name and identify the four variable properties of a gas.
- To describe the effects discussed in the five experiments in the text.
- To define pressure and standard pressure.

Focus

Review text Table 5.1 for the physical properties of gases.

A gas is characterized by four variable properties: **volume**, V, **pressure**, P, **temperature**, T, and **quantity** of gas. If three of the variables are fixed, then the fourth must also be fixed. In studying the physical properties of gases, we generally fix two of the variables and gather data on the relationship between the other two.

To summarize the experiments in the text:

Experiment 1: As temperature increases, volume increases.
Experiment 2: As pressure decreases, volume increases and vice versa.
Experiment 3: As the quantity of gas increases, volume increases.
Experiment 4: Each gas at a fixed P, V, and T has its own mass; different gases have different masses and different densities.
Experiment 5: Lighter gases diffuse (escape) faster than heavier gases.

Pressure is the force that a substance exerts per unit area. Force is weight. Everyone exerts force on the earth's surface; the area of your feet is the one area over which the force is applied.

A gas exerts a force per unit area against the walls of a container.

$$\text{standard pressure} = 1 \text{ atm} = 760 \text{ mmHg} = 760 \text{ torr}$$

at 0°C. Also from Appendix A-4, 1 atm = 14.7 lb/in^2(psi) = 101.3 kPa = 1.013 bars.

Example 5.1 What is the pressure in torr of 0.770 atm?

Answer 585 torr

Solution Once again use dimensional analysis in solving problems.

Given quantity and units: 0.770 atm
Desired quantity and units: pressure (torr)
Conversion factor: 1 atm = 760 torr

$$\text{pressure (torr)} = 0.770\ \text{atm} \times \frac{760\ \text{torr}}{\text{atm}} = 585\ \text{torr}$$

SECTION 5.2 BOYLE'S LAW

Objective

- To express mathematically the relationship between changes in pressure and volume on a given quantity of gas at constant temperature.

Focus

As the volume of a gas increases (container becomes larger), the pressure it exerts against the wall of the container decreases. As the volume of a gas decreases, the pressure increases proportionally. Expressed mathematically,

$$P \propto \frac{1}{V}, \quad \text{or} \quad P \times V = \text{constant}$$

where $\propto$ means "is proportional to."

In performing any calculations involving pressure and volume changes, it is to your advantage to use the conceptual *common sense* method described in the text. This method encourages you to develop a physical association for changes in pressure and volume rather than a purely mathematical one.

Example 5.2 An air-fuel mixture in an internal combustion engine occupies 500 cm^3 at a pressure of 764 torr.† If the piston moves so that the volume of the mixture is reduced to 59.8 cm^3, what is the resulting pressure of the mixture if there is no change in temperature?

Answer 6500 torr

Solution

Given quantities: P_1 = 764 torr
V_1 = 500 cm^3
V_2 = 58.8 cm^3
Desired quantity: P_2 (torr)

According to Boyle's law, a decrease in the volume of a gas produces a proportional increase in pressure by the volume ratio 500 cm^3/58.8 cm^3.

† Recall from Chapter 4 that all zeros in a measured quantity are significant.

$$P_{new} = P_{original} \times \text{volume correction}$$

$$P_{new}(\text{torr}) = 764 \text{ torr} \times \frac{500 \text{ cm}^3}{58.8 \text{ cm}^3} = 6500 \text{ torr} \qquad \text{(three significant figures)}$$

Although the units of pressure and volume are arbitrary, make sure they are consistent.

SECTION 5.3 CHARLES' LAW: KELVIN TEMPERATURE

Objectives

- To calculate changes in temperature and volume of a given sample of gas at constant pressure.
- To define absolute zero and calculate the Kelvin temperature.
- To define standard temperature and pressure (STP) for a gas.

Focus

At constant pressure a rise in temperature increases proportionally the volume of a gas. Expressed mathematically,

$$\boldsymbol{V \propto T}, \quad \text{or} \quad \boldsymbol{V = \textbf{constant} \times T}, \quad \text{or} \quad \boldsymbol{\frac{V}{T} = \textbf{constant}}$$

This proportion is valid only when the temperature of the gas is stated in kelvins (Celsius temperature + 273.15). From Chapter 1,

$$T\text{ (K)} = t(°\text{C}) + 273.15$$

STP defines the mass and volume of a gas at standard temperature and pressure: 1 atm (or 760 torr) and 0°C (or 273.15 K).

Learn Charles' volume-temperature relationship according to the common sense method. Note that units for volume must be consistent and those for temperature must *always* be in kelvins.

Example 5.3 A 150 L sample of nitrogen gas is heated at constant pressure from 21°C to 42°C. What is the new volume of the nitrogen gas?

Answer 161 L

Solution

Given quantity: $V_1 = 150$ L

$T_1 = 21°\text{C} + 273 = 294$ K

$T_2 = 42°\text{C} + 273 = 315$ K

Desired quantity: V_2 (L)

Since the change in volume of a gas is directly proportional to its change in Kelvin temperature, the volume increases by a factor equal to the ratio of the two temperatures, which must be greater than 1. The correct factor is therefore 315 K/294 K.

$$V_{new} = V_{original} \times \text{temperature correction}$$

$$V_{new}\,(L) = 150\ L \times \frac{315\,\cancel{K}}{294\,\cancel{K}} = 161\ L$$

SECTION 5.4 THE LAW OF COMBINING VOLUMES AND THE AVOGADRO LAW

Objectives

- To explain the relationship between the volume of a gas and the number of its molecules in the system.
- To calculate the atomic or molecular weight of a gas using Avogadro's law.
- To define the volume occupied by 1 mole of gas at STP.

Focus

As the number of molecules of a gas increases in a system, the volume also increases if the pressure and temperature remain constant.

Expressed mathematically,

$$N \propto V, \quad \text{or} \quad N = \textbf{constant} \times V, \quad \text{or} \quad \frac{N}{V} = \textbf{constant}$$

This statement suggests that at a constant pressure, temperature, *and* volume, the number of gas molecules must also be constant and independent of molecular weight or size. *Equal volumes of gases at the same T and P contain the same number of molecules.*

Therefore, if the number of molecules of two gases is the same in two different systems which have the same temperature, pressure, and volume, common sense tells us that the *weights* of the two gases are not the same because the molecular weights of the two gases are not the same.

Example 5.4 A sample of hydrogen gas occupying a volume of 2.76 L exerts a pressure of 627 torr at 27°C.

(a) If the quantity of hydrogen in the container is tripled, what is the new volume if the pressure and temperature remain unchanged?

(b) If the quantity of hydrogen is reduced by one-half in the same 2.76 L container, what is the pressure exerted by the remaining gas at 27°C?

Answer (a) 8.28 L; (b) 314 torr.

Solution

(a) Given quantity: $V_1 = 2.76\ V$
$N_1 = N$ (number of molecules)
$N_2 = 3N$
Desired quantity: $V_2\,(L)$

An increase in the number of molecules produces a proportional increase in the volume by a factor equal to the ratio of the number of molecules in the system, $3\,N/N$,

$$V_{new}\ (L) = V_{original} \times \text{correction for number of molecules}$$

$$V_{new} = 2.76\ L \times \frac{3\cancel{N}}{\cancel{N}} = 8.28\ L$$

(b) Given quantities: $P_1 = 627$ torr
$N_1 = N$
$N_2 = \frac{1}{2}N$
Desired quantity: P_2 (torr)

A decrease in the number of molecules results in a proportional decrease in pressure by a factor equal to the ratio of the number of molecules in the system, $\frac{1}{2}N/1N$

$$P_{new}\ (\text{torr}) = P_{original} \times \text{correction for number of molecules}$$

$$P_{new} = 627\ \text{torr} \times \frac{\frac{1}{2}\cancel{N}}{\cancel{N}} = 314\ \text{torr}$$

Example 5.5 A 600 mL flask is filled with N_2 gas to a pressure of 718 torr at 23°C. The mass of N_2, determined by weighing the flask before and after evacuation, is 0.650 g. Then the flask is filled with an unknown pure gas to a pressure of 718 torr at 23°C. The mass of this gas is 1.02 g. If the molecular weight of N_2 is 28.0, what is the molecular weight of the unknown gas?

Answer 44.0

Solution From Avogadro's law, since V, P, and T remain unchanged, the number of molecules in the two systems is the same. Therefore the ratio of molecular weights of the two molecules is the same as the ratio of the weights of the two gases or

$$\frac{1.02\ g}{0.65\ g} = 1.57$$

If the molecular weight of N_2 is 28.0, the molecular weight of the unknown is 28.0 × 1.57, or 44.0.

The **molar volume** of a gas, the volume occupied by 1 mole of gas at STP, is 22.4 L; this volume contains 6.02×10^{23} molecules.

Example 5.6

(a) Calculate the volume occupied by 3.00 moles O_2 at STP.
(b) How many molecules (N) of O_2 are present?

Answer (a) 67.2 L; (b) 1.81×10^{24} molecules

Solution (a) Since 1.0 mole of any gas occupies 22.4 liters at STP, then for 3.00 moles:

$$V_{new} = 3.00\ \cancel{\text{mol}} \times \frac{22.4\ L}{\cancel{\text{mol}}} = 67.2\ L$$

(b) If 1.0 mole of gas contains 6.02×10^{23} molecules at STP, then for 3.00 moles:

$$N = 3.00\ \cancel{\text{mol}} \times \frac{6.02 \times 10^{23}\ \text{molecules}}{\cancel{\text{mol}}}$$

$$= 1.81 \times 10^{24}\ \text{molecules}$$

SECTION 5.5 THE IDEAL GAS LAW

Objectives

- To apply the ideal gas law to gaseous systems.
- To apply a combination of Boyle's law, Charles' law, and Avogadro's law to changes in T, P, V, and n on a gas sample.

Focus

A combination of Boyle's law, Charles' law, and Avogadro's law produces the **ideal gas law**,

$$PV = nRT$$

where n equals the number of moles of gas.

Several values of R are listed in the text, Appendix E-2, but the value most commonly used is $R = 0.0821$ (L atm)/(mole K). The units of P, V, n, and T must be the same as those that appear in the value for R (L, atm, mol, K). This is where mistakes often occur in calculations.

Example 5.7 Calculate the volume occupied by 6.71 mol H_2 at 2130 torr pressure and 26°C.

Answer 58.8 L

Solution

Given quantities: $P = 2130$ torr
$n = 6.71$ mol
$T = 26°C$
Desired quantity: V (L)

Since R has units of (L atm/mole K), pressure must be expressed in atmospheres and temperature in kelvins. The answer is in liters.

$$P\text{ (atm)} = 2130\ \cancel{\text{torr}} \times \frac{1\text{ atm}}{760\ \cancel{\text{torr}}} = 2.80\text{ atm}$$

$$T\text{ (K)} = 26°\text{C} + 273 = 299\text{ K}$$

Rearranging the ideal gas law equation:

$$V(\text{L}) = \frac{nRT}{P} = \frac{(6.71\ \cancel{\text{mol}})\left(\dfrac{0.0821\ \text{L}\ \cancel{\text{atm}}}{\cancel{\text{mol}}\ \cancel{\text{K}}}\right)(299\ \cancel{\text{K}})}{2.80\ \cancel{\text{atm}}}$$

$$V = 58.8\text{ L}$$

If P, T, or V for a fixed number of moles of gas changes, then a combination of Boyle's law and Charles' law can be used to calculate the change.

Example 5.8 A total of 3.70 moles of krypton gas exerts a pressure of 823 torr in a 750 mL flask at 38°C. The gas expands into a larger flask in which its pressure is 514 torr and its temperature is 27°C. Calculate the final volume occupied by the krypton gas.

Answer 1160 mL

Solution

Given quantities: $P_1 = 823$ torr $\quad P_2 = 514$ torr
$V_1 = 750$ mL
$T_1 = 38°C \quad T_2 = 27°C$
$n_1 = 3.70$ mol $\quad n_2 = 3.70$ mol
Desired quantity: V_2 (mL)

The conditions of P, V, and T change for a fixed number of moles of krypton gas, so we use both Boyle's law and Charles' law for the calculation.

When the pressure decreases, the volume increases proportionally by a ratio of the pressures that is greater than one (Boyle's law), 823 torr/514 torr. The decrease in temperature decreases the volume proportionally by a ratio of the Kelvin temperatures (K) less than one (Charles' law),

$$\frac{(273 + 27)\text{K}}{(273 + 38)\text{K}}$$

Accordingly,

$$V_{\text{new}} = V_{\text{original}} \times \left(\begin{array}{c}\text{pressure correction}\\\text{on the original}\\\text{volume}\end{array}\right) \times \left(\begin{array}{c}\text{temperature correction}\\\text{on the original}\\\text{volume}\end{array}\right)$$

$$V_{\text{new}}\ (\text{mL}) = 750\ \text{mL} \times \frac{823\ \cancel{\text{torr}}}{514\ \cancel{\text{torr}}} \times \frac{(273 + 27)\cancel{\text{K}}}{(273 + 38)\cancel{\text{K}}} = 1160\ \text{mL}$$

SECTION 5.6 MOLECULAR WEIGHTS OF GASES

Objective

- To use molecular weight data in making ideal gas law calculations.

Focus

For a gas in which three of the four variables (P, V, T, and n) are known, the ideal gas law can employ molecular weight (molar mass) data to calculate the fourth value.

Example 5.9 A krypton sample has a density of 3.74 g/L. If the pressure gauge on the cylinder has a reading of 138 kPa, what is the temperature of the gas?

Answer 371 K

Solution The density of Kr gas represents 3.74 g Kr in 1 liter.

Given quantities: (L, atm, and mol)

$$P = 138\ \cancel{\text{kPa}} \times \frac{\text{atm}}{101.3\ \cancel{\text{kPa}}} = 1.36\ \text{atm}$$

$$V = 1.00\ \text{L}$$

$$n = 3.74\ \cancel{\text{g}} \times \frac{\text{mol}}{83.8\ \cancel{\text{g}}} = 0.0446\ \text{mol}$$

Desired quantity: T (K)

Using the ideal gas law,

$$T\text{ (K)} = \frac{PV}{nR} = \frac{(1.36\text{ atm})(1.00\text{ L})}{(0.0446\text{ mol})\left(\frac{0.0821\text{ L atm}}{\text{mol K}}\right)}$$

$$= 371\text{K}$$

SECTION 5.7 CHEMICAL REACTIONS INVOLVING GASES

Objective

- To employ gas law data in calculating quantities of gases that react or are produced in a chemical reaction.

Focus

Balanced equations read in terms of moles. The moles of gas are calculated by using the ideal gas law and/or the fact that 1 mole of gas occupies 22.4 L at STP.

Example 5.10 Calculate the volume of O_2 gas, collected at 748 torr and 22°C, produced from the thermal decomposition of 2.33 g $KClO_3$.

$$2KClO_3(c) \rightarrow 2KCl(c) + 3O_2(g)$$

Answer 0.701 L

Solution We can calculate the volume of O_2 collected at 748 torr and 22°C in two steps. First, calculate the mol O_2, n, produced from 2.33 g $KClO_3$.

Given quantity and units: 2.33 g $KClO_3$
Desired quantity and unit: mol O_2
Conversion factors:

$$1\text{ mol } KClO_3 = 122.6\text{ g } KClO_3$$

$$2\text{ mol } KClO_3 \text{ yields } 3\text{ mol } O_2$$

$$n = 2.33\text{ g } KClO_3 \times \frac{\text{mol } KClO_3}{122.6\text{ g } KClO_3} \times \frac{3\text{ mol } O_2}{2\text{ mol } KClO_3}$$

$$= 0.0285\text{ mol } O_2$$

Second, use the ideal gas law to calculate volume O_2.

Given quantities: (atm, mol, and K) $P = 748\text{ torr} \times \frac{1\text{ atm}}{760\text{ torr}} = 0.984\text{ atm}$

$$n = 0.0285\text{ mol } O_2$$

$$t = 273 + 22°\text{C} = 295\text{ K}$$

Desired quantity: V (L)

$$V\,(\text{L}) = \frac{nRT}{P} = \frac{(0.0285\,\text{mol})\left(\dfrac{0.0821\ \text{L atm}}{\text{mol K}}\right)(295\,\text{K})}{0.984\ \text{atm}}$$

$$= 0.701\ \text{L}$$

SECTION 5.8 DALTON'S LAW OF PARTIAL PRESSURES

Objectives

- To explain that the total pressure of a mixture of gases is equal to the sum of the pressures exerted by the gases.
- To calculate the amount of gas collected over a volatile liquid.

Focus

If two or more nonreacting gases are placed in a closed system, each gas occupies the container's entire volume and exerts its own pressure against the walls of the container. The *total pressure equals the sum of the partial pressures* ($P_{tot} = p_1 + p_2 + p_3 + \cdots$) *exerted by each gas.* The partial pressure of a gas is the pressure it would exert if it were the only gas in the container.

Example 5.11 A 500 mL combustion chamber is filled with methane gas, CH_4, to a pressure of 0.840 atm at 30°C. A total of 0.0400 mole of oxygen gas, O_2, is added to the chamber at 30°C. Calculate the total pressure of gas before combustion.

Answer 2.83 atm

Solution

$$P_{tot} = p_{CH_4} + p_{O_2}$$

From the ideal gas law calculate the pressure each gas exerts

$$p_{O_2} = \frac{nRT}{V} = \frac{(0.0400\,\text{mol})\left(\dfrac{0.0821\,\text{L atm}}{\text{mol K}}\right)(303\,\text{K})}{0.500\,\text{L}} = 1.99\ \text{atm}$$

p_{CH_4} = 0.840 atm (given quantity)
P_{tot} = 0.840 atm + 1.99 atm = 2.83 atm

SECTION 5.9 IDEAL GASES: THE KINETIC MOLECULAR THEORY

Objective

- To explain the effect of changes in pressure, volume, temperature, and moles of gases in terms of the kinetic molecular theory.

Focus

A knowledge of the postulates in the kinetic molecular theory is absolutely necessary. According to the theory, ideal gases can be interpreted as having the following properties:

1. Ideal gas molecules have no volume. Therefore gases can be easily compressed by reducing only empty space.
2. Ideal gas molecules have no intermolecular attractive or repulsive forces.
3. Pressure is caused by molecules hitting the walls of the container. If the volume of the container decreases, the frequency of collisions with the walls increases; the result is an increase in pressure (Boyle's law)
4. Temperature also affects pressure: the higher the temperature, the greater is the speed, u, of the molecules and therefore the force and the frequency of the wall collisions. To maintain a constant pressure at higher temperatures, the volume of gas must increase (Charles' law).
5. The average kinetic energy, E_K, of the molecules in gases, liquids, and solids is constant at constant temperature:

$$E_K = \tfrac{1}{2}mu^2 \propto T(K)$$

Perhaps this is the most important interpretation resulting from the kinetic molecular theory.

Example 5.12 Explain each of the following in terms of the kinetic molecular theory:

(a) Pressure is a result of ____________________.

(b) As the volume of a gas increases, at constant temperature, the pressure of the gas __________ because ____________________.

(c) As the temperature decreases at constant pressure, the volume must __________ because ____________________.

(d) The pressure of a gas __________ as a result of a temperature increase because ____________________.

(e) The frequency of wall collisions __________ as a result of added gas molecules at constant volume and temperature.

(f) If the gas is confined to a rigid container, what effect does the application of an external pressure have on the internal pressure of a gas?

(g) What can be done to a gaseous system that will increase the frequency of wall collisions by gas molecules?

(i) ____________________.

(ii) ____________________.

(h) Samples of xenon gas, Xe, and helium gas, He, are each placed in identical containers at the same temperature and pressure.

(i) Comment on the relative number of gaseous atoms in each system.

(ii) Comment on the relative average kinetic energy of the gaseous atoms in the two systems.

(iii) Comment on the relative average speed of the gaseous atoms.

Answer (a) *An increased number of collisions by the molecules with the walls of the container*: the frequency with which molecules hit each other has no effect.

(b) *Decreases: the frequency with which molecules hit the walls decreases*.

(c) *Decreases: as the temperature is lowered the frequency with which molecules hit the walls decreases*. To maintain the same number of collisions, the walls must be moved in closer, which means that the volume must be decreased.

(d) *Increases: at a higher temperature the force and the frequency with which molecules hit the wall increases*.

(e) *Increases*: the more molecules in a system, the greater frequency of collisions with the wall.

(f) No effect. If the container is rigid, the internal molecules still hit the walls with the same force and frequency.

(g) (i) *Increase the temperature.*

(ii) *Decrease the volume.*

(h) (i) Same number of atoms in each. Avogadro's law: If P, V, and T are the same, then n is the same.

(ii) If the temperatures are the same, the average kinetic energies are the same because $E_K \propto T$ (K)

(iii) At the same temperature:

$$E_K\,(\mathrm{He}) = E_K\,(\mathrm{Xe})$$
$$\tfrac{1}{2}mu_{\mathrm{He}}^2 = \tfrac{1}{2}mu_{\mathrm{Xe}}^2$$

Since $m_{\mathrm{Xe}} > m_{\mathrm{He}}$, then $u_{\mathrm{He}} > u_{\mathrm{Xe}}$ for the equality to be valid.

5.10 GRAHAM'S LAW

Objective

- To learn how the rate of effusion (or diffusion) of a gas is related to its density and molecular weight.

Focus

Effusion is the passage of gas molecules through an opening (orifice), whereas diffusion is the mere dispersal of molecules. Graham's law states that the rate of effusion of a gas molecule is proportional to the inverse of the square root of its molar mass or density.

$$\textbf{rate} \propto \frac{1}{\sqrt{\textbf{density}}} \propto \frac{1}{\sqrt{\textbf{molar mass}}}$$

Rearranging, we have

$$\text{rate} \times \sqrt{\text{density}} = \text{constant}$$

or

$$\text{rate} \times \sqrt{\text{molar mass}} = \text{constant}$$

for all gases.

From these expressions the greater the molar mass or density of a substance, the slower is its rate of effusion or diffusion. In other words, heavyweights move slower than lightweights.

Example 5.13 The two gases propane, C_3H_8, and butane, C_4H_{10}, are separated by the effusion process. The density of propane is 1.96 g/L and that of butane is 2.59 g/L at standard conditions. Which gas effuses faster? By what factor?

Answer C_3H_8; 1.15 times faster.

Solution Because C_3H_8 is a lighter gas, it effuses faster by a factor equal to the inverse ratio of the square roots of their densities.

$$\frac{\text{rate } C_3H_8}{\text{rate } C_4H_{10}} = \sqrt{\frac{2.59 \text{ g/L}}{1.96 \text{ g/L}}} = 1.15$$

Rate C_3H_8 = 1.15 rate C_4H_{10}; that is, if butane's rate of effusion is multiplied by a factor of 1.15, it then equals propane's rate of effusion.

SECTION 5.11 DISTRIBUTION OF MOLECULAR SPEEDS

Objective

- To explain how molecular collisions and temperature affect the speed of molecules.

Focus

The speed of *individual* molecules, at constant temperature, varies significantly within a system because of individual differences in their intermolecular collisions. The average speed for all the molecules in a system, however, remains constant, with some moving faster and others moving slower than average.

Any shift in the distribution of molecular speeds is due to temperature effects. According to the kinetic molecular theory

$$E_K = \tfrac{1}{2}mu^2 \propto T(\text{K})$$

The temperature is proportional to the square of the speed for a given molecule (or molecular system).

Example 5.14 If the typical average molecular speed is about 400 m/sec for nitrogen at 0°C (see text, Section 5.10 marginal note), what is the average speed of nitrogen molecules at 23°C?

Answer 417 m/sec

Solution Temperature (K) is proportional to the square of the molecular speed:

$$\frac{T_1}{T_2} = \frac{u_1^2}{u_2^2}$$

Rearranging,

$$u_2 = \sqrt{\frac{u_1^2 T_2}{T_1}} = \sqrt{\frac{(400\text{ m/sec})^2(273 + 23)\cancel{K}}{(273 + 0)\cancel{K}}} = 417\text{ m/sec}$$

SECTION 5.12 DEVIATION FROM IDEAL BEHAVIOR

Objectives

- To explain the difference between ideal and real gases in terms of the kinetic molecular theory.
- To define conditions for "near" ideal gas behavior.

Focus

Two postulates of the kinetic molecular theory do not apply to real molecules in a gaseous system.

1. Real gas molecules *do* attract each other. This fact is observed when gases liquefy. Since intermolecular forces do exist, "sticky collisions" occur between real molecules. (See text, Figure 5.13b.) This effectively reduces the number of molecules in the system. A smaller number of molecules in the system lessens the frequency of molecular collisions with the walls; therefore the pressure is lower than that of an ideal gas. This effect is significant only at low pressures. The result is a negative deviation from ideal gas behavior. (See text, Figure 5.16.)
2. Real gas molecules *do* occupy volume. For this reason the real gas exerts a pressure that is greater than that of an ideal gas. Because molecules in a real gas system have less space in which to move (see text, Figure 5.15), they hit the walls more frequently, thus causing a greater pressure. This effect is more significant at very high pressures where the molecules are more crowded. This results in a positive deviation from ideal behavior. (See text, Figure 5.16.)

It must be noted, however, that at one atmosphere pressure or less and at room temperature or greater, the deviation from ideal behavior for a real gas is less than 1% for most gases. Nearly ideal behavior occurs for real gases at high temperatures and low pressures. Under these conditions, the molecules spread out leaving more space between individual molecules. Thus there are fewer "sticky collisions" and the size of molecules becomes insignificant.

QUESTIONS

Section 5.1

5.1 The superduper weatherman says the barometer reading is 29.9 inches Hg and rising. Convert this reading to atmospheres.

5.2 The pressure of nitrogen gas in a steel cylinder reads 83.5 kPa. What pressure is this in units of torr?

5.3 What is the minimum length, in meters, of glass tubing for a barometer that is required to measure 1 atm of pressure when water (density of 1 g/cm^3) is substituted for mercury (density of 13.6 g/cm^3)?

Section 5.2

5.4 Specify the conditions under which Boyle's law is valid.

5.5 Fluorine gas occupies 200 mL at a pressure of 1478 torr. What pressure (in atmospheres) does the fluorine gas exert if it expands into a 6.00 L reaction vessel?

5.6 A combustible gas is stored in a 1200 mL vessel at 40°C and a pressure of 200 kPa. What is the volume of the reaction vessel if the gas pressure is reduced to 0.33 kPa at 40°C before combustion?

5.7 A natural gas well with an approximate volume of 1.7×10^9 L has a reservoir pressure of 3.2 atm. If all the gas is stored in a steel tank at a pressure of 5.1 atm, to store the natural gas what must be the minimum volume of the tank?

Section 5.3

5.8 State the conditions in which Charles' law is valid.

5.9 Define the standard conditions of a gas.

5.10 Propane gas is confined to a 200 mL cylinder equipped with a movable piston. If the temperature of the gas is changed from 18°C to 300°C at constant pressure, what is the adjusted volume occupied by the gas?

5.11 Chlorine gas is stored in a cylinder at −19°C at a pressure of 600 torr. The gas is then transferred to a 7.0 L flask and heated to 107°C where again the pressure is recorded to be 600 torr. What is the original volume of the cylinder?

5.12 Two cubic meters of acetylene at −48°C expand into a 12 m^3 combustion chamber. Assuming no change in pressure, what is the temperature of the acetylene gas in the combustion chamber?

Section 5.4

5.13 A 62.1 mL flask has a total of 1.67×10^{21} molecules of methane, CH_4, at STP. Calculate the volume occupied by 9.77×10^{23} molecules of CH_4 at STP.

5.14 A total of 6.46×10^{23} molecules of oxygen, O_2, occupy a 4.00 L flask at 6.00 atm and 273 K. Calculate the number of molecules in a 34 L reaction vessel that exerts a pressure of 6.00 atm at 273 K.

5.15 Two gases are separately confined to 1 L flasks at STP. The mass of gas A is 1.43 g and that of gas B is 0.179 g. If the molecular weight of gas A is 32, what is the molecular weight of gas B?

5.16 For separate flasks the following is recorded:
Flask A: 2.63 mol O_2
Flask B: 3.18×10^{24} molecules CH_4
Flask C: 63.1 L C_3H_8 at standard conditions
Calculate the total volume the three gases occupy at STP.

Section 5.5

5.17 A total of 0.583 g xenon gas occupies a volume of 400 mL at 23°C. What pressure does the gas exert?

5.18 (a) Carbon monoxide gas, CO, is purchased in a 0.425 L lecture bottle at a pressure of 5.00 atm at 23°C. What volume does the CO occupy when the pressure is changed to 6000 torr at a temperature of −12°C?

(b) What mass, in grams, of CO is purchased?

Section 5.6

5.19 A 21.3 g sample of a known carcinogen is confined to a 7.73 L vessel at a 0.880 atm and 30°C. Calculate the molecular weight of the carcinogen.

Section 5.7

5.20 A volume of 600 L C_4H_{10} at 150°C and 2.33 atm is burned with excess oxygen.

$$2C_4H_{10}(g) + 13O_2(g) \rightarrow 8CO_2(g) + 10H_2O(g)$$

(a) Calculate the mass CO_2 released to the atmosphere.

(b) What volume does the CO_2 occupy if released to the atmosphere at a pressure of 705 torr and 16°C?

Section 5.8

5.21 A volume of 25.1 mL of oxygen gas, O_2, is collected over water at 19°C. The pressure of the collected gas is adjusted to that of atmospheric pressure, 749.0 torr. The vapor pressure of H_2O at 19°C is 16.5 torr (see Figure 5.1).

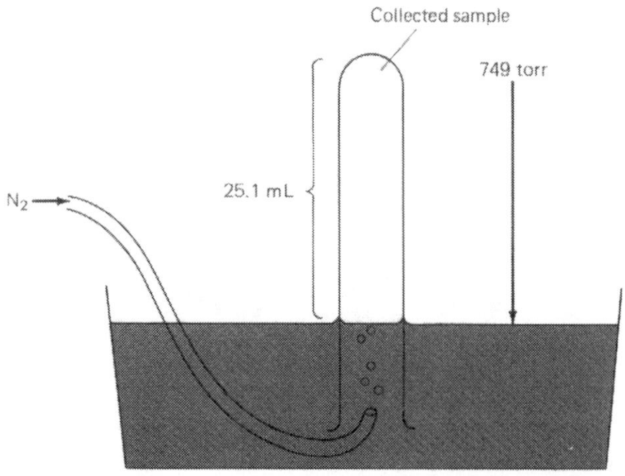

Figure 5.1 Collection of O_2 gas over water.

(a) What substances are present in the collected sample?

________________ and ________________.

(b) What volume does the O_2 occupy?

(c) What volume does the water vapor occupy?

(d) What pressure does the water vapor exert?

(e) What is the partial pressure of O_2 in the collected sample?

(f) What volume does the O_2 alone (if passed through a drying tube to remove the water) occupy at STP?

(g) How many moles of O_2 are collected?

Section 5.10

5.22 A volume of 48.7 mL of carbon dioxide, CO_2, passes through an orifice in 5.70 minutes. A second gas requires 13.4 minutes for 52.5 mL of gas to pass through the same one. If the density of CO_2 is 1.96 g/L (STP), what is the density of the unknown gas?

5.23 A total mass of 0.797 g krypton passes through an orifice in 16.3 minutes.
(a) If the same number of moles of an unknown gas passes through the same orifice in 10.7 minutes, what is the total mass of the unknown gas sample? (*Hint*: the rate of passage of molecules through an orifice is expressed in units of "per time.")
(b) What is the molecular weight of the second gas?

Section 5.11

5.24 The peak of the distribution curve (see text, Figure 5.6) refers to the "most probable speed"; the "average speed" lies to the right of the peak. Explain.

5.25 How does the area under the distribution curve vary with temperature?

Section 5.12

5.26 A 1 L vessel contains 1 mole of an ideal gas at 273 K. Another 1 L vessel contains 1 mole of methane, CH_4, at 273 K.

(a) Which vessel has the most free space between molecules? ________

(b) Which molecular system exerts the greatest pressure? ________

5.27 When an ideal gas undergoes a rapid expansion, there is no temperature change. On the other hand, most real gases cool when they expand. Explain.

ANSWERS

5.1 0.999 atm

5.2 626 torr

5.3 10.3 m

5.4 Boyle's law is valid for an ideal gas at constant T and n

5.5 0.0648 atm

5.6 7.27×10^5 mL

5.7 1.07×10^9 L

5.8 Charles' law is valid for an ideal gas at constant P and n

5.9 Standard conditions at 0°C (273K) and 1 atm pressure

5.10 394 mL

5.11 4.7 L

5.12 1350 K, 1077°C

5.13 3.63×10^4 mL

5.14 5.49×10^{24} molecules

5.15 4.00

5.16 240 L

5.17 0.270 atm

5.18 (a) 0.237 L CO
(b) 2.45 g CO

5.19 77.8 g/mol

5.20 (a) 7.09 kg CO_2
(b) 4.12×10^3 L

5.21 (a) N_2 gas and H_2O vapor
(b) 25.1 mL
(c) 25.1 mL
(d) 16.5 torr
(e) 732.5 torr
(f) 22.6 mL
(g) 1.01×10^{-3} mol O_2

5.22 9.30 g/L

5.23 (a) 0.343 g
(b) 36.1 g/mol

5.24 The "average speed" lies to the right because some molecules have a very high speed; it is the average speed of all molecules from zero to infinity. The "most probable speed" is the speed for the greatest number of molecules.

5.25 The area under the curve does not vary with temperature, there is merely a greater distribution of speeds.

5.26 (a) the ideal gas
(b) the ideal gas

5.27 Real molecules absorb energy and thereby cool the system when they expand because energy is required to overcome the intermolecular forces that are keeping the molecules attracted to each other.

Chapter 6

THERMOCHEMISTRY

SECTION 6.1 INTRODUCTION

Objective

- To explain how energy can be evolved or absorbed in a chemical reaction.

Focus

In a chemical reaction, two things must happen: bonds between atoms must break, requiring an absorption of energy, and bonds must form, resulting in a release of energy. The process with the greater energy change determines whether heat is evolved (**exothermic**) or absorbed (**endothermic**). The resultant heat change is the heat of reaction.

SECTION 6.2 HEAT MEASUREMENTS

Objectives

- To explain the concept of heat flow.
- To calculate heat and temperature changes associated with the specific heat of a substance.

Focus

Heat, expressed in units of joules or calories, flows from a warmer object to a cooler one. Heat flows; cold does not flow. Technically, cold is the absence of heat.

The **specific heat** of a substance is the amount of heat required to change the temperature of 1 gram by 1°C.

$$\text{sp ht} = \frac{\text{heat (J)}}{\text{mass (g)} \times \Delta t(°\text{C})}$$

Example 6.1 Calculate the mass of copper that can be heated from 22°C to 42°C if 60 kJ are absorbed.

Answer 7.9 kg

Solution From Table 6.1, the specific heat of copper is 0.38 J/(g °C). The quantity of heat, the temperature change, and the specific heat determine the mass of Cu heated. Use dimensional analysis.

Given quantity and units:

$$\Delta t = t_2 - t_1 = 42°\text{C} - 22°\text{C} = 20°\text{C}$$

$$q = 60 \text{ kJ}$$

$$\text{sp ht} = 0.38 \text{ J/(g °C)}$$

Desired quantity and units: g Cu
Conversion factor:

$$10^3\text{J} = \text{kJ}$$

$$\text{mass Cu} = 60\,\cancel{\text{kJ}} \times \frac{10^3\cancel{\text{J}}}{\cancel{\text{kJ}}} \times \frac{\text{g} \times \cancel{°\text{C}}}{0.38\,\cancel{\text{J}}} \times \frac{1}{20\cancel{°\text{C}}} = 7.9 \times 10^3 \text{ g} = 7.9 \text{ kg}$$

or using the formula,

$$m = \frac{q}{\text{sp ht} \times \Delta t} = \frac{6.0 \times 10^4\cancel{\text{J}}}{\left(\frac{0.38\,\cancel{\text{J}}}{\text{g}\,\cancel{°\text{C}}}\right)(20\cancel{°\text{C}})} = 7.9 \times 10^3 \text{ g} = 7.9 \text{ kg}$$

Example 6.2 Suppose the heat in Example 6.1 is supplied from a 5.0 gal hot water bath, initially at 100°C. What is the final temperature of the water after the 60 kJ of heat are supplied? Assume the density of water is 0.97 g/mL at this temperature.

Answer 99.2°C

Solution The water loses heat; therefore

$$q = -60 \text{ kJ} = -6.0 \times 10^4 \text{ J}$$

The mass of 5.0 gal water with a density of 0.97g/mL is

$$\text{mass } H_2O = 5.0\,\cancel{\text{gal}} \times \frac{4\cancel{\text{qt}}}{\cancel{\text{gal}}} \times \frac{\cancel{\text{L}}}{1.06\,\cancel{\text{qt}}} \times \frac{1\,\cancel{\text{mL}}}{10^{-3}\cancel{\text{L}}} \times \frac{0.97 \text{ g}}{1\,\cancel{\text{mL}}} = 1.8 \times 10^4 \text{ g}$$

Note the use of dimensional analysis in converting gal H_2O to g H_2O. Since q = sp ht × mass × Δt, then

$$-6.0 \times 10^4\text{J} = \frac{4.184 \text{ J}}{\cancel{\text{g}}\,°\text{C}} \times 1.8 \times 10^4\cancel{\text{g}} \times (y - 100)\cancel{°\text{C}}$$

$$-0.80°\text{C} = (y - 100)°\text{C}$$

$$y = 99.2°\text{C}$$

SECTION 6.3 MEASUREMENTS OF HEAT OF REACTION: THERMOCHEMICAL EQUATIONS

Objectives

- To define the heat capacity of a calorimeter.
- To calculate the molar heat of combustion for a reaction from experimental data.
- To explain the difference between ΔE and ΔH.
- To write and explain thermochemical equations.

Focus

The **heat capacity** of a calorimeter is the amount of energy required to raise the temperature of the calorimeter 1°C,

$$\text{heat capacity} = \text{J/°C}$$

The **molar heat of combustion**, the heat evolved in the combustion of 1 mole of a substance, can be measured with a constant volume calorimeter. At constant temperature and volume this energy change is written as

$$\Delta E = E_{\text{products}} - E_{\text{reactants}} = q_V.$$

At constant temperature and pressure, the heat evolved or absorbed in the reaction is expressed by its change in enthalpy, ΔH

$$\Delta H = H_{\text{products}} - H_{\text{reactants}} = q_p$$

For exothermic reactions, the values of ΔE and ΔH have negative values because the energy or the enthalpy of the products is less than that for the reactants.

For endothermic reactions, ΔE and ΔH have positive values because the energy or enthalpy of the products is greater than that for the reactants.

When writing thermochemical equations, note the physical state of each substance; obviously, the energy in 1 mole of steam at 100°C is greater than that in 1 mole of liquid water at 100°C.

Example 6.3 A mass of 10.0 g $B_5H_9(g)$ reacts spontaneously with O_2 in a constant pressure calorimeter that has a heat capacity of 817 kJ/°C. The temperature rise in the calorimeter is 0.875°C. Calculate ΔH for the reaction

$$2B_5H_9(g) + 12O_2(g) \rightarrow 5B_2O_3(c) + 9H_2O(\ell) \qquad \Delta H = ?$$

Answer $\Delta H = -9.0 \times 10^3$ kJ

Solution Because the temperature increases, the reaction is exothermic. The molecular weight of B_5H_9 is 63.0 and the calorimeter absorbs 817 kJ per each 1°C temperature rise. For a 0.875°C temperature change, the calorimeter absorbs

$$\frac{817\ \text{kJ}}{\cancel{°C}} \times 0.875\cancel{°C} = 715\ \text{kJ}$$

Therefore, 10.0 g B_5H_9 produce 715 kJ. The heat evolved per mole B_5H_9 is

$$-\frac{715\ \text{kJ}}{10.0\ \text{g}} \times \frac{63.0\ \text{g}}{\text{mol}} = -4.50 \times 10^3\ \frac{\text{kJ}}{\text{mol } B_5H_9}$$

In the reaction 2 mol B_5H_9 react; the ΔH for the reaction is therefore

$$2\ \text{mol } B_5H_9 \times \left(-\frac{4.50 \times 10^3\ \text{kJ}}{\text{mol } B_5H_9}\right) = -9.0 \times 10^3\ \text{kJ}$$

SECTION 6.4 HESS'S LAW

Objective

- To calculate the heat of reaction, ΔH, using data from other thermochemical equations (Hess's law).

Focus

By rearranging and algebraically adding thermochemical equations, the ΔH for a reaction can be calculated. Experimentally, this procedure is valid because reactants may form products in one step or in a series of steps. What we observe experimentally is the total energy (or enthalpy) change for a reaction, not the change for each step.

Example 6.4 From the following thermochemical equations:

1. $C(graphite) + O_2(g) \rightarrow CO_2(g) \qquad \Delta H = -394\ \text{kJ}$
2. $C_3H_7OH(\ell) + \frac{9}{2} O_2(g) \rightarrow 3CO_2(g) + 4H_2O(\ell) \qquad \Delta H = -2010\ \text{kJ}$
3. $H_2O(\ell) \rightarrow H_2(g) + \frac{1}{2} O_2(g) \qquad \Delta H = +286\ \text{kJ}$

determine the heat of reaction for

$$3C(graphite) + 4H_2(g) + \frac{1}{2} O_2(g) \rightarrow C_3H_7OH(\ell) \qquad \Delta H = ?$$

Answer $\Delta H = -316$ kJ

Solution Arrange the three thermochemical equations so that they add up to the desired equation.

(a) We need 3C(*graphite*) on the left, so multiply Equation 1 by 3. This means we must also multiply the ΔH value by 3.

$$3C(c) + 3O_2(g) \rightarrow 3CO_2(g) \qquad \Delta H = 3(-394\ \text{kJ})$$

(b) We need $C_3H_7OH(\ell)$ on the right, so reverse Equation 2 and change the sign of ΔH.

$$3CO_2(g) + 4H_2O(\ell) \rightarrow \frac{9}{2} O_2(g) + C_3H_7OH(\ell) \qquad \Delta H = +2010\ \text{kJ}$$

(c) Notice that $4H_2(g)$ must be on the left; so reverse Equation 3 and multiply by 4:

$$4H_2(g) + 2O_2(g) \rightarrow 4H_2O(\ell) \qquad \Delta H = 4(-286\ \text{kJ})$$

Add the three equations algebraically and cancel like moles of substances appearing on both sides of the equation.

$$3C(graphite) + \cancel{3O_2(g)} + \cancel{3CO_2(g)} + \cancel{4H_2O(\ell)} + 4H_2(g) + \overset{\frac{1}{2}O_2(g)}{\cancel{2O_2(g)}} \rightarrow$$

$$\cancel{3CO_2(g)} + \cancel{\tfrac{9}{2}O_2(g)} + C_3H_7OH(\ell) + \cancel{4H_2O(\ell)}$$

$$3C(graphite) + 4H_2(g) + \tfrac{1}{2}O_2(g) \rightarrow C_3H_7OH(\ell)$$

Add the ΔH for each of the reactions to obtain ΔH for the desired reaction:

$$\begin{aligned}\Delta H_{\text{reaction}} &= 3(-394 \text{ kJ}) + (2010 \text{ kJ}) + 4(-286 \text{ kJ}) \\ &= -316 \text{ kJ}\end{aligned}$$

SECTION 6.5 HEATS OF FORMATION

Objective

- To calculate the heat of reaction, ΔH, from the heats of formation, ΔH_f, for the substances appearing in the balanced equation.

Focus

The **heat of reaction**, ΔH, is calculated by subtracting the sum of the heats of formation of the reactants from the sum of the heats of formation of the products.

$$\Delta H_{\text{reaction}} = \Sigma\, \Delta H_f\,(\text{products}) - \Sigma\, \Delta H_f\,(\text{reactants})$$

where Σ represents a summation sign.

For any substance its **heat of formation**, ΔH_f, is the heat absorbed or released when a compound is formed from its elements, each being in its normal physical (or standard) state at 25°C and 1 atm pressure. ΔH_f for all *elements* in their normal physical state at 25°C and 1 atm pressure equals zero. In Example 6.4 we calculated the ΔH_f for $C_3H_7OH(\ell)$ at 25°C and 1 atm pressure.

Example 6.5 Using the text, Appendix D, calculate the heat of reaction for

$$4NH_3(g) + 7O_2(g) \rightarrow 4NO_2(g) + 6H_2O(\ell) \qquad \Delta H = ?$$

Answer $\Delta H = -1397.82$ kJ

Solution

$$\begin{aligned}\Delta H_{\text{reaction}} &= \Sigma\, \Delta H_f\,(\text{products}) - \Sigma\, \Delta H_f\,(\text{reactants}) \\ \Delta H_{\text{reaction}} &= [4\Delta H_{f,\,NO_2\,(g)} + 6\Delta H_{f,\,H_2O(\ell)}] - [4\Delta H_{f,\,NH_3\,(g)} + 7\Delta H_{f,\,O_2\,(g)}] \\ &= [4(+33.13 \text{ kJ}) + 6(-285.836 \text{ kJ})] - [4(-46.11 \text{ kJ}) + 4(0)]\end{aligned}$$

Watch the signs!

$$\begin{aligned}\Delta H_{\text{reaction}} &= -1582.26 \text{ kJ} - [-184.44 \text{ kJ}] \\ &= -1397.82 \text{ kJ}\end{aligned}$$

SECTION 6.6 BOND ENERGY

Objectives

- To calculate a bond energy.
- To employ bond energies in calculating a heat of reaction.

Focus

Bond energy is the energy required to break one bond in a gaseous molecule into separated gaseous atoms: the larger the bond energy, the stronger and more stable the bond is. Bond energies may be determined experimentally or by using Hess's law.

Example 6.6 From the following data, calculate the H—Cl bond energy.

	ΔH_f
HCl(*g*)	−92.307 kJ
H(*g*)	+217.965 kJ
Cl(*g*)	+121.39 kJ

Answer $\Delta H = +431.66$ kJ

Solution We want to calculate ΔH for

$$HCl(g) \rightarrow H(g) + Cl(g)$$

$$\begin{aligned} \Delta H_{\text{reaction}} &= \Sigma\, \Delta H_f\,(\text{products}) - \Sigma\, \Delta H_f\,(\text{reactants}) \\ &= [\Delta H_{f,\,H(g)} + \Delta H_{f,\,Cl(g)}] - [\Delta H_{f,\,HCl(g)}] \\ &= [217.965\text{ kJ} + 121.39\text{ kJ}] - [-92.307\text{ kJ}] \\ &= +431.66\text{ kJ} \end{aligned}$$

The heat of reaction, ΔH, is also calculated from the bond energies involved in the breaking and making of bonds.

Example 6.7 Calculate the heat of formation of gaseous cyclohexane from the bond energy data in the text, Table 6.2, Appendix D, and the equation

$$C(\textit{graphite}) \rightarrow C(g) \qquad \Delta H = +718.4\text{ kJ}$$

The formula for cyclohexane is

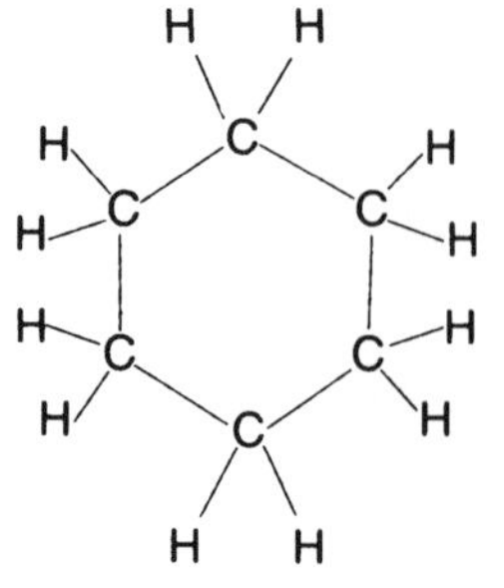

Answer $\Delta H = -100$ kJ

Solution The heat of formation is the energy released (exothermic) when cyclohexane is formed from its elements at 25°C and 1 atm pressure. We want to calculate ΔH for the equation

$$6C(\mathit{graphite}) + 6H_2(g) \rightarrow C_6H_{12}(g)$$

But bond energy data give ΔH for the reaction

$$6C(g) + 12H(g) \rightarrow C_6H_{12}(g)$$

In this reaction 6 C — C bonds form. From Table 6.2, the C — C bond energy is 339 kJ; the energy released is

$$6 \times (-339 \text{ kJ}) = -2034 \text{ kJ}$$

The C — H bond energy is 416 kJ/mol. The formation of 12 C — H bonds releases

$$12 \times (-416 \text{ kJ}) = -4992 \text{ kJ}$$

Addition of these enthalpy changes gives ΔH for

$$6C(g) + 12H(g) \rightarrow C_6H_{12}(g) \qquad \Delta H = -7026 \text{ kJ}$$

Now change C(*graphite*) to C(g) and $H_2(g)$ to H(g):

$$6C(\mathit{graphite}) \rightarrow 6C(g) \qquad \Delta H = 6(718.4 \text{ kJ}) = +4310 \text{ kJ}$$

$$6H_2(g) \rightarrow 12H(g) \qquad \Delta H = 12(218 \text{ kJ}) = +2616 \text{ kJ}$$

Add the last three equations and the corresponding ΔH's:

$$6C(\mathit{graphite}) + 6H_2(g) \rightarrow C_6H_{12}(g)$$

$$\Delta H = -7026 \text{ kJ} + 4310 \text{ kJ} + 2616 \text{ kJ}$$

$$= -100 \text{ kJ}$$

SECTION 6.7 THE FIRST LAW OF THERMODYNAMICS

Objectives

- To explain the first law of thermodynamics: the change in energy of a system as a result of heat and work changes.
- To calculate how the total energy of a system is affected by the addition or removal of heat and by the amount of work that is done on or by the system.

Focus

The first law of thermodynamics stated mathematically is: $\mathbf{\Delta E = q - w}$

$+q$ = heat absorbed (energy gained) by the system
$-q$ = heat evolved (energy lost) by the system
$+w$ = work done by (energy lost) the system
$-w$ = work done on (energy gained) the system
ΔE = change in the energy resulting from heat gained ($+q$) or lost ($-q$) and the work done by ($+w$) or on ($-w$) the system. The system is merely the reaction being studied.

Example 6.8 Calculate w for a system that absorbs 260 kJ of heat and ΔE is 157 kJ. Is work done on or by the system?

Answer $w = +103$ kJ; work is done by the system.

Solution

$$\begin{aligned} \Delta E &= q - w \\ q &= +260 \text{ kJ; energy added} \\ 157 \text{ kJ} &= 260 \text{ kJ} - w \\ w &= +103 \text{ kJ.} \end{aligned}$$

Since w is positive, work is done *by* the system.

Example 6.9 In a combustion cylinder, 9000 kJ of heat are produced and, in addition, a piston compresses the gases to an equivalency of 700 kJ of work. What is the total energy change of the system?

Answer $\Delta E = 9700$ kJ

Solution Since 700 kJ of work is done on the system, w is negative:

$$\begin{aligned} w &= -700 \text{ kJ} \\ q &= +9000 \text{ kJ} \quad \text{(Heat is added to the system)} \\ \Delta E &= q - w \\ &= 9000 \text{ kJ} - (-700 \text{ kJ}) = 9700 \text{ kJ} \end{aligned}$$

ΔE is dependent only on the initial and final states and is independent of the pathway between the two, q and w are processes by which the system goes from the initial to the final state. For example, if $\Delta E = -75$ kJ for the system, the total heat absorbed or evolved by the system and the total work done by or on the system may be any amount (even up to 100,000,000's of kilojoules), but the value of $q - w$ is nevertheless equal to -75 kJ.

SECTION 6.8 THE DIFFERENCES BETWEEN ΔE AND ΔH

Objective

- To explain the differences between ΔE and ΔH.

Focus

ΔE is a heat of reaction measured in a constant volume container (usually in a bomb calorimeter shown in the text, Figure 6.1) at constant temperature.

ΔH is a heat of reaction measured at constant temperature and pressure, generally atmospheric pressure.

$\Delta E = \Delta H$ for reactions in which only liquids or solids are reactants and products and when the number of moles of gaseous reactants equals the number of moles of gaseous products.

However, when a change in the number of moles of gas in a reaction occurs, ΔE does not equal ΔH because of the work, w, that is done *by* the gas in expanding against the atmosphere. Remember the first law: $\Delta E = q - w$.

For an expanding system, that is, moles of gaseous products are greater than moles of gaseous reactants, ΔE is less than ΔH, because the system does work against the atmosphere; w is positive and therefore $\Delta E = q - (+w)$ is less than ΔH.

For a contracting system, that is, moles of gaseous products are less than moles of gaseous reactants, ΔE is greater than ΔH, because the atmosphere does work *on* the system. In this case, w is negative and therefore $\Delta E = q-(-w)$ is greater than ΔH.

QUESTIONS

Section 6.2

6.1 Calculate the quantity of heat, in kJ, required to heat 250 mL of water from 18°C to 100°C. Assume the density of the water to be 1.00 g/mL.

6.2 What is the final temperature of a 500 g iron rod at an initial temperature of 20.0°C if it absorbs 200 kJ of energy?

Section 6.3

6.3 The known molar heat of combustion of solid benzoic acid, C_6H_5COOH, is 3227 kJ. The combustion of 0.4339 g C_6H_5COOH (molecular weight = 122.12) increases the temperature of a calorimeter from 22.635° to 24.122°C. What is the heat capacity of the calorimeter?

6.4 A mass of 1.170 g naphthalene, $C_{10}H_8$ (molecular weight = 128.2), is oxidized in a constant volume (bomb) calorimeter that has a heat capacity of 24.3 kJ/°C. The temperature of the calorimeter increases from 17.73°C to 19.66°C. Calculate ΔE for the oxidation of 1 mole of naphthalene.

Section 6.4

6.5 From the following data, calculate ΔH for

$$CS_2(\ell) + 3O_2(g) \rightarrow CO_2(g) + 2SO_2(g) \qquad \Delta H = ?$$
$$C(graphite) + O_2(g) \rightarrow CO_2(g) \qquad \Delta H = -393.5 \text{ kJ}$$
$$S(c) + O_2(g) \rightarrow SO_2(g) \qquad \Delta H = -296.8 \text{ kJ}$$
$$C(graphite) + 2S(c) \rightarrow CS_2(\ell) \qquad \Delta H = +\ 87.9 \text{ kJ}$$

6.6 Calculate ΔH for the reaction

$$C(graphite) + H_2O(g) \rightarrow CO(g) + H_2(g) \qquad \Delta H = ?$$

from the following thermochemical data:

$$H_2(g) + \frac{1}{2} O_2(g) \rightarrow H_2O(\ell) \qquad \Delta H = -285.8 \text{ kJ}$$

$$CO(g) + \frac{1}{2} O_2(g) \rightarrow CO_2(g) \qquad \Delta H = -283.0 \text{ kJ}$$

$$C(graphite) + O_2(g) \rightarrow CO_2(g) \qquad \Delta H = -393.5 \text{ kJ}$$

$$H_2O(g) \rightarrow H_2O(\ell) \qquad \Delta H = -44.0 \text{ kJ}$$

Section 6.5

6.7 For the reaction

$$4HNO_3(\ell) + 5N_2H_4(\ell) \rightarrow 7N_2(g) + 12H_2O(g)$$

calculate the ΔH, given the standard heats of formation:

$HNO_3(\ell)$, −172 kJ $N_2H_4(\ell)$, +50.2 kJ $H_2O(g)$, −242 kJ

6.8 (a) Calculate ΔH for the reaction

$$3OF_2(\ell) + 4NH_3(\ell) \rightarrow 2N_2(g) + 3H_2O(g) + 6HF(g) \qquad \Delta H = ?$$

The standard ΔH_f's are:

$OF_2(\ell)$ = +12.6 kJ; $NH_3(\ell)$ = −66.9 kJ; $H_2O(g)$ = −242 kJ; and $HF(g)$ = −271 kJ.

(b) How much $OF_2(\ell)$, in grams, is consumed in the production of 1000 kJ energy?

Section 6.6

6.9 Determine ΔH_f for C_2H_5SH, 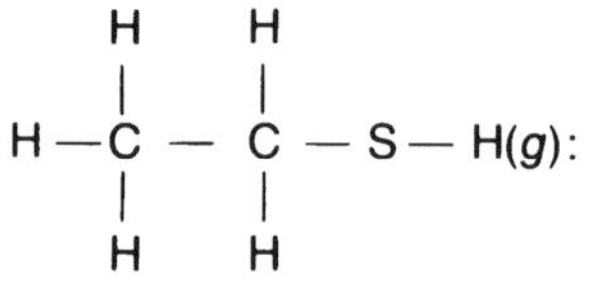

$$2C\ (graphite) + S(c) + 3H_2(g) \rightarrow C_2H_5SH(g)$$

using the bond energy values in the text, Table 6.2, the bond energy values for C — S (+259 kJ) and S — H (+339 kJ), and the following ΔH_f values:

$C(graphite) \rightarrow C(g)$	+720 kJ
$S(c) \rightarrow S(g)$	+223 kJ
$\frac{1}{2}\ H_2(g) \rightarrow H(g)$	+218 kJ

Remember a bond energy is the energy required to separate two bonded *gaseous* atoms, not atoms in their standard state.

6.10 (a) Using the bond energy values (see text, Table 6.2), estimate ΔH for the reaction

$$CH_4(g) + 2Cl_2(g) \rightarrow CH_2Cl_2(g) + 2HCl(g)$$

(b) Calculate ΔH for the reaction of 2.00 g CH_4.

6.11 Using the bond energy values (see text, Table 6.2), calculate ΔH for the reaction

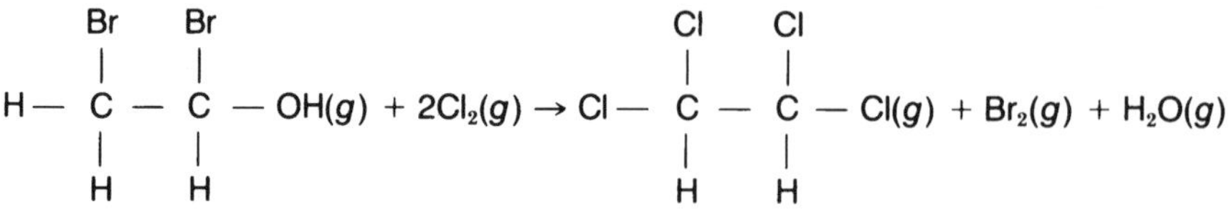

6.12 Use the reactions

$Sn(c) + 2Cl_2(g) \rightarrow SnCl_4(\ell)$	$\Delta H_f = -534$ kJ
$Sn(c) \rightarrow Sn(g)$	$\Delta H = -301$ kJ
$Cl_2(g) \rightarrow 2Cl(g)$	$\Delta H = +244$ kJ
$SnCl_4(\ell) \rightarrow SnCl_4(g)$	$\Delta H + +133$ kJ

to assign a bond energy to the Sn — Cl bond.

Section 6.7

6.13 One mole CH_4 is burned in a cylinder to produce CO_2 and H_2O:

$$CH_4(g) + 2O_2(g) \rightarrow CO_2(g) + 2H_2O(g) \qquad \Delta E = -892.4 \text{ kJ}$$

If the piston in the cylinder performs 317.0 kJ of work, how much heat is wasted? (See Figure 6.1).

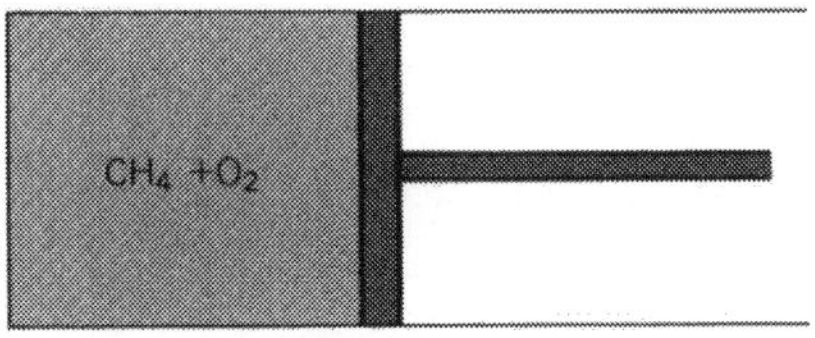

Figure 6.1 Combustion of CH_4.

6.14 One mole gas absorbs 400 kJ of heat, but in turn is compressed by an energy equivalent of 275 kJ. Calculate the energy change, ΔE, of the gas sample.

6.15 In a combustion cylinder the total energy change produced from the burning of'a fuel is −2000 kJ (exothermic). The cooling system absorbs 875 kJ. What is the amount of work, w, that can be done by the system?

ANSWERS

6.1 +85.7 kJ

6.2 890°C

6.3 +7.711 kJ/°C

6.4 −5140 kJ/mol

6.5 −1075.0 kJ

6.6 +131.3 kJ

6.7 −2467 kJ

6.8 (a) −2122 kJ
(b) 76.3 g

6.9 −46 kJ

6.10 (a) −194 kJ
(b) −24.3 kJ

6.11 −151 kJ

6.12 147 kJ

6.13 −575.4 kJ

6.14 +675 kJ

6.15 +1125 kJ

Chapter 7

THE ELECTRONIC STRUCTURE OF ATOMS

SECTION 7.1 THE NATURE OF LIGHT

Objectives

- To define the wavelength, frequency, and speed of light.
- To characterize a photon in terms of its particlelike properties.

Focus

The **wavelength** of light is the length, generally expressed in nanometers, between two adjacent crests or troughs in a wave. The **frequency** of a wave is defined as the number of waves that pass a given point per second (sec^{-1}) or the number of times a single crest (or trough) passes a given point per second.

The product of the wavelength, λ, and the frequency, ν, equals the speed, c, at which electromagnetic radiation travels:

$$\lambda(\text{m}) \times \nu(\text{sec}^{-1}) = c(\text{m sec}^{-1})$$

The speed, c, of all electromagnetic radiation equals 2.998×10^8 m/sec.

Light has properties that are dualistic: it has wavelike and particlelike properties. Like other waves, light produces diffraction patterns.

The energy properties of light are explained when we consider light's particle properties. For example, when light strikes the surface of a metal, some electrons may be ejected from that surface. The energy of the ejected electrons depends upon the energy of the light and not upon its intensity (brightness). The higher the energy of the light, the greater the kinetic energy of the ejected electrons. Because of this effect (called the photoelectric effect), light is said to be quantized, meaning that it consists of packets of energy called **photons**. A photon must have a certain amount of energy to be able to "knock off" an electron. Any energy in excess of this amount is carried by the ejected electron in the form of kinetic energy.

$$E_{\text{photon}} = E_{\text{to remove } e^-} + \frac{1}{2}\text{mv}^2$$

If the energy of the photon is too low, then it cannot cause ejection of an electron no matter how many photons strike the surface. The energy of the photon, E, is proportional to the frequency:

$$E = h\nu \quad \text{or} \quad E = \frac{hc}{\lambda}$$

$$\text{where } h(\text{Planck's constant}) = 6.63 \times 10^{-34} \frac{\text{J sec}}{\text{photon}}$$

Example 7.1 The energy of a photon needed to remove an electron from a copper surface is 7.29×10^{-19} J. What is the wavelength (nm) of this light (assuming zero kinetic energy)?

Answer 272 nm

Solution The energy that a photon needs to remove the electron is *at least* 7.29×10^{-19} J.
The wavelength of the light corresponding to this energy is

$$\lambda = \frac{hc}{E} = \frac{\left(6.63 \times 10^{-34} \frac{\text{J sec}}{\text{photon}}\right)\left(3.00 \times 10^{8} \frac{\text{m}}{\text{sec}}\right)\left(\frac{\text{nm}}{10^{-9}\text{m}}\right)}{7.29 \times 10^{-19} \frac{\text{J}}{\text{photon}}}$$

$$\lambda = 272 \text{ nm}$$

This wavelength is in the ultraviolet region of the electromagnetic spectrum.

Example 7.1 illustrates the particlelike (it can knock an electron off the surface) and wavelike properties (it has a wavelength) of a photon. If the wavelength is less than 272 nm (greater energy), the photon has more than enough energy to remove an electron from the copper surface. The extra energy gives the electron some kinetic energy. If the wavelength is greater than 272 nm, the photons, regardless of their number, lack the energy necessary to remove an electron.

SECTION 7.2 MATTER WAVES

Objective

- To explain the wavelike properties of matter.

Focus

The last section established the fact that light has both wavelike and particlelike properties. In this section you will see that subatomic particles, such as neutrons and electrons, also have a dual nature. The diffraction patterns shown by subatomic particles are evidence that they act as waves as well as particles. The wavelength of a particle is inversely proportional to its velocity and mass: $\lambda = h/mv$

Example 7.2 The mass of a baseball is 145 grams. At what velocity must it be thrown to have a wavelength of 1.00 mm?

$$1\text{J} = \frac{1\text{ kg m}^2}{\text{sec}^2}$$

Answer 4.57×10^{-35} m/sec

Solution

$$\nu = \frac{h}{m\lambda} = \frac{\left(6.63 \times 10^{-34} \frac{\cancel{\text{J}}\,\cancel{\text{sec}}}{\cancel{\text{particle}}} \times \frac{\cancel{\text{kg}}\text{ m}^{\cancel{2}}}{\cancel{\text{J}}\text{ sec}^{\cancel{2}}}\right)}{\left(\frac{0.145\ \cancel{\text{kg}}}{\cancel{\text{particle}}} \times 0.00100\cancel{\text{m}}\right)} = 4.57 \times 10^{-30}\text{ m/sec}$$

Obviously, a baseball cannot be thrown at this velocity and thus the naked eye cannot see such a wavelength. However, an electron with a wavelength of 1 mm and a mass of 9.11×10^{-31} kg has a velocity of

$$\nu = \frac{h}{m\lambda} = \frac{\left(6.63 \times 10^{-34} \frac{\cancel{\text{J}}\,\cancel{\text{sec}}}{\cancel{\text{particle}}} \times \frac{\cancel{\text{kg}}\text{ m}^{\cancel{2}}}{\cancel{\text{J}}\text{ sec}^{\cancel{2}}}\right)}{\left(\frac{9.11 \times 10^{-31}\ \cancel{\text{kg}}}{\cancel{\text{particle}}} \times 0.001\ \cancel{\text{m}}\right)} = 0.728\text{ m/sec}$$

This is a velocity we can appreciate.

SECTION 7.3 SPECTRA OF ELEMENTS

Objectives

- To distinguish between an emission and an absorption spectrum.
- To distinguish between a continuous spectrum and a line spectrum.
- To demonstrate that a line spectrum is characteristic of one element.
- To realize that the energy of light absorbed or emitted by atoms is quantized.

Focus

When light from an excited atom passes through a prism, an **emission spectrum**, characteristic of the sample, is obtained.

If white light (light containing all wavelengths of light) passes through a sample, the atoms are excited by the absorption of some photons. If the photons that are *not* absorbed pass through a prism, the entire color spectrum is seen with the exception of the photons absorbed by the sample. This is called an **absorption spectrum**—all wavelengths of the spectrum except those the sample absorbs.

White light, produced by heating a solid to incandescence, gives a **continuous spectrum**, which is one that shows all wavelengths of light in the visible region.

A **line spectrum** consists only of certain wavelengths appearing on a black background. Line spectra are emission spectra that originate from the deexcitation of atoms. Each element produces a characteristic line spectrum upon deexcitation. This is caused by the emission of a photon from an excited atom where each photon has a quantized amount of energy. Each spectral line arises from photons of a definite energy. Likewise an atom *absorbs* only quantized amounts of energy. Therefore, the line spectrum for an element is its fingerprint.

The frequencies of the photons in the line spectrum of atomic hydrogen are related to whole numbers (a and b) by the equation:

$$\nu = \frac{c}{\lambda} = 3.2898 \times 10^{15}\ \text{sec}^{-1} \left(\frac{1}{a^2} - \frac{1}{b^2}\right)$$

b is always greater than a.

Example 7.3 Calculate the energy of the photon emitted by the hydrogen atom when $b = 4$ and $a = 2$.

Answer $E = 4.09 \times 10^{-19}$ J

Solution Substitute $b = 4$ and $a = 2$ into the equation and solve for ν:

$$\nu = 3.2898 \times 10^{-15}\ \text{sec}^{-1} \left(\frac{1}{2^2} - \frac{1}{4^2}\right)$$

$$= 3.2898 \times 10^{-15}\ \text{sec}^{-1} \left(\frac{16 - 4}{4(16)}\right)$$

$$= 6.168 \times 10^{-14}\ \text{sec}^{-1}$$

$$E = h\nu = \left(6.63 \times 10^{-34}\ \frac{\text{J}\ \cancel{\text{sec}}}{\text{photon}}\right)(6.168 \times 10^{14}\ \cancel{\text{sec}^{-1}})$$

$$= 4.09 \times 10^{-19}\ \text{J/photon}$$

This frequency corresponds to the 481.6 nm line of the hydrogen spectrum. (See text, Figure 7.11.)

SECTION 7.4 QUANTIZATION OF THE ENERGY OF AN ELECTRON IN AN ATOM

Objectives

- To summarize the concept of an electron functioning as a wave and having only quantized energy values.
- To identify electron energy levels and the principal quantum number.
- To calculate the energy for an electron in the hydrogen atom at any quantized energy level.

Focus

Erwin Schrödinger developed a mathematical model for the hydrogen atom based on the concept that an electron has wavelike properties. He said that an electron behaves as a standing wave because the limits of electron wave propagation must be within the confines of an atom. Since wavelengths for a standing wave must be quantized (see Figure 7.1), Schrödinger showed that an electron's allowed energies are also quantized.

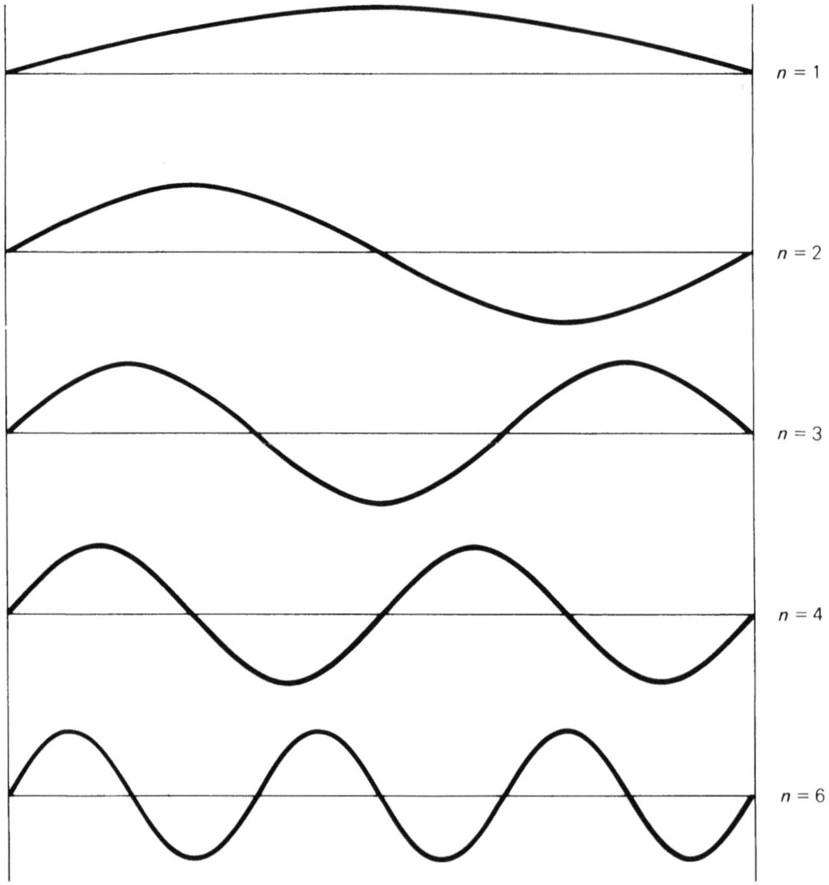

Figure 7.1 Wavelengths for a standing wave.

These quantized energies are called *energy levels*. The level also corresponds to the number of "loops" in the standing-wave electron in Schrödinger's mathematical model.

One to four dimensionless numbers designate the electron's energy levels in the atom. In the hydrogen atom, only one number, the **principal quantum number**, n, determines the energy of the electron. The most fundamental standing electron wave between the confined restraints of the atom is 1 loop, or $n = 1$. This is the lowest energy level for an electron in an atom (ground state for the hydrogen atom.) The second highest energy level is $n = 2$, and so on.

The energies of an electron (the energy levels) in the hydrogen atom are calculated from

$$E_n = \frac{-2.179 \times 10^{-18}\ \text{J/atom}}{n^2}$$

See the text for an explanation of the negative sign in the equation.

Example 7.4 Calculate the energy of the electron in hydrogen for $n = 4$ and for $n = 2$. Calculate the difference between these two energy levels and compare it with Example 7.3

Answer $E(n = 4) = -1.362 \times 10^{-19}$ J
$E(n = 2) = -5.448 \times 10^{-19}$ J
$E(n = 4) - E(n = 2) = 4.086 \times 10^{-19}$ J

Solution

$$E(n = 4) = \frac{-2.179 \times 10^{-18}\ \text{J}}{n^2} = \frac{-2.179 \times 10^{-18}\ \text{J}}{4^2}$$

$$= -1.362 \times 10^{-19}\ \text{J}$$

$$E(n = 2) = \frac{-2.179 \times 10^{-18}\ \text{J}}{2^2}$$

$$= -5.448 \times 10^{-19}\ \text{J}$$

$$E(n = 4) - E(n = 2) = -1.362 \times 10^{-19}\ \text{J} - (-5.448 \times 10^{-19}\ \text{J})$$

$$= 4.086 \times 10^{-19}\ \text{J}$$

This compares favorably with the 4.08×10^{-19} J in Example 7.3

SECTION 7.5 ORIGIN OF SPECTRAL LINES

Objective

- To account for the line spectra of elements in terms of the quantized energy levels for electrons in an atom.

Focus

The appearance of a line in an emission line spectrum for an atom results from an electron dropping from a higher energy level to a lower one. The emitted energy appears as a photon. The difference in the two energy levels is the exact energy of the photon.

Conversely, when an atom absorbs a photon, an electron is promoted from a lower to a higher energy level. The photon is equal in energy to the difference between the two levels. Accordingly, in an absorption spectrum only certain energies (wavelengths) of light do *not* appear in the spectrum.

Example 7.5 (a) Calculate the energy of a photon required to excite an electron in the H atom from the $n = 1$ to the $n = 7$ energy level. (b) Calculate the wavelength of the absorbed photon.

Answer (a) 2.135×10^{-18} J; (b) 93.04 nm

Solution (a) From the equation in Section 7.4, the energy for any electron in an H atom is calculated from

$$E_n = \frac{-2.179 \times 10^{-18}\ \text{J/atom}}{n^2}$$

For $n = 1$,

$$E_1 = \frac{-2.179 \times 10^{-18}\ \text{J/atom}}{1^2} = -2.179 \times 10^{-18}\ \frac{\text{J}}{\text{atom}}$$

For $n = 7$,

$$E_7 = \frac{-2.179 \times 10^{-18}\ \text{J/atom}}{7^2} = -4.447 \times 10^{-20}\ \frac{\text{J}}{\text{atom}}$$

The energy of the photon needed for the excitation is

$$\Delta E = E_{\text{final}} - E_{\text{initial}} = E_7 - E_1$$

or

$$\Delta E = \left(-4.447 \times 10^{-20}\ \frac{\text{J}}{\text{atom}}\right) - \left(-2.179 \times 10^{-18}\ \frac{\text{J}}{\text{atom}}\right)$$

$$= +2.135 \times 10^{-18}\ \frac{\text{J}}{\text{atom}} \quad \left(\text{or}\ \frac{\text{J}}{\text{photon}}\right)$$

The "+" sign indicates that energy is absorbed for the excitation of the electron.

(b) From Planck's equation, $E = hc/\lambda$ and

$$\lambda = \frac{hc}{E} = \frac{(6.626 \times 10^{-34}\ \cancel{\text{J}}\ \cancel{\text{sec}}/\cancel{\text{photon}})\ (2.998 \times 10^{8}\ \text{m}/\cancel{\text{sec}})}{(2.135 \times 10^{-18}\ \cancel{\text{J}}/\cancel{\text{photon}})}$$

$$= 9.304 \times 10^{-8}\ \text{m}$$

$$= 93.04\ \text{nm}$$

This wavelength is in the ultraviolet region of the electromagnetic spectrum.

SECTION 7.6 THE ELUSIVE ELECTRON AND ATOMIC ORBITALS

Objectives

- To represent the probability location of an electron by means of an electron cloud, the high probability region being called an atomic orbital.
- To express the energy of an electron using the four quantum numbers: n, ℓ, m_ℓ, and m_s.
- To explain the relationship between n, ℓ, and m_ℓ.
- To account for an atom's magnetism.

Focus

The Schrödinger equation for electrons tells us something about the energy of an electron and its most probable location in the electron cloud.

Specifically, it tells us that:

1. When we plug a set of related whole numbers (quantum numbers), designated by n, ℓ, and m_ℓ, into his wave equation, ψ, we can calculate a set of energy levels for electrons. The differences between these levels are in agreement with the line spectrum of the hydrogen atom.
2. The square of the wave equation, ψ^2, produces a probability location for an electron at various energies. These probability locations are called **atomic orbitals** and have the shapes shown in the text, Figure 7.13. The size, shape, and position of these atomic orbitals are fixed by the values of n, ℓ, and m_ℓ. (The word "orbital" is a poor choice of words in that an electron's calculated probability location has no resemblance to an orbit.) The orbital shape is not a pathway by which the electron moves in the atom; it only represents a 90% to 95% probability location of the electron.

The significance of the quantum numbers, n, ℓ, and m_ℓ is very important in this section. This set of interrelated integers identifies the energy for the electrons, their probability location (an orbital), and a spatial orientation of each orbital for each electron.

The **quantum numbers** are as follows:

The principal quantum number, n, represents the allowed electronic levels in the atom. n may have values of 1, 2, 3, 4, . . . ∞; the larger the value of n, the higher is the energy of the electron. All electrons having the same value of n belong to the same shell.

The quantum number, ℓ, determines the shape of the orbital for the electron and the number of subshells for a given n. The number of subshells in a given shell equals n. The subshells have values of $\ell = 0, 1, 2, 3, \ldots (n - 1)$ and $s, p, d, f, \ldots$ orbitals represent them.

An $\ell = 0$ subshell is an s orbital (see text, Figure 7.14).
$\ell = 1$ subshell is a p orbital.
$\ell = 2$ subshell is a d orbital.
$\ell = 3$ subshell is an f orbital.
$\ell = 4$ subshell is a g orbital, and so on.

The larger the value of ℓ, the higher is the energy of the electron. An n, ℓ designation identifies an electron's relative energy in a specific orbital of an atom. For example, an electron with $n = 4$ and $\ell = 2$, a $4d$ electron, has a higher energy than one with $n = 4$ and $\ell = 1$, a $4p$ electron.

The magnetic quantum number, m_ℓ, indicates the number of orbitals that make up each subshell in an atom. The number of values of m_ℓ determines the number of orbitals for a given ℓ subshell. All orbitals with a given m_ℓ have the same energy. The values of m_ℓ are all integral values from $+\ell$ to $-\ell$ including zero for a total of $2\ell + 1$ values or $2\ell + 1$ orbitals.

m_ℓ also determines the number of spatial orientations of each orbital (see text, Figure 7.14).

Therefore, each atomic orbital is defined by n, ℓ, and m_ℓ. Table 7.1 provides a summary.

Table 7.1†

n (energy level)	ℓ (subshell)	$n + \ell$ (orbital)	m_ℓ	Number of orbitals
1	0	1s	0	One 1s orbital
2	0	2s	0	One 2s orbital
	1	2p	+1	
			0	Three 2p orbitals
			−1	
3	0	3s	0	One 3s orbital
	1	3p	+1	
			0	Three 3p orbitals
			−1	
	2	3d	+2	
			+1	
			0	Five 3d orbitals
			−1	
			−2	
4	0	4s	0	One 4s orbital
	1	4p	+1	
			0	Three 4p orbitals
			−1	
	2	4d	+2	
			+1	
			0	Five 4d orbitals
			−1	
			−2	
	3	4f	+3	
			+2	
			+1	
			0	Seven 4f orbitals
			−1	
			−2	
			−3	

† To summarize the table, we say that the $n = 1$ shell has one orbital, so the $n = 2$ shell has one s and three p orbitals. The $n = 3$ shell has one s, three p, and five d orbitals. The $n = 4$ shell has one s, three p, five d, and seven f orbitals.

Example 7.6 What is the minimum value of n for $\ell = 2$, a d orbital?

Answer $n = 3$

Solution The values of ℓ are restricted by n.

$$\ell = 0, 1, 2, \ldots (n - 1)$$

Since $\ell = 2$, n is 3 or more.

Example 7.7 List the values of m_ℓ for the d orbitals. How many d orbitals are there?

Answer 2, 1, 0, −1, −2; five

Solution For a d orbital, $\ell = 2$. Since m_ℓ has all integral values from $+\ell$ to $-\ell$ including zero, then

$$m_\ell = 2, 1, 0, -1, -2$$

Then there are five d orbitals in an atom which has more than three energy shells, $n = 3, 4, 5, \ldots$.

Example 7.8 Which of the following sets of quantum numbers *do not* represent acceptable solutions to the Schrödinger equation?

	n	ℓ	m_ℓ
(a)	3	0	−1
(b)	2	1	0
(c)	4	3	−1
(d)	1	1	1

Answer (a) and (d)

Solution (a) For $n = 3$, $\ell = 0, 1, 2$; therefore, $\ell = 0$ is allowed. When $\ell = 0$, $m_\ell = 0$ only; therefore, $m_\ell = -1$ is not allowed.

(b) For $n = 2$, $\ell = 0, 1$; therefore, $\ell = 1$ is allowed. For $\ell = 1$, $m_\ell = +1, 0, -1$; therefore, $m_\ell = 0$ is allowed.

(c) For $n = 4$, $\ell = 0, 1, 2, 3$; therefore, $\ell = 3$ is allowed. For $\ell = 3$, $m_\ell = +3, +2, +1, 0, -1, -2, -3$; therefore, $m_\ell = -1$ is allowed.

(d) For $n = 1$, $\ell = 0$; therefore, $\ell = 1$ is not allowed. Since $\ell = 1$ is not allowed, then $m_\ell = 1$ cannot be allowed. For $\ell = 0$, m_ℓ can only equal 0.

Example 7.9 How many orbitals have the quantum numbers $n = 6$ and $\ell = 0$?

Answer One

Solution Since $n = 6$, $\ell = 0, 1, 2, 3, 4, 5$. When $\ell = 0$, $m_\ell = 0$, which is *one* value or orbital.

Example 7.10 Determine the number of subshells for $n = 5$.

Answer Five

Solution For $n = 5$, $\ell = 0, 1, 2, 3, 4$. There are five allowed ℓ values for $n = 5$, corresponding to five subshells: 5*s*, 5*p*, 5*d*, 5*f*, and 5*g*.

Example 7.11 How many spatial orientations are there for a *g* orbital?

Answer Nine

Solution $\ell = 0, 1, 2, 3, 4$
s p d f g

For a *g* subshell, $\ell = 4$. The m_ℓ values are +4, +3, +2, +1, 0, −1, −2, −3, −4, which is a total of nine values corresponding to nine spatial orientations.

For a summary of the n, ℓ, and m_ℓ quantum numbers, see Table 7.2.

The spin quantum number, m_s, completes the description of the energy for an electron.

An electron may spin either in a clockwise or counterclockwise direction; thus only two quantum values of m_s are allowed for an electron: $+\frac{1}{2}$ and $-\frac{1}{2}$. The magnetism of an atom is produced by an electron spinning on its axis, causing a "mini" magnetic field.

Table 7.2

Quantum number	Allowed values	Significance
n	1, 2, 3, . . . (called principal energy level or shells)	Allowed energy levels for electrons in an atom based on the standing electron wave concept from the Schrödinger wave equation
ℓ	0, 1, 2, . . . ($n - 1$) (called subshells)	1. Allowed electronic subshells belonging to each principal energy level in an atom. 2. Indicates a 90 to 95% probability location for the electron, which represents the shape of the *orbital*
m_ℓ	All integral values from $+\ell$ to $-\ell$ including zero (called orbitals), designated by $s, p, d, f, g, h, \ldots$	1. Allowed number of orbitals for a given subshell 2. Indicates the number of orientations of the orbitals for a given ℓ quantum number

SECTION 7.7 DISTRIBUTION OF ELECTRONS IN ATOMS

Objectives

- To organize the relative energies of the subshells according to the $n + \ell$ rule.
- To predict the electron configuration of an atom.
- To determine the number of unpaired electrons in an atom.

Focus

The set of four quantum numbers n, ℓ, m_ℓ, and m_s defines the energy for each electron in an atom.

The $(n + \ell)$ rule determines the relative energies of the subshells.

1. The lower the $(n + \ell)$ value for a subshell, the lower its energy in the atom.
2. If the $(n + \ell)$ sum is the same for two subshells, then the one with the lower n value is of lower energy.

Example 7.12 Arrange the following orbitals with the following n, ℓ notations in order of increasing energy: 3*s*, 4*f*, 2*p*, 3*p*, 5*d*.

Answer 2*p*, 3*s*, 3*p*, 4*f*, 5*d*

Solution Determine the $(n + \ell)$ sum of each orbital:

For 3*s*: $n = 3, \ell = 0, (n + \ell) = 3$
4*f*: $n = 4, \ell = 3, (n + \ell) = 7$
2*p*: $n = 2, \ell = 1, (n + \ell) = 3$
3*p*: $n = 3, \ell = 1, (n + \ell) = 4$
5*d*: $n = 5, \ell = 2, (n + \ell) = 7$

The $(n + \ell)$ value for the 3*s* and 2*p* are the same, but because $n = 2$ is of lower energy than $n = 3$, the 2*p* is of lower energy than the 3*s*.

For the same reason, the 4*f* has a lower energy than the 5*d*.

Before placing electrons into orbitals, we must consider the **Pauli exclusion principle**: No two electrons in an atom can have the same set of four quantum numbers.

Example 7.13 For a given value of m_ℓ, determine the number of electrons in the orbital.

Answer 2 electrons $+\frac{1}{2}$, $-\frac{1}{2}$

Solution A given m_ℓ means that we are discussing a specific *orbital* in a given ℓ subshell. Since only two orientations or spin quantum values for electrons exist, $m_s = +\frac{1}{2}$ and $-\frac{1}{2}$ and since no two electrons can have the same set of four quantum numbers (Pauli), there can only be two electrons in a given orbital, m_ℓ.

Let's consider a specific 3*p* orbital, say the one in which $m_\ell = -1$. If there are only two possible values for the electrons in that orbital, $m_s = +\frac{1}{2}$ and $-\frac{1}{2}$, then each electron has its own set of four quantum numbers and only two electrons can occupy that orbital. The sets of four quantum numbers for the two electrons in this 3*p* orbital are

n	ℓ	m_ℓ	m_s
3	1	−1	$+\frac{1}{2}$
3	1	−1	$-\frac{1}{2}$

Example 7.14 Determine the number of electrons that can have the following quantum numbers:

(a) $n = 4, \ell = 2, m_\ell = 0$
(b) $n = 3, \ell = 2$
(c) $n = 5, \ell = 1, m_\ell = 0, m_s = +\frac{1}{2}$
(d) $n = 2$

Answer

(a) 2
(b) 10
(c) 1
(d) 8

Solution (a) The $m_\ell = 0$ indicates that we are considering only one orbital. Therefore, there can be only 2 electrons in this orbital, one with $m_s = +\frac{1}{2}$ and the other with $m_s = -\frac{1}{2}$. There are only two electrons with the quantum numbers $n = 4$, $\ell = 2$ and $m_\ell = 0$.

The sets of four quantum numbers for these two electrons are:

n	ℓ	m_ℓ	m_s
4	2	0	$+\frac{1}{2}$
4	2	0	$-\frac{1}{2}$

(b) Two of the four quantum numbers are given. For $\ell = 2$, $m_\ell = +2, 1, 0, -1, -2$, or five values. Each m_ℓ orbital has two electrons corresponding to $m_s = +\frac{1}{2}$ and $-\frac{1}{2}$. Therefore, the five orbitals have a total of 10 electrons.

(c) All four quantum numbers are given. Only one electron in the atom has this set.

(d) Only one quantum number is given. For $n = 2$, $\ell = 0, 1$. There are two subshells, the 2*s* and 2*p*. Each subshell has a number of orbitals that corresponds to the number of m_ℓ values.

$\ell = 0$ $m_\ell = 0$, one *s* orbital
$\ell = 1$ $m_\ell = 1, 0, -1$, three *p* orbitals

There are four m_ℓ orbitals. With two values of m_s for each m_ℓ orbital, a total of 8 electrons have the quantum number $n = 2$.

Example 7.15 Account for the following table.

Subshell	Number of orbitals	Maximum Number of electrons
$\ell = 0$, *s* subshell	1	2
$\ell = 1$, *p* subshell	3	6
$\ell = 2$, *d* subshell	5	10
$\ell = 3$, *f* subshell	7	14

Solution For $\ell = 0$ (*s* subshell), there is only one value of m_ℓ ($m_\ell = 0$). Since $m_s = +\frac{1}{2}$ and $-\frac{1}{2}$ for the electrons in each m_ℓ orbital, there are only two electrons.

For $\ell = 1$ (*p* subshell), $m_\ell = +1, 0, -1$ or three *p* orbitals with two electrons ($m_s = +\frac{1}{2}$ and $-\frac{1}{2}$) per orbital—a total of 6 electrons.

For $\ell = 2$ (*d* subshell), $m_\ell = +2, +1, 0, -1, -2$ or five *d* orbitals with two electrons per orbital—a total of 10 electrons.

For $\ell = 3$ (*f* subshell), $m_\ell = +3, +2, +1, 0, -1, -2, -3$ or seven *f* orbitals with two electrons per orbital—a total of 14 electrons.

Knowing the relationship between the four quantum numbers and knowing that each corresponds to an energy state for an electron, we can build an energy-level diagram using the $(n + \ell)$ rule for a "general" atom. (See text, Figure 7.17.) We can then fill these energy levels with electrons, with each added electron occupying the lowest available energy level. This gives the most stable arrangement of electrons in the atom —its ground state electron configuration.

The ground state electron configuration is derived from the following rules:

1. The number of electrons in an atom equals the atomic number.
2. Each added electron enters the orbitals in order of increasing energy (Figure 7.17).
3. **Pauli exclusion principle.** No two electrons in a given atom have the same set of four quantum numbers; an orbital can hold only two electrons.
4. **Hund's rule.** When electrons enter a subshell consisting of more than one orbital, they remain unpaired with parallel spins in the orbitals until each orbital has at least one electron.

Example 7.16 Write the electron configuration for $_{14}Si$. How many unpaired electrons are in the $_{14}Si$ atom?

Answer $1s^22s^22p^63s^23p^2$ or $[Ne]3s^23p^2$; two unpaired electrons.

Solution $_{14}Si$ has 14 electrons. We fill the orbitals in order of increasing energy (see text, Figure 7.17) as follows:

- 2 in the 1*s* orbital
- 2 in the 2*s* orbital
- 6 in the 2*p* orbital
- 2 in the 3*s* orbital and
- 2 in the 3*p* orbital (according to Hund's rule)

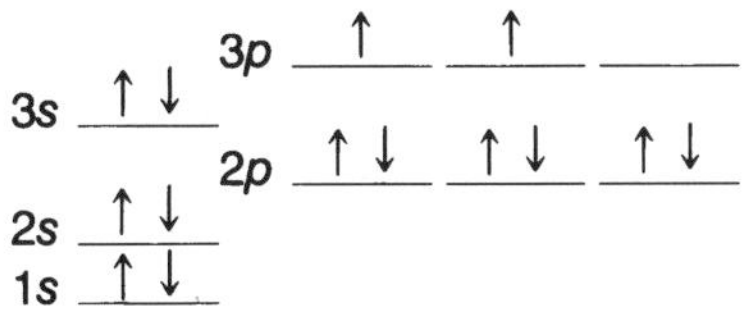

Example 7.17 Write the electron configuration for $_{34}Se$.

Answer $1s^22s^22p^63s^23p^64s^23d^{10}4p^4$ or $[Ar]4s^23d^{10}4p^4$

Solution We fill the orbitals with 34 electrons as follows:

- 2 in the 1*s* orbital
- 2 in the 2*s* orbital
- 6 in the 2*p* orbitals
- 2 in the 3*s* orbital
- 6 in the 3*p* orbitals
- 2 in the 4*s* orbital
- 10 in the 3*d* orbitals
- 4 in the 4*p* orbitals (according to Hund's rule)

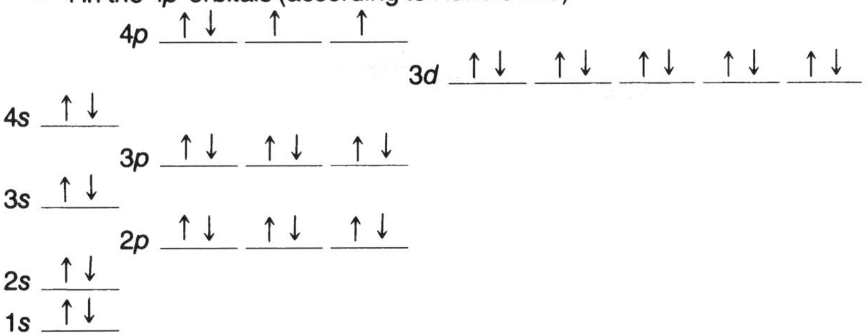

Because properties of the elements depend on electron configurations, it is important that you develop the ability to write them quickly. The two easiest ways to remember the sequence in which the orbitals are filled with electrons are to consult either Figure 7.17 or the periodic table.

Although the periodic table is not discussed until Chapter 8 of the text, its use in writing electron configurations is presented now for convenience. Only a casual familiarity with the periodic table is necessary to determine an atom's electron configuration. You may choose to skip the remaining material in this section until Chapter 8 (consult your instructor).

The following relationships between the periodic table and the energy-level diagram in Figure 7.17 in the text are:

1. For elements in Groups 1A and 2A, the final two electrons added to the levels in Figure 7.17 are placed in an s orbital; the value of n for the electrons is the same as the period (horizontal row) number of the element in the periodic table. For example, the Group IA element, $_{11}$Na, ends with $3s^1$. One electron is added to an s orbital in the third shell ($n = 3$) (see Study Guide, Figure 7.2).

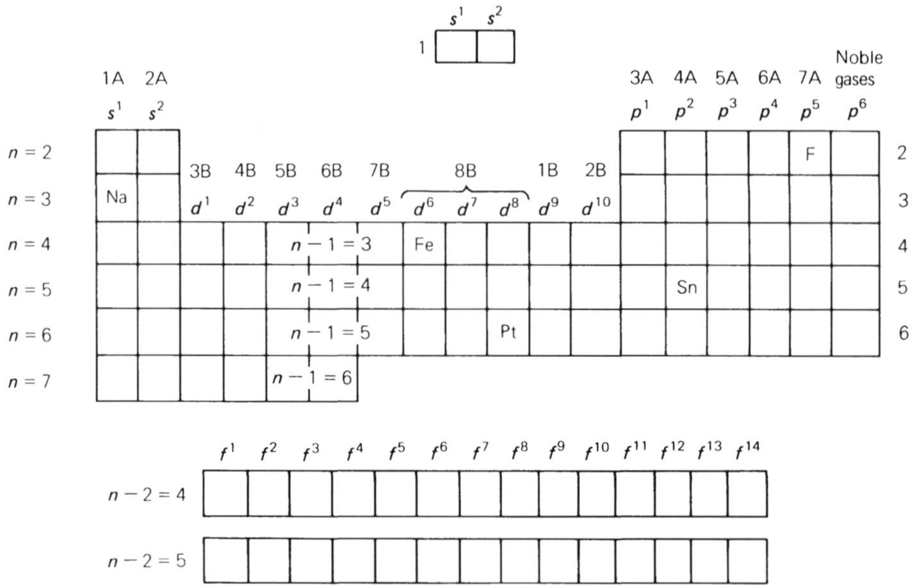

Figure 7.2 Periodic table of the elements.

2. For elements in Group 3A through the noble gases, the last six electrons are added to the p orbitals. There are six columns, corresponding to six successive electrons in the three p orbitals with a given n. The value of n for these p orbitals is the same as the period number for the element in the periodic table. For example, the electron configuration of F ends with $2p^5$. Notice that $n = 2$ and F is in period 2. Also notice that F, which is in Group 7A, is in the fifth column of this section of the periodic table and has five p electrons as do all the elements in that group. (see Figure 7.2).
3. For the Group B elements, the last electrons are added to d orbitals and the value of n is *one less* than the period number in the periodic table. For example, the electron configuration of $_{26}$Fe, ends with $3d^6$. Notice that $n = 3$ and Fe is in period 4. Also notice that Fe, which is in the sixth column of the Group B elements, has six d electrons as do all elements in the Fe group (see Figure 7.2).

4. Finally, the electron configurations of the elements in the two rows at the bottom of the table are completed by electrons in *f* orbitals with a value of *n*, which is *two less* than the period number.

To summarize: In determining electron configurations using the periodic table, begin by filling an *s* orbital (with 2 electrons) in Groups 1A and 2A; the three *p* orbitals (6 electrons) Groups 3A through the noble gases; the five *d* orbitals (10 electrons) in the B groups; and the seven *f* orbitals (14 electrons) in the bottom two rows of elements.

Example 7.18 Write the electron configuration for $_{50}Sn$ (see Figure 7.2).

Answer $1s^2 2s^2\ 2p^6 3s^2\ 3p^6 4s^2\ 3d^{10} 4p^6 5s^2 4d^{10} 5p^2$

$$\underset{\text{He}}{\uparrow}\ \ \underset{\text{Ne}}{\uparrow}\ \ \underset{\text{Ar}}{\uparrow}\ \ \underset{\text{Kr}}{\uparrow}$$

Solution $_{50}Sn$ is in the fifth period ($n = 5$), and has 2 electrons in the *p* orbitals (Group 4A). The electron configuration of $_{50}Sn$ therefore ends with $5p^2$. The rest of the electron configuration for $_{50}Sn$ is determined by going back and forth across the periodic table and writing each part down. Starting with the previous noble gas, we find the electron configuration for $_{50}Sn$ is $[Kr]5s^2 4d^{10} 5p^2$.

Example 7.19 Write the electron configuration for $_{78}Pt$ (see Figure 7.2).

Answer $[Xe]6s^2 4f^{14} 5d^8$

Solution Pt is a Group B element in the sixth period, so that its electron configuration ends with a $5d$. Being the 8th element in the $5d$ series, the electron configuration of Pt must end in a $5d^8$. After the previous noble gas, the electron configuration for $_{78}Pt$ is: $[Xe]6s^2 4f^{14} 5d^8$.

When *positive* ions are formed from the element, the electrons with the highest $(n + \ell)$ value (or the highest *n* in case of a tie) are lost first. For example, the electron configuration for Fe is $[Ar]4s^2 3d^6$, but for Fe^{2+} it is $[Ar]3d^6$; $_{50}Sn^{2+}$ is $[Kr]5s^2 4d^{10}$ and $_{50}Sn^{4+}$ is $[Kr]4d^{10}$.

SECTION 7.8 PARAMAGNETISM

Objectives

- To define diamagnetism and paramagnetism.
- To determine the degree of paramagnetism of an atom.

Focus

Substances whose atoms have only paired electrons are **diamagnetic**, but those that have one or more unpaired electrons are **paramagnetic**.

Paramagnetic substances are attracted to a magnetic field, but diamagnetic substances are slightly repelled. The greater the number of unpaired electrons, the more strongly the substance is attracted to a magnetic field.

Example 7.20 Which of the following atoms are paramagnetic? If the atom is paramagnetic, indicate the number of unpaired electrons.

$$_{15}P;\ _{20}Ca;\ _{30}Zn;\ _{13}Al;\ _{35}Br$$

Answer $_{15}P$, 3 unpaired e^-; $_{13}Al$, 1 unpaired e^-; $_{35}Br$, 1 unpaired e^-

Solution Write the electron configuration of each atom:

$_{15}P$: $[Ne]3s^23p^3$. The 3s orbital is filled with 2 electrons. The 3 electrons in the 3p orbitals occupy separate orbitals according to Hund's rule. 3p $\underline{\uparrow}\ \underline{\uparrow}\ \underline{\uparrow}$

$_{20}Ca$: $[Ar]4s^2$. All electrons are paired.

$_{30}Zn$: $[Ar]4s^23d^{10}$. All electrons are paired.

$_{13}Al$: $[Ne]3s^23p^1$. The 2 electrons in the 3s orbital are paired. The lone electron in the 3p orbital is unpaired.

$_{35}Br$: $[Ar]4s^23d^{10}4p^5$. The 2 electrons in the 4s are paired and the 10 electrons in the five 3d orbitals are paired. But the 5 electrons in the three 4p orbitals are arranged as 4p $\underline{\uparrow\downarrow}\ \underline{\uparrow\downarrow}\ \underline{\uparrow}$. There is one lone, unpaired electron.

QUESTIONS

Section 7.1

7.1 The energy required to remove an electron from a sodium atom is 3.68×10^{-19} J. Calculate the kinetic energy of an ejected electron, if light having a wavelength of 272 nm removes an electron from the sodium atom.

7.2 (a) Calculate the energy of a photon having a wavelength of 5.00×10^{-5} cm. This λ appears as a green color in the visible spectrum.

(b) Calculate the energy of 1 mole of these photons.

(c) The bond energy for the C—I bond is 213 kJ/mol. Will green light cause this bond to break?

7.3 The first ionization energy of Na, ($Na(g) \rightarrow Na(g)^+ + e^-$), is 495.8 kJ/mole.

(a) What frequency of a photon (Hz) is necessary for the ionization of *one* Na atom? 1 Hz = 1 sec^{-1}.

(b) Calculate the wavelength of the photon.

Section 7.2

7.4 What is the wavelength of an O_2 molecule at 27°C when it has a velocity of 480 m/sec? The mass of an O_2 molecule is calculated using the formula weight of O_2 and Avogadro's number.

7.5 What is the velocity of a neutron (mass = 1.7×10^{-28} kg) that has a wavelength of 1.0×10^{-10} m?

Section 7.3

7.6 Calculate the wavelength in nm corresponding to the line in the atomic hydrogen spectrum in which a = 2 and b = 6.

Section 7.4

7.7 (a) Calculate the energy of the electron in the hydrogen atom for $n = 2$ and $n = 6$.

(b) Calculate the difference in energy of these two energy levels.

(c) Using the equation $E = hc/\lambda$, calculate the wavelength that corresponds to this energy. Compare your answer to the answer in Question 7.6.

Section 7.5

7.8 Calculate the energy of the photon necessary to excite an electron from the $n = 2$ to $n = 6$ energy level in the H atom.

7.9 Calculate the energy required to remove an electron ($n = \infty$) from the H atom when the electron is in the $n = 2$ energy level.

7.10 How much energy in kJ is released in the deexcitation 1 *mole* of H atoms from the $n = 4$ energy level to the ground state ($n = 1$)? Is this emitted energy sufficient to break the H — H bond, 435.9 kJ/mole?

Section 7.6

7.11 For the $\ell = 1$ subshell, representing *p* orbitals, describe the spatial orientation of the three *p* orbitals.

7.12 Describe the shape of an *s* orbital.

7.13 (a) Which quantum number characterizes the shape of an atomic orbital?

____

(b) What subshell does an electron with quantum numbers $n = 3$ and $\ell = 1$ occupy? ________

(c) What is the lowest energy shell in which the *d* subshell exists? ________

(d) How many values of m_ℓ are possible for an electron in a 4*d* orbital? ________

(e) How many values of ℓ are possible for $n = 6$? ________

(f) How many values of m_ℓ are there for a 3*p* orbital? ________

(g) Is the set of quantum numbers $n = 4$, $\ell = 2$, $m_\ell = -2$ allowed for an electron in an atom? Explain.

(h) How many subshells are allowed in the $n = 4$ energy level? ________

(i) How many spatial orientations are there for an *h* orbital? ________

(j) How many orbitals have the 4*d* designation? ________ the 5*f* designation?

____

Section 7.7

7.14 (a) What is the maximum number of electrons which can have the $n = 5$, $\ell = 2$ designation? ________

(b) How many 4*p* electrons are in an $_{51}$Sb atom? ________

(c) How many unpaired electrons are in the $_{26}$Fe atom? ________

(d) How many *p* electrons are in the $_{15}$P atom? ________

(e) What is the maximum number of electrons in an atom that can have the $n = 3$ quantum number? ________

(f) List all the elements that have a [noble gas]ns^2 electronic configuration?

________, ________, ________, ________, ________, ________.

(g) What is the maximum number of electrons in an atom that have a 5*g* designation? ________

(h) How many unpaired electrons are in each of the noble gases? ________

7.15 Write the electron configurations for $_{30}Zn$, $_{53}I$, and $_{16}S$.

7.16 Use the shorthand method for writing the electron configuration for $_{56}Ba$, $_{49}In$, and $_{27}Co$.

7.17 Use the periodic table to determine the electron configuration for $_{33}As$, $_{55}Cs$, and $_{77}Ir$.

7.18 Write the electron configuration of the following elements by using the periodic table or Figure 7.17 as a guide:

(a) $_{40}Zr$ (f) $_{28}Ni^{2+}$†
(b) $_{32}Ge$ (g) $_{12}Mg$
(c) $_{19}K$ (h) $_{80}Hg$
(d) $_{19}K^+$ (i) $_{51}Sb$
(e) $_{28}Ni$ (j) $_{51}Sb^{3+}$

Section 7.8

7.19 Determine the atoms or ions that are paramagnetic. If paramagnetic, determine the number of unpaired electrons.

(a) $_{34}Se$ (f) $_{37}Rb^+$
(b) $_{23}V^{3+}$† (g) $_{26}Fe^{2+}$†
(c) $_{80}Hg$ (h) $_{14}Si$
(d) $_{27}Co^{3+}$† (i) $_{16}S$
(e) $_{18}Ar$ (j) $_{53}I^-$

ANSWERS

7.1 $\frac{1}{2}mv^2 = 3.63 \times 10^{-19}$ J

7.2 (a) 3.98×10^{-19} J/photon
(b) 2.40×10^5J/mol
(c) Yes, 240 kJ/mol > 213 kJ/mol

7.3 (a) 1.24×10^{15} Hz
(b) 242 nm

7.4 $\lambda = 2.60 \times 10^{-11}$ m

7.5 $v = 3.9 \times 10^4$ m/sec

7.6 $\lambda = 410.1$ nm

† The 3*d* subshell is more stable than the 4*s*.

7.7 (a) $E_2 = -5.448 \times 10^{-19}$ J
$E_6 = -6.053 \times 10^{-20}$ J

(b) $E_2 - E_6 = 4.843 \times 10^{-19}$ J

(c) $\lambda = 410.2$ nm; agrees very well with value in 7.6.

7.8 $E_2 - E_6 = +4.843 \times 10^{-19}$ J

7.9 $+5.448 \times 10^{-19}$ J

7.10 (a) $E_1 - E_4 = -2.043 \times 10^{-19}$ J/atom
$= -1.23 \times 10^3$ kJ/mol

(b) Yes; 1.23×10^3 kJ/mol > 435.9 kJ/mol

7.11 Each *p* orbital is dumbbell-shaped and the three are located along the x-, y-, and z-axes.

7.12 The *s* orbital is spherical.

7.13 (a) ℓ

(b) 3*p*

(c) $n = 3$

(d) 5; +2, +1, 0, −1, −2

(e) 6; $\ell = 0, 1, 2, 3, 4, 5$

(f) 3; +1, 0, −1

(g) Yes. $n = 4$; $\ell = 0, 1, 2, 3$, so $\ell = 2$ is allowed. $\ell = 2$, $m_\ell = +2, +1, 0, -1, -2$, so $m_\ell = -2$ is allowed.

(h) 4; $\ell = 0, 1, 2, 3$

(i) 11; $\ell = 5$ *h* orbital, m_ℓ has 11 values

(j) 5, 7

7.14 (a) 10

(b) 3

(c) 4

(d) 9

(e) 18

(f) Be, Mg, Ca, Sr, Ba, Ra

(g) 18

(h) 0

7.15 $_{30}$Zn: $1s^22s^22p^63s^23p^64s^23d^{10}$
$_{53}$I: $1s^22s^22p^63s^23p^64s^23d^{10}4p^65s^24d^{10}5p^5$
$_{16}$S: $1s^22s^22p^63s^23p^4$

7.16 $_{56}$Ba: [Xe]$6s^2$
$_{42}$In: [Kr]$5s^24d^{10}5p^1$
$_{27}$Co: [Ar]$4s^23d^7$

7.17 $_{33}$As: [Ar]$4s^23d^{10}4p^3$
$_{55}$Cs: [Xe]$6s^1$
$_{77}$Ir: [Xe]$6s^24f^{14}5d^7$

7.18 (a) $_{40}$Zr: [Kr]$5s^24d^2$

(b) $_{32}$Ge: [Ar]$4s^23d^{10}4p^2$

(c) $_{19}K$: $[Ar]4s^1$

(d) $_{19}K^+$: $[Ar]$

(e) $_{28}Ni$: $[Ar]4s^2 3d^8$

(f) $_{28}Ni^{2+}$: $[Ar]3d^8$

(g) $_{12}Mg$: $[Ne]3s^2$

(h) $_{80}Hg$: $[Xe]6s^2 4f^{14} 5d^{10}$

(i) $_{51}Sb$: $[Kr]5s^2 4d^{10} 5p^3$

(j) $_{51}Sb^{3+}$: $[Kr]5s^2 4d^{10}$

7.19 (a) 2 (f) 0

(b) 2 (g) 4

(c) 0 (h) 2

(d) 4 (i) 2

(e) 0 (j) 0

Chapter 8

CHEMICAL PERIODICITY

SECTION 8.1 FROM DOBERIENER TO MENDELEEV

Objective

- To recognize scientists for their contributions to the modern-day version of the periodic table.

Focus

Four early scientists generally associated with the organization of the periodic table are:

- Dobereiner—lists of "triads"
- Newlands—law of octaves
- Mendeleev and Meyer—when the elements were arranged in order of increasing atomic weights, elements having similar chemical and physical properties were grouped together (with a few minor exceptions)

Mendeleev gained prominence for his organization of the elements because he maintained that undiscovered elements would fit into his organization. In addition, his predicted properties were so near the experimental properties of those discovered later, he is generally recognized as the founder of the periodic table.

SECTION 8.2 THE PERIODIC LAW AND PERIODIC TABLE; TYPES OF ELEMENTS

Objectives

- To locate and identify various specific sections of the periodic table.
- To use the periodic table for writing the electron configuration of an atom or ion.

Focus

If you are able to identify and interpret the various sections of the periodic table, you have a valuable source of information. Become familiar with each of the following and know the corresponding electron configuration data.

- **Group A Elements**
  The **representative** elements: the *s* subshell is filling in Groups 1A and 2A; and the *p* subshell in Groups 3A through 8A. A marked change in chemical and physical properties occurs across the period.

- **Group B Elements**
  The **transition** elements: the *d* subshell is filling in Groups 3B through 1B. There is less distinction in properties between neighboring transition elements than there is in neighboring representative elements.

- **Lanthanoids and Actinoids**
  The **inner transition elements:** the *f* subshell is filling in the last two rows of 14 elements (see text, Figure 8.3). The properties are more similar between these neighboring elements than they are between neighboring transition elements.

More specifically, you should be able to locate the alkali metals, the alkaline-earth metals, the iron triad, the coinage metals, the rare earth metals, the noble gases, and the halogens.

Another statement of the periodic law (slightly different than what appears in the text) is **when the elements are listed in order of increasing atomic number, there is a periodic (regular) arrangement of elements which have similar chemical and physical properties**. Each group, then, has similar properties: all Group 1A elements react vigorously with water, all Group 7A elements are diatomic, all Group 2B elements commonly form the 2+ ion, and so on. This is because all elements of a group have the same outer shell electron configuration.

Example 8.1 Determine the number of 3*d* and 4*p* electrons in an atom of $_{32}Ge$.

Answer Ten 3*d* electrons and two 4*p* electrons

Solution Ge is the second element of the *p* subshell group in period 4 ($n = 4$). Hence it has two 4*p* electrons; the 3*d* orbitals, being lower in energy than the 4*p* orbitals, are therefore filled. At 2 electrons per orbital, they have a total of 10 electrons.

Using the periodic table, let's review Section 7.8 (Study Guide), that is, the determination of an electron configuration of an element.

Example 8.2 Write the electron configurations for $_{73}Ta$ and $_{49}In$ after the previous noble gas.

Answer $_{73}Ta$: $[Xe]6s^2 4f^{14} 5d^3$
$_{49}In$: $[Kr]5s^2 4d^{10} 5p^1$

Solution $_{73}Ta$ appears as the third member of the 5*d* series of transition elements; therefore it has three 5*d* electrons. Working backward to the previous noble gas, xenon, the 4*f* series (14 electrons) and the 6*s* series (two electrons) appear. Hence the configuration of $_{73}Ta$ after the xenon core, [Xe], is $6s^2 4f^{14} 5d^3$.

$_{49}In$ is the first member of the 5*p* series, having one 5*p* electron. The 4*d* series (10 electrons) and the 5*s* series (two electrons) appear after the previous noble gas, krypton. Hence, for $_{49}In$: $[Kr]5s^2 4d^{10} 5p^1$

SECTION 8.3 THE PERIODICITY OF VALENCE

Objective

- To determine the valence of each representative element.

Focus

Valence is an element's combining capacity when compared to hydrogen which has a valence of 1. Chlorine combines with hydrogen in a 1:1 atomic ratio to form HCl; therefore the valence of chlorine is also 1. All Group 7A elements combine similarly with hydrogen and therefore have a valence of 1.

One oxygen atom combines with two hydrogen atoms to form H_2O; therefore the valence of oxygen is 2 as are all other Group 6A elements.

One Group 2A atom combines with two Group 7A atoms; therefore the valence of all Group 2A elements is 2.

To generalize,

For Groups 1A to 4A, common valence = Group A number
For Groups 5A to 8A, common valence = 8-Group A number

Example 8.3 Determine the valence for the following representative elements.

(a) Al	(d) Cl	(g) Sn
(b) Sr	(e) S	(h) N
(c) K	(f) As	(i) Ar

Answer From the stated rules the valence of the respective element is

(a) 3	(d) 1	(g) 4
(b) 2	(e) 2	(h) 3
(c) 1	(f) 3	(i) 0

SECTION 8.4 SIZES OF ATOMS AND IONS

Objective

- To explain the trends in covalent and ionic radii of the atoms within a given group or period of elements.

Focus

Five trends involving covalent and ionic radii of elements are discussed in terms of the magnitude of the nuclear charge versus electron-electron repulsions. An increased nuclear charge causes the electron cloud to contract; an increased number of electrons causes it to expand.

1. Covalent radii decrease from left to right across a period.
   With each successive element, one electron is added to the existing shell. The effect of the increased nuclear charge more than offsets the increased electron-electron repulsion; therefore, the contracting effect dominates. An exception occurs in the Group 6A elements where the fourth *p* electron pairs up with an electron already in a *p* orbital, causing an additional electron-electron repulsion effect.

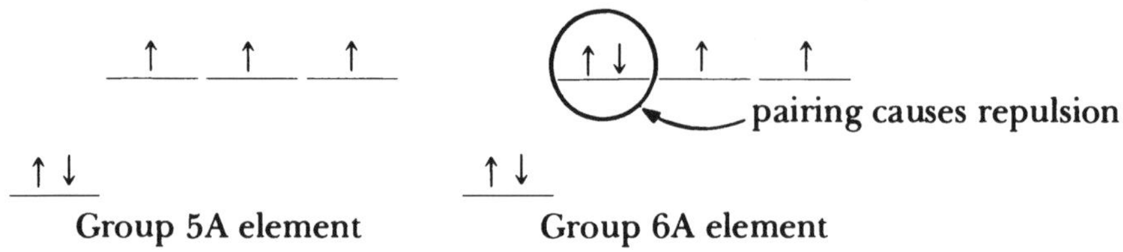

2. Covalent radii increase from top to bottom within a given group of elements.
   With each successive element within the group, an electron is added to a new shell. Because electrons in successive elements occupy higher shells, the electrons are moved farther from the nucleus, and the increased nuclear charge does not offset the electron-electron repulsion.
3. Positive ionic radii are smaller than the neutral atom.
   For a positive ion, one or more electrons have been removed, reducing the magnitude of the electron-electron repulsion. However, the nuclear charge remains the same.
4. Negative ionic radii are larger than the neutral atom.
   To form a negative ion, one or more electrons are added to an atom. This increases the electron-electron repulsion, making the negative ion larger.
5. For atoms and ions that are isoelectronic (the same ground-state electronic configuration) but have different nuclear charges, the atom or ion with the greatest nuclear charge has the smallest radius.

Example 8.4 Select the smaller atom or ion from each of the following pairs:

(a) Na or Si
(b) Cl^- or S^{2-}
(c) Ca^{2+} or K^+
(d) Cl^- or Cl
(e) Se or Te
(f) Ba or Mg

Answer

(a) Si
(b) Cl^-
(c) Ca^{2+}
(d) Cl
(e) Se
(f) Mg

Example 8.5 Circle the sets of isoelectronic species.

(a) Ca^{2+}, Sc^{3+}	(d) N^{3-}, Be^{2+}
(b) Na^{+}, Cl^{-}	(e) Br^{-}, Cl^{-}
(c) S^{2-}, Ar	(f) Sr^{2+}, Br^{-}

Answer (a), (c), (f)

SECTION 8.5 IONIZATION ENERGY

Objectives

- To define the ionization energy of an atom.
- To predict the relative ionization energies for the atoms within a given group or period of elements.

Focus

The first **ionization energy** of an atom is the energy required to remove the least tightly bound electron from a *gaseous* atom.

1. This energy decreases from top to bottom within a group. The electron is removed more easily from the larger atoms at the bottom of the group because it is in a higher energy shell.
2. Ionization energies tend to increase from left to right across a period. This corresponds to the general decrease in covalent radii. The two exceptions are:
   (a) From Group 2A to 3A. Group 3A elements have lower ionization energies because the electron is removed from a *p* subshell. Slightly less energy is required to remove the electron from a *p* subshell than from the *s* subshell in the Group 2A elements.

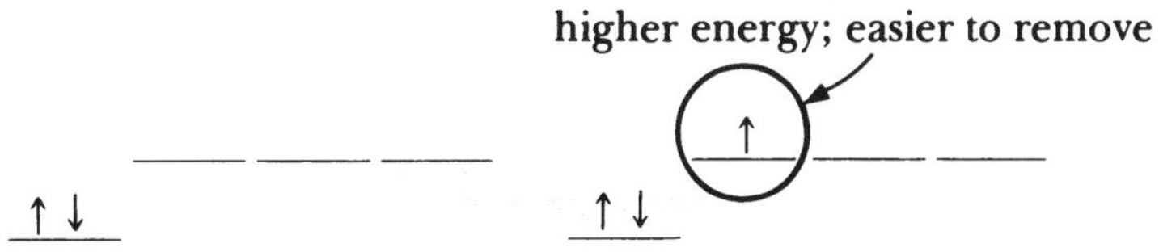

Group 2A element Group 3A element

   (b) From Group 5A to 6A. The mutual repulsions between paired electrons in a given orbital are greater than between single electrons in separate orbitals. Therefore, it is easier to remove an electron from a *p* subshell that has paired electrons (Group 6A) than from a *p* subshell that does not (Group 5A).

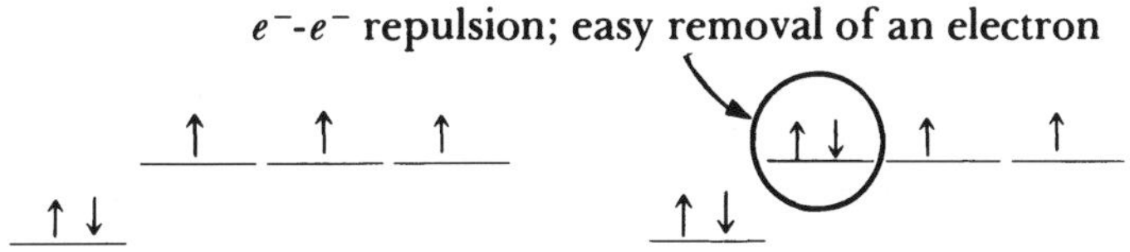

Group 5A element Group 6A element

Example 8.6 Which element in each pair has the higher first ionization energy (requires the greater amount of energy to remove the first electron)?

(a) Ge versus As (d) F versus I
(b) As versus Se (e) Si versus Cl
(c) Mg versus Al (f) Sn versus Pb

Answer

(a) As—trend within period (d) F—trend within group
(b) As—exception 2b above (e) Cl—trend within period
(c) Mg—exception 2a above (f) Sn—trend within group

SECTION 8.6 ELECTRON AFFINITY

Objectives

- To define electron affinity
- To explain endothermic and exothermic energy values.

Focus

Electron affinity is the energy change accompanying the addition of an electron to a gaseous atom. When a gaseous atom readily gains an electron, energy is *released* because the attraction between the positive nucleus and the electron is greater than the resulting electron-electron repulsion within the electron cloud. The process is exothermic. Nonmetal atoms and smaller atoms have larger electron affinities (release more energy when an electron is accepted), although this is not universally true.

A gaseous negative ion is reluctant to accept a second electron (second electron affinity) and therefore the process requires energy, an endothermic process.

SECTION 8.7 THE ELEMENTS

Objective

- To distinguish between the various groups of elements in the periodic table by means of ground-state electron configurations, valence, and characteristic chemical and physical properties.

Focus

We can make predictions about the chemical properties of an element, based upon its ground-state electron configuration. Some properties of the representative elements are listed below. Read this section very carefully to gain some understanding of an element's properties. Many elements in Groups 4A, 5A, and 6A exist in allotropic forms (see the text, Table 8.9);

that is, the element exists in more than one form or the atoms are bonded in more than one way. The two allotropic forms of oxygen are O_2 and O_3. These two forms are comprised of only O atoms but are bonded in different ways.

Elements that border the "stairstep" line in Tables 8.3 and 8.9 have some properties characteristic of both metals and nonmetals. These are called semimetals (metalloids).

Table 8.1 summarizes the properties of the representative elements.

The transition elements are metals having the predicted properties of metals: hard, lustrous, conduct electric current and heat, malleable. Most compounds of the transition elements and even some of the metals have color. The colors of gold, copper, and silver metals, and the salts of nickel, copper, iron, chromium, and manganese make a chemistry stockroom very colorful.

Little work has been done with the inner transition metals. Separation of the lanthanoids is difficult because of the low natural abundance and the high degree of similarity in their chemical and physical properties.

The actinoid metals are radioactive, and the elements after uranium are synthetic (man-made).

Table 8.1 Summary of Properties of Representative Elements

Group	Electron configuration	Common valence	Chemical properties
Hydrogen	$1s^1$	1	Less dense than air, burns with O_2; very reactive.
1A—alkali metals	ns^1	1	Very reactive with air or water, produces oxides. Heavier metals react with more oxygen, producing peroxides (O_2^{2-}) and superoxides (O_2^-). All react vigorously with H_2O, producing the hydroxide and H_2. Stored in inert liquid. Soft, lustrous metals.
2A—alkaline earth metals	ns^2	2	Harder and less reactive than 1A elements. Increasing reactivity with H_2O from Ca to Ba; Ba and Sr are similar to alkali metals in their reactivity. Be is nonreactive with H_2O. All form the oxide and react with water (except Be), forming the hydroxide and H_2.
3A	ns^2np^1	3	B is the only nonmetal. Al quickly forms impervious Al_2O_3 layer when exposed to air. The production of Al metal consumes large amounts of electrical energy for the electrolytic breakdown of Al_2O_3. Others form the oxide at elevated temperatures, but all are unreactive in H_2O. Ga, In, and Tl have limited use.

(continued)

Group	Electron configuration	Common valence	Chemical properties
4A	ns^2np^2	4	C is the only nonmetal. Si and Ge are considered semimetals (metalloids). All form oxides at elevated temperatures. C has two allotropic forms: diamond and graphite. Si comprises 27% of mass of Earth's crust. "Pure" Si and Ge are used in making semiconductors. Silicates and silica (SiO_2) are used widely in construction and in glass making. Sn and Pb are common in alloys. Pb compounds are generally insoluble. Pb is used primarily in lead storage batteries and for radioactive shielding. Pb is toxic.
5A	ns^2np^3	3	N_2 is relatively inert but reacts with O_2 in internal combustion engines, producing NO and NO_2, constituents of photochemical smog. The white allotropic form of P_4 is very reactive with O_2. Large quantities of phosphates (PO_4^{3-}) are in detergents and fertilizers. As and its compounds are very toxic.
6A	ns^2np^4	2	O_2 reacts with most elements. O_2 and O_3 are allotropic forms of oxygen. O_3, a product of photochemical smog, is very reactive and toxic. S readily forms SO_2 and SO_3 for H_2SO_4 and fertilizer production. Most metals react with S forming the sulfide. Se is used to make yellow glass for stoplights. H_2Te is probably the most foul-smelling inorganic compound.
7A—halogens	ns^2np^5	1	F_2 is a corrosive toxic gas; Cl_2 is a poisonous gas; Br_2 is a toxic liquid and I_2 is a blue-black solid. All halogens form the binary acids, HF, HCl, HBr, HI, and react with metals to form the corresponding halide.
8A	ns^2np^6	0	All noble gases are resistant to chemical attack.

QUESTIONS

Section 8.2

8.1 Categorize the following as a representative, transition, or inner-transition element. If the element belongs to a special group, such as the alkali metals, the coinage metals, the halogens, and so on, label it accordingly.

(a) $_{40}Zr$	(e) $_{92}U$	(i) $_{58}Ce$
(b) $_{67}Ho$	(f) $_{32}S$	(j) $_{36}Kr$
(c) $_{29}Cu$	(g) $_{82}Pb$	(k) $_{11}Na$
(d) $_{56}Ba$	(h) $_{17}Cl$	(l) $_{62}Sm$
		(m) $_{26}Fe$

8.2 How many 2*p* and 3*d* electrons does the atom $_{23}V$ have?

8.3 Using the periodic table as a guide, write the electron configuration for $_{92}U$, $_{27}Co$, $_{62}Sm$, and $_{83}Bi$.

8.4 Which group in the periodic table has the following *general* ground-state electron configurations?

(a) ns^2np^3	(d) ns^2np^6
(b) ns^1	(e) $ns^2(n-1)d^1$
(c) $ns^2(n-1)d^{10}$	(f) ns^2np^5

Section 8.4

8.5 Differentiate between an ionic and a covalent radius.

8.6 Explain why the oxygen atom is slightly larger than the nitrogen atom.

8.7 Select the smaller from each of the following pairs:

(a) Ca, Ge	(d) Ca^{2+}, Sc^{3+}	(g) Na^+, K^+
(b) O, F	(e) Te^{2-}, I^-	(h) In, In^{3+}
(c) S^{2-}, S	(f) Ba^{2+}, Ba	(i) C^{4+}, C^{4-}

Section 8.5

8.8 Select the element which has the higher ionization energy in each of the following pairs:

(a) Sb or Sn	(d) Sr or Rb
(b) Al or P	(e) Ba or Na
(c) S or Te	(f) Ca or Ga

Section 8.6

8.9 Which of the following has the greater electron affinity?

(a) S or As
(b) Na or Cl
(c) O or Te

Section 8.7

8.10 List the representative elements that are radioactive.

8.11 Characterize the chemical properties of the two common allotropes of phosphorus: white and red.

8.12 List all the elements that are gases at room temperature and pressure.

8.13 What is the difference between an allotrope and an isotope?

8.14 Write the four balanced equations for the reactions that occur between H_2O and Ba, Sr, Na, and K.

ANSWERS

8.1 (a) transition
(b) inner-transition; rare-earth or lanthanoid
(c) transition; coinage
(d) representative; alkaline-earth
(e) inner transition; heavy rare-earth or actinoid
(f) representative
(g) representative
(h) representative; halogen
(i) inner transition; rare-earth or lanthanoid
(j) representative; noble gas
(k) representative; alkali
(l) inner transition; rare-earth or lanthanoid
(m) transition

8.2 Six 2*p* electrons and three 3*d* electrons

8.3 ${}_{92}U$: $[Rn]7s^2 5f^4$
${}_{27}Co$: $[Ar]4s^2 3d^7$
${}_{62}Sm$: $[Xe]6s^2 4f^6$
${}_{83}Bi$: $[Xe]6s^2 4f^{14} 5d^{10} 6p^3$

8.4 (a) Group 5A (d) Group 8A
(b) Group 1A (e) Group 3B
(c) Group 2B (f) Group 7A

8.5 A covalent radius is the radius of a neutral atom whereas an ionic radius is that of a charged (positive or negative) atom.

8.6 In the nitrogen atom each 2*p* orbital has a single electron. But the fourth electron in the 2*p* orbitals of the oxygen atom results in a pairing of electrons in one 2*p* orbital. This electron-electron repulsion causes the oxygen atom to be slightly larger.

8.7 (a) Ge (d) Sc^{3+} (g) Na^+
(b) F (e) I^- (h) In^{3+}
(c) S (f) Ba^{2+} (i) C^{4+}

8.8 (a) Sb (d) Sr
(b) P (e) P
(c) S (f) Ca

8.9 (a) S
(b) Cl
(c) O
(d) C

8.10 Fr, Ra, Po, At, Rn

8.11 The vapor of white phosphorus ignites spontaneously in air. It is stored under water. Red phosphorus is fairly stable and not as dangerous to handle. Both allotropes are toxic.

8.12 H_2, N_2, O_2, F_2, Cl_2, He, Ne, Ar, Kr, Xe, Rn

8.13 An allotrope is a number of atoms of the *same* element bonded in two or more ways, whereas an isotope is an atom of a given element that has a different atomic mass.

8.14 $Ba(c) + 2H_2O(\ell) \rightarrow H_2(g) + Ba(OH)_2(aq)$
$Sr(c) + 2H_2O(\ell) \rightarrow H_2(g) + Sr(OH)_2(aq)$
$2Na(c) + 2H_2O(\ell) \rightarrow H_2(g) + 2NaOH(aq)$
$2K(c) + 2H_2O(\ell) \rightarrow H_2(g) + 2KOH(aq)$

Chapter 9

TYPES OF CHEMICAL BONDS

SECTION 9.1 BONDING TYPES

Objective

• To explain the differences among the four different types of bonding between atoms.

Focus

The four types of bonding and the general properties they impart to substances are:

1. **Metallic** bonding is only between metal atoms. These substances are good conductors of electricity.
2. **Ionic** bonding is electrostatic attraction between ions. These substances are good conductors of electricity in molten and dissolved states and have high melting points.
3. **Molecular covalent** bonding is intramolecular bonding between atoms. These substances are nonconductors of electricity and have low melting points.
4. **Network covalent** bonding is intramolecular bonding between atoms. These substances are nonconductors of electricity and have very high melting points.

SECTION 9.2 LEWIS THEORY OF BONDING

Objectives

• To write Lewis symbols for the representative elements and their ions.

• To define the octet rule and its relationship to noble gas electronic configuration.

Focus

The **Lewis symbol** represents an element's number of valence electrons by placing a respective number of dots around the symbol for the element. For representative elements, the number of dots is the same as the Group number.

Example 9.1 Write the Lewis symbols for Pb, Ca, and Cl.

Answer $\cdot\dot{Pb}:$, $Ca:$, $\cdot\underset{\cdot\cdot}{\ddot{Cl}}:$

Solution Pb is a member of Group IVA: $\cdot\dot{Pb}:$
Ca is a member of Group IIA: $Ca:$
Cl is a member of Group VIIA: $:\underset{\cdot\cdot}{\ddot{Cl}}\cdot$

Representative elements tend to gain, lose, or share valence electrons when forming a compound so that they become isoelectronic with a noble gas. Since all noble gases, except He, have eight valence electrons, this is often known as the **octet rule**.

Example 9.2 Write the Lewis symbols for Ba^{2+} and Cl^-. With which noble gas is each isoelectronic?

Answer Ba^{2+}, isoelectronic with Xe; $:\underset{\cdot\cdot}{\ddot{Cl}}:^-$, isoelectronic with Ar.

Solution Ba^{2+} has two *less* electrons than $Ba:$, leaving only Ba^{2+} as the Lewis formula. Ba^{2+} is isoelectronic with Xe.

Cl^- has one *more* electron than $:\underset{\cdot\cdot}{\dot{Cl}}:$, producing $:\underset{\cdot\cdot}{\ddot{Cl}}:^-$; $:\underset{\cdot\cdot}{\ddot{Cl}}:^-$ is isoelectronic with Ar.

SECTION 9.3 IONIC BOND

Objective

- To show the formation of an ionic bond using Lewis symbols and formulas.

Focus

The metallic representative elements tend to lose valence electrons and become isoelectronic with a noble gas in the formation of a bond. The nonmetal, which combines with the metal, tends to gain electrons and also be isoelectronic with a noble gas.

Example 9.3 Write the Lewis formula for the ionic compounds (a) barium chloride and (b) potassium sulfide.

Answer (a) $Ba^{2+} + 2:\underset{\cdot\cdot}{\ddot{Cl}}:^-$ ($BaCl_2$); (b) $2K^+ + :\underset{\cdot\cdot}{\ddot{S}}:^{2-}$ (K_2S)

Solution (a) From the periodic table, the Ba atom (Group 2A), [Xe]$6s^2$, has two valence electrons that it loses in ionic bond formation to form Ba^{2+} and become isoelectronic with Xe.

The Cl atom, [Ne]$3s^23p^5$, has seven valence electrons and needs one electron to form $:\ddot{Cl}:^-$ and become isoelectronic with Ar. Therefore two Cl atoms must be available per Ba atom for ionic bond formation and for each atom to be isoelectronic with a noble gas.

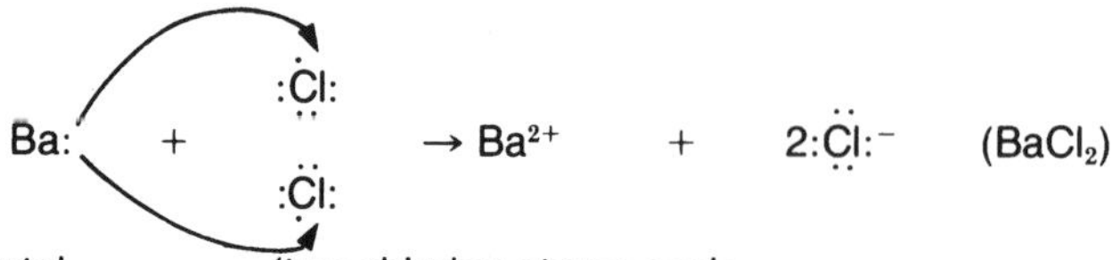

(a barium atom loses a total of 2 electrons)

(two chlorine atoms each gain 1 electron for a total of 2 electrons)

(b) K·, [Ar]$4s^1$, has one valence electron to lose (a metal) to be isoelectronic with Ar (a noble gas) :Ṣ:, [Ne]$3s^23p^4$, has to gain two valence electrons (a nonmetal combining with a metal) to be isoelectronic with Ar (a noble gas). Therefore two K·'s are needed for each :Ṣ:

(two K's lose a total of 2 electrons)

(S gains a total of 2 electrons)

The Ba and K atoms lose electrons in the formation of ionic compounds; this loss of electrons is called **oxidation**. The Cl and S atoms gain electrons in the formation of ionic compounds; this gain of electrons is called **reduction**. The reaction to form an ionic bond, in which a substance gains electrons and another substance loses electrons, is called an **oxidation–reduction (redox) reaction**.

SECTION 9.4 FORMATION OF IONIC SOLIDS

Objectives

- To explain and calculate ΔH_f for ionic compounds, given the various energy factors involved in their formation.
- To explain the effects of ionization energy, electron affinity, and crystal lattice energy on the stability of an ionic compound.

Focus

The transformation of solid K and liquid Br_2 into ionic KBr according to the equation

$$K(c) + \tfrac{1}{2}Br_2\ (\ell) \rightarrow KBr(c)$$

can be pictured as a cycle of energy steps (the Born-Haber cycle) as shown in Figure 9.1.

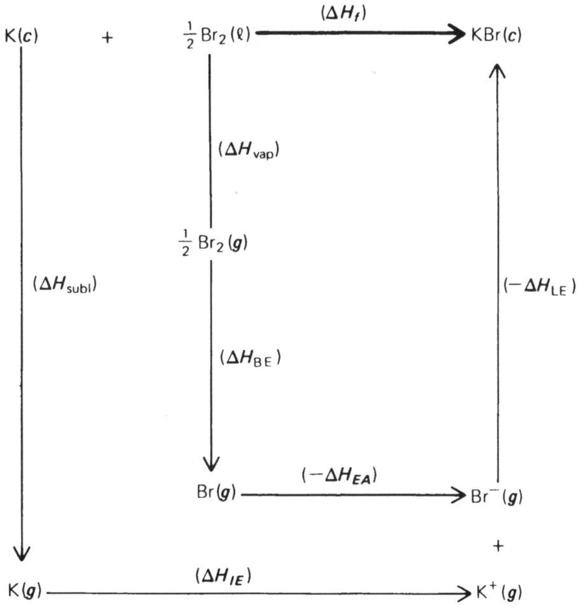

Figure 9.1 The Born-Haber cycle:

$$\Delta H_f = \underbrace{\Delta H_{subl} + \Delta H_{vap} + \Delta H_{BE} + \Delta H_{IE}}_{\text{endothermic}} - \underbrace{\Delta H_{EA} - \Delta H_{LE}}_{\text{exothermic}}$$

where *vap* is vaporization, *LE* is lattice energy, *subl* is sublimation, *BE* is bond energy, *IE* is ionization energy, and *EA* is energy affinity.

For gaseous nonmetals such as F_2, Cl_2, and O_2, $\Delta H_{vap} = 0$ because they are already in the gaseous state.

The three major factors explaining the stability of ionic compounds are:

1. ΔH_{IE}—an endothermic quantity
2. ΔH_{EA}—an exothermic quantity
3. ΔH_{LE}—an exothermic quantity

The more negative (exothermic) the sum of these three values, the more stable is the ionic bond; therefore, a low IE and a large EA, coupled with a large LE, result in a more stable ionic bond. ΔH_{IE} values, for the formation of positive ions, increase from left to right within a period and decrease from top to bottom within a group of elements. ΔH_{EA} values, for the formation of negative ions, increase (more exothermic) from left to right within a period and decrease from top to bottom within a group of elements. ΔH_{LE} values increase (more exothermic) when the charge on the positive and negative ions is large. This larger charge causes a greater ionic attraction in the gaseous state and the formation of a more stable salt.

SECTION 9.5 COVALENT BOND: COVALENCY

Objectives

- To determine the covalency of a nonmetallic representative element.
- To write the Lewis formula of a covalent compound by using the covalency requirement for each element in the compound.

Focus

A **covalent bond** is a sharing of electrons between two nonmetals in order to achieve greater stability.

The **covalency** of a nonmetallic representative element is the number of covalent bonds it forms. Therefore, its covalency is equal to the number of electrons it requires to form a neutral molecule and become usually isoelectronic with a noble gas.

Example 9.4 Determine the covalency of C, P, Sn, Cl.

Answer C—4, P—3, Sn—4, Cl—1

Solution C has 4 valence electrons (IVA) and needs to share 4 more from other atoms. Each C atom needs to form 4 covalent bonds to achieve an octet of electrons.

P has 5 valence electrons and needs 3 more by sharing through 3 covalent bonds—a covalency of 3.

Sn has 4 valence electrons and needs 4 more to be isoelectronic with a noble gas. The formation of 4 covalent bonds is needed—a covalency of 4.

Cl has 7 valence electrons and needs 1 more. The formation of 1 covalent bond is needed—a covalency of 1.

If the covalency does not occur (the required number of covalent bonds are not formed), yet the atom is isoelectronic with a noble gas, then the atomic group has an electrical charge. A group of covalently bonded atoms having a charge is called a *polyatomic ion*. For example, in OH^-, the covalency of oxygen is 2, but only one covalent bond is formed.

For oxygen to complete its octet, an additional electron is needed, giving the OH group a negative charge.

$$\cdot\ddot{\underset{\cdot\cdot}{O}}{:}H + e^- \rightarrow {:}\ddot{\underset{\cdot\cdot}{O}}{:}H^- \quad \text{or} \quad -\ddot{\underset{\cdot\cdot}{O}}-H^-$$

Example 9.5 Write the Lewis formulas for SiH_4, PCl_3, and H_2O_2 consistent with the octet rule and covalency requirements.

Answer

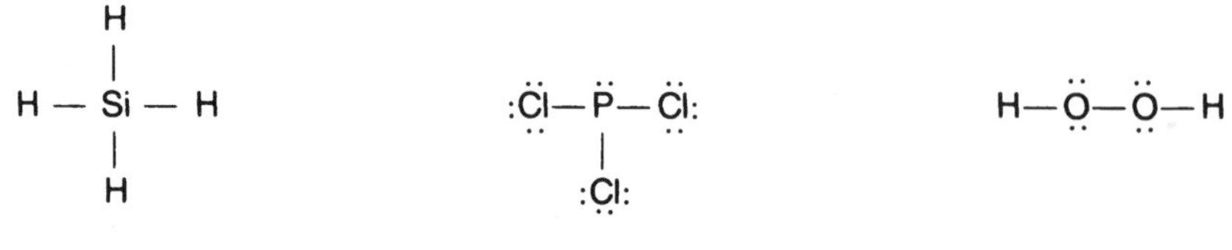

Solution The covalency requirement for Si is 4, —Si—, and for H is 1, —H.

Since there are 4 H atoms, each one bonds to the Si atom:

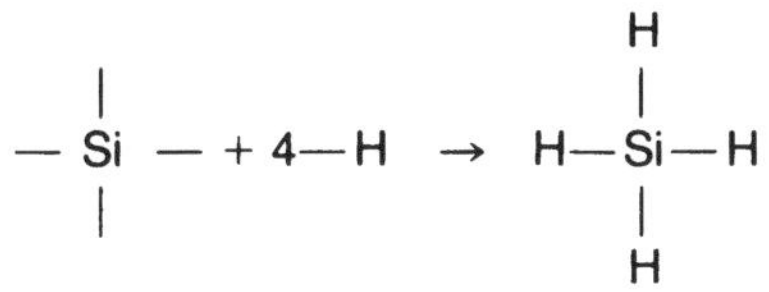

Each "—" in the *formula* of SiH_4 represents a pair of bonding electrons.

The covalency requirement for P is 3, —P—, and for Cl is 1, —Cl:. The 3 Cl atoms therefore bond to the P atom:

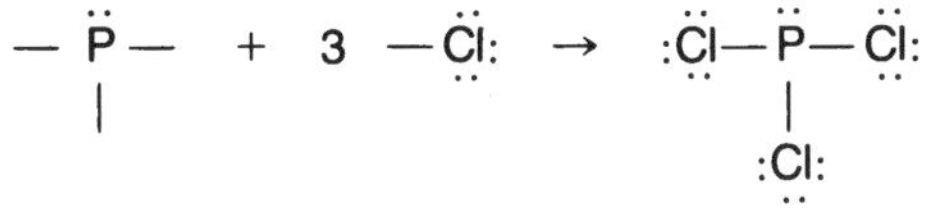

The covalency requirement for each oxygen atom is 2, —O—, and each H atom is 1, — H. If the two H atoms bond to a single O atom, H_2O forms and the other —O— is omitted; therefore the two oxygen atoms must bond together, or

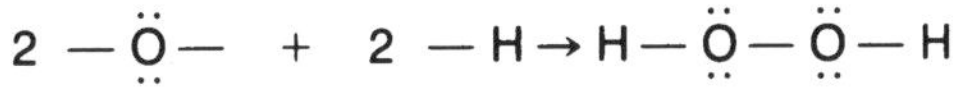

Notice that each atom, except H, in the three formulas follow the octet rule and each H atom is isoelectronic with the He atom.

When undecided on which atom to select for building the Lewis formula, start with the one having the highest covalency. If two or more atoms have the same highest covalency, they will probably share a common bond.

SECTION 9.6 POLAR COVALENT BONDS: ELECTRONEGATIVITY

Objective

• To predict the relative polarity of a bond based upon the electronegativity values for the elements in that bond.

Focus

The **electronegativity** of an element is the relative attraction an atom has for electrons in a chemical bond: the greater the difference in electronegativity between the bonded atoms, ΔEn, the greater its polarity.

Example 9.6 From the text, Table 9.3, arrange the bonds between the following pairs of atoms in order of increasing polarity: Sb, Cl; Ge, Br; C, I; and P, F.

Answer C — I < Ge — Br < Sb — Cl < P — F

Solution Determine Δ*En* for each pair.

	Sb:Cl	Ge:Br	C:I	P:F
En	1.8:2.8	2.0:2.7	2.5:2.2	2.1:4.1
Δ*En*	1.0	0.7	0.3	2.0

Therefore the C — I bond is the least polar and the P — F bond the most.

SECTION 9.7 MULTIPLE BONDS

Objective

- To write Lewis formulas for molecules with multiple bonds.

Focus

In some cases, especially for the elements C, N, O, and S, the covalency and octet requirement for an atom is possible only through the formation of multiple bonds. For example, in carbon dioxide, CO_2, the covalency for C is 4, as drawn in the following forms:

$$(-\underset{|}{\overset{|}{\mathrm{C}}}- \quad \text{or} \quad =\mathrm{C}= \quad \text{or} \quad -\overset{|}{\mathrm{C}}= \quad \text{or} \quad -\mathrm{C}\equiv)$$

and for each oxygen it is

$$2\,(-\underset{..}{\ddot{\mathrm{O}}}- \quad \text{or} \quad =\underset{..}{\ddot{\mathrm{O}}})$$

For CO_2, C is $=\mathrm{C}=$ and O is $=\underset{..}{\ddot{\mathrm{O}}}$; therefore, the Lewis formula is

$$=\mathrm{C}= \; + \; 2\; =\underset{..}{\ddot{\mathrm{O}}} \rightarrow \underset{..}{\ddot{\mathrm{O}}}=\mathrm{C}=\underset{..}{\ddot{\mathrm{O}}}$$

Example 9.7 Write the Lewis formula consistent with the octet rule and covalency requirements for C_2F_4 and HNCO.

Answer

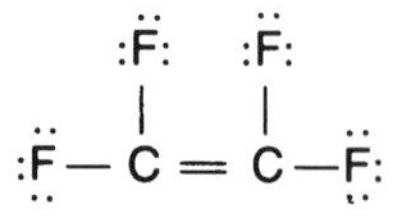

$$\underset{..}{\ddot{\mathrm{O}}}=\mathrm{C}=\ddot{\mathrm{N}}-\mathrm{H} \quad \text{or} \quad \mathrm{H}-\underset{..}{\ddot{\mathrm{O}}}-\mathrm{C}\equiv\ddot{\mathrm{N}}$$

Solution 2C, covalency is 4, $-\underset{|}{\overset{|}{\mathrm{C}}}- \quad \text{or} \quad -\overset{|}{\mathrm{C}}= \quad \text{or} \quad =\mathrm{C}= \quad \text{or} \quad -\mathrm{C}\equiv$

4F, covalency is 1, $-\underset{..}{\ddot{\mathrm{F}}}:$

Remember, in writing the Lewis formula first write the atom having the highest covalency requirement; if two have the highest, they are most likely bonded to one another. The two C atoms must be bonded to one another.

$$2\,{-}\overset{|}{C}{=} + 4\,{-}\ddot{\underset{..}{F}}{:} \rightarrow {:}\ddot{F}{-}\overset{:\ddot{F}:}{\overset{|}{C}}{=}\overset{:\ddot{F}:}{\overset{|}{C}}{-}\ddot{F}{:}$$

1 H: covalency requirement is 1: — H

1 N: covalency requirement is 3: $-\ddot{N}-$ (with a single bond below) or $-\ddot{N}=$ or $\ddot{N}\equiv$

1 C: covalency requirement is 4: $-\overset{|}{\underset{|}{C}}-$ or $-\overset{|}{C}=$ or $=C=$ or $-C\equiv$

1 O: covalency requirement is 2: $-\ddot{\underset{..}{O}}-$ or $=\ddot{\underset{..}{O}}$

Starting with C, several structures are possible.

(a) Single bonds to C, $-\overset{|}{\underset{|}{C}}-$; none

(b) Double bonds to C, $-\overset{|}{C}{=}$; none

Double bonds to C, $=C=$;

$$\ddot{\underset{..}{O}}{=} + {=}C{=} + {=}\ddot{N}{-} + {-}H \rightarrow \ddot{\underset{..}{O}}{=}C{=}\ddot{N}{-}H$$

(c) Triple bond to C, $-C\equiv$;

$$H{-} + {-}\ddot{\underset{..}{O}}{-} + {-}C{\equiv} + {\equiv}\ddot{\underset{..}{N}} \rightarrow H{-}\ddot{\underset{..}{O}}{-}C{\equiv}\ddot{N}$$

SECTION 9.8 EXCEPTIONS TO THE OCTET RULE

Objective

- To write a Lewis formula for compounds that are exceptions to the covalency requirement and octet rule.

Focus

In writing Lewis formulas for molecules or ions that are exceptions to the covalency requirement and octet rule, the major difficulty is recognizing the exception. How do you know whether a particular molecule or ion conforms? When do you insert a double bond so that the substance conforms?

The text lists five exceptions to the covalency requirement and the octet rule. Here we list some rules to help you recognize these exceptions.

1. Generally, C, N, O, and S (the elements in the first complete row in the periodic table) form *double* or *triple* bonds in addition to *single bonds* in order to conform to "the rules."

2. All other elements in the table *usually* form only *single* bonds. These bonded atoms may or may not conform to "the rules." The atom that most often adjusts to a new covalency requirement is the central atom. It is usually the lone atom in the molecule or ion having the lowest electronegativity and the highest covalency.
3. If a molecule has an odd number of valence electrons (that is, NO_2 has $17e^-$) or if it contains a noble gas atom (that is, XeF_4), the molecule cannot conform to the rule of covalency and octet formation.

Example 9.8 SF_4 exceeds its covalency requirement and the octet rule. How can this be recognized and what is the Lewis formula?

Answer We recognize the exception because the covalency requirement of S (=2) cannot meet the covalency requirement for 4 F atoms. No double bonds form. Therefore, S exceeds its covalency requirement to accommodate the F atoms.

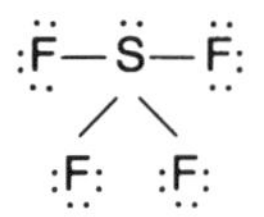

Solution S has a covalency requirement of 2, —S̈— and F have a covalency requirement of 1, —F:

S, the lone atom, has the highest covalency requirement and the lowest electronegativity. Therefore, S serves as the central atom and accomodates the covalency requirement of F. As a result, S expands its availability of electrons to accomodate the F atoms: ·S· or —S—.

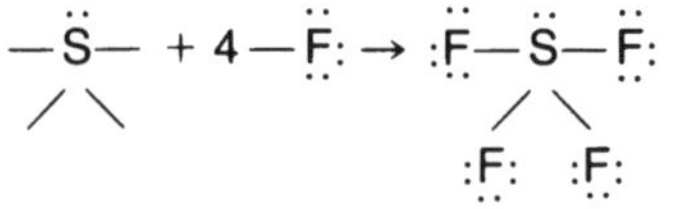

Thus the S atom accommodates 10 valence electrons.

Example 9.9 Write the Lewis formula for $XeOF_4$.

Answer The O atom accepts 2 electrons from Xe to complete its octet.

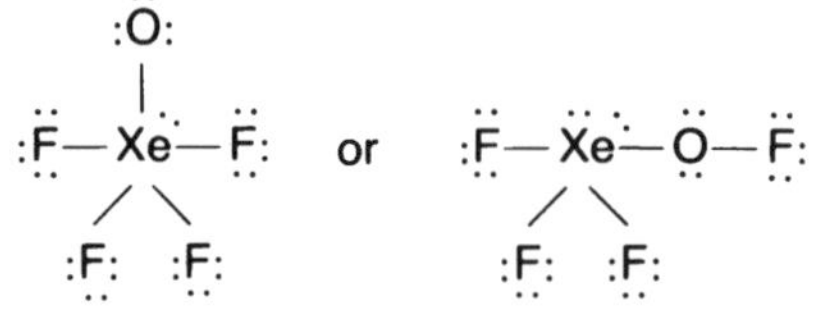

Solution We recognize immediately an exception to the octet rule with the presence of Xe. Xe and O are lone atoms in the molecule; since Xe is the least electronegative, it is the central atom in the Lewis formula. Xe increases its covalency for bonding to oxygen, —O—, and the 4 F atoms, —F:.

Xe prefers to form only single bonds (rule 2). Two Lewis formulas are possible with Xe arranging its 8 electrons: The O atom, :O:, accepts 2 electrons from Xe to complete its octet.

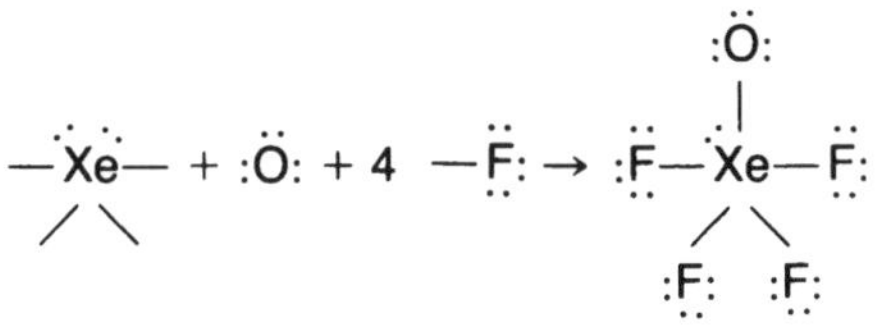

(This configuration of atoms is the accepted arrangement)

or the O atom accepts 1 electron from Xe and 1 electron from F

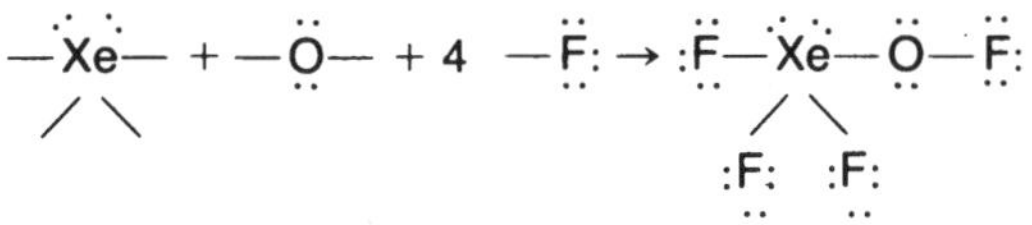

SECTION 9.9 OXIDATION NUMBER

Objective

- To assign an oxidation number to, or determine the oxidation state of, an element in covalently bonded molecules and polyatomic ions.

Focus

Assigning an **oxidation number** to an atom is a bookkeeping device used to account for its valence electrons when a bond forms with another element.

Memorize the four rules that are listed in the text for assigning oxidation numbers.

Example 9.10 Assign the appropriate oxidation state to P in the compounds H_3PO_4 and H_3PO_3.

Answer H_3PO_3, P is 3+; H_3PO_4, P is 5+.

Solution According to rule 3, H is 1+ and O is 2−.

H_3PO_3	H_3PO_4
3 H atoms, 1+ for each = 3+ 3 O atoms, 2− for each = 6−	3 H atoms, 1+ for each = 3+ 4 O atoms, 2− for each = 8−

According to rule 4, the sum of the oxidation states for a molecule must equal zero; therefore

$$(3+) + (P) + (6-) = 0 \qquad (3+) + P + (8-) = 0$$
$$P = 3+ \qquad P = 5+$$

A change in oxidation state shows that an oxidation-reduction reaction has occurred. **Reduction** refers to a decrease in oxidation number of an atom and **oxidation**, an increase in oxidation number.

Example 9.11 In the reaction, $Br_2 + SO_2 + 2H_2O \rightarrow 2HBr + H_2SO_4$ determine which elements change oxidation state and whether they are oxidized or reduced.

Answer Br—reduced; S—oxidized

Solution Determine the oxidation number of each element in each compound of the equation.

Br_2:	Br	= 0 (rule 1)
SO_2:	O	= 2− (rule 3); therefore
	S	= 4+ (rule 4)
H_2O:	O	= 2−, H = 1+ (rule 3)
HBr:	H	= 1+ (rule 3); therefore
	Br	= 1− (rule 4)
H_2SO_4:	O	= 2− (rule 3), H = 1+ (rule 3); therefore
	S	= 6+ (rule 4)

Br = 0 → Br = 1—, a decrease in oxidation state; reduction
S = 4+ → S = 6+, an increase in oxidation state; oxidation

SECTION 9.10 MORE ON LEWIS FORMULAS

Objective

- To write more systematic Lewis structures for more difficult structures.

Focus

Additional rules for writing Lewis structures are listed in the Study Guide, Section 9.8. Review and practice the rules using more examples. A fourth rule can be added for writing formulas for polyatomic ions.

Rule 4. When writing the Lewis formula for a polyatomic ion, add (for negative ions) or remove (for positive ions) an appropriate number of electrons from the central atom.

Example 9.12 Write the Lewis formula for ICl_4^- and PF_4^+.

Answer

Solution I, the lone atom, has the lower electronegativity. Therefore, as the central atom, it adjusts its covalency to accomodate the covalency requirement of the 4 Cl atoms. In addition, 1e⁻ is added to the I atom, accounting for the 1− charge on the ion:

$$:\ddot{\underset{..}{I}}\cdot + e^- \rightarrow [:\ddot{\underset{..}{I}}:]^-$$

$[:\ddot{\underset{..}{I}}:]^-$ must adjust its covalency to accomodate the 4 Cl atoms or

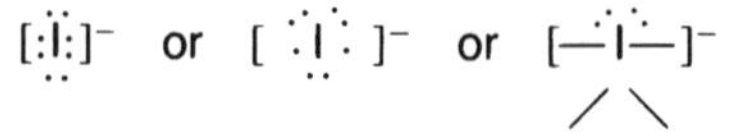

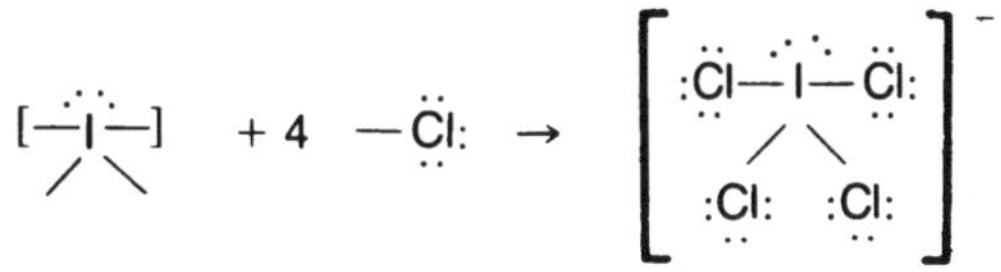

To make the PF_4^+ ion, an electron is removed from the total number of valence electrons. Here we remove it from the central P atom.

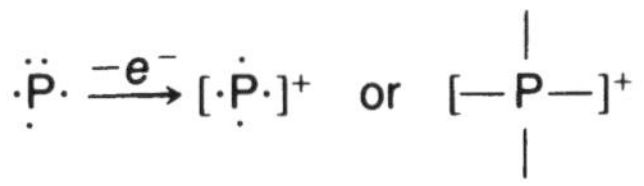

Now bond the 4 F atoms, each with a covalency requirement of 1, to the P^+ ion

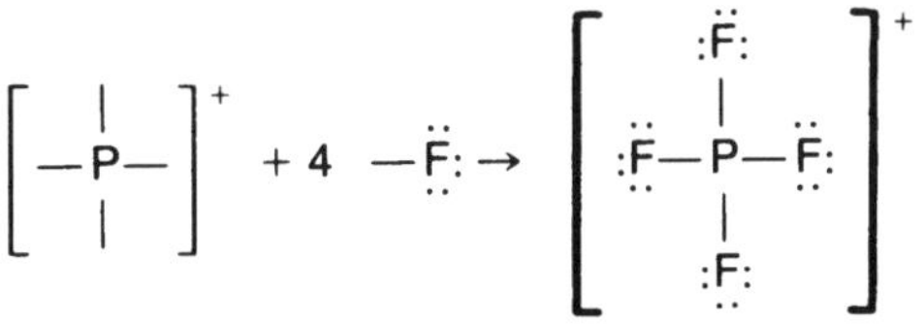

SECTION 9.11 PROPERTIES OF COVALENT BONDS

Objectives

- To explain the relationship between bond distance, bond polarity, and bond strength.
- To define a bond angle.

Focus

Shorter bond lengths and greater bond polarities produce stronger bonds. Multiple bonds are stronger than single ones.

Example 9.13 Select the stronger bond from each.

(a) :C≡O: or :Ö=C=Ö:

(b) :B̤̈r—F̤̈: or :C̤̈l—F̤̈:

Answer (a) :C≡O:

(b) :B̤̈r—F̤̈:

Solution (a) The triply bonded CO in CO is stronger and has a shorter bond length than the doubly bonded CO in CO_2.

(b) The Br—F is more polar, having a greater bond strength than Cl—F.

Bond angle is a property of molecules having three or more atoms. However, there is no relationship between bond angle and bond length, bond strength, or bond polarity.

SECTION 9.12 RESONANCE

Objectives

- To define resonance structures.
- To write contributing Lewis structures for a molecule or ion, when a single structure does not correctly represent the location of the electron pairs.

Focus

If several correct Lewis formulas can be written for a single molecule or ion, then none represents its actual structure. The correct structure has characteristics of each of these Lewis formulas. For each formula,

1. The skeleton arrangement of the atoms stays the same.
2. The number of electron pairs in the formula remains the same.
3. The most stable resonance structure is one in which each atom has an octet of electrons and shows its common covalency.

Example 9.14 Write all the Lewis formulas for SO_2 and N_3^-.

Answer

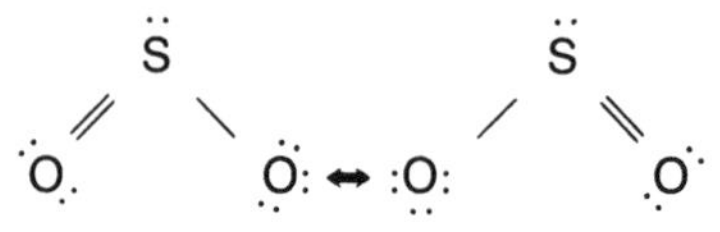

$$[:N=N=N:]^- \leftrightarrow [:N\equiv N-N:]^- \rightarrow [:N-N\equiv N:]^-$$

Solution S, the lone atom and the least electronegative atom, is the central atom with a covalency requirement of 2, —S— or =S. Each O atom also has a covalency requirement of 2, —O— or O: or =O.

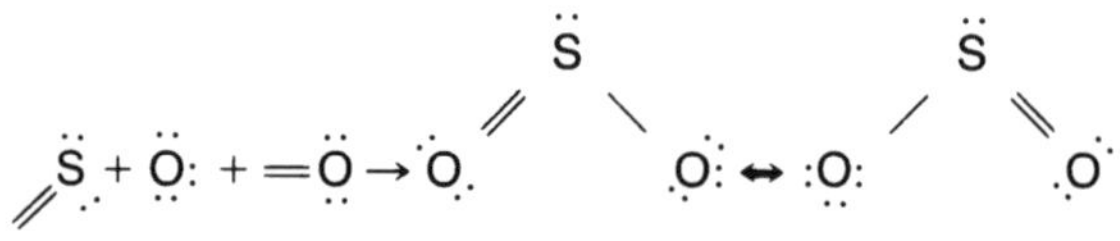

The actual SO_2 structure is neither one nor the other but somewhere in between. For N_3^-, we must add one electron to the central nitrogen atom $\cdot N\cdot + e^- \rightarrow [\cdot N\cdot]^-$.

$$[:N=N=N:]^- \leftrightarrow [:N\equiv N-N:]^- \leftrightarrow [:N-N\equiv N:]^-$$

SECTION 9.13 SHAPES OF MOLECULES

Objective

- To predict the shape of a molecule or ion based upon the number of pairs of electrons in the valence shell of the central atom.

Focus

The shape of molecules and ions can be predicted by positioning the electron pairs in the valence shell of the central atom in such a way as to reduce electron pair-electron pair repulsions.

The five basic shapes of molecules and ions and their corresponding number of electron pairs where there are only single bonds is summarized in Table 9.1.

Table 9.1 Electron Pairs and Shapes of Molecules

Electron pairs	Bonding pairs	Unshared pairs	Orientation of e^- pairs	Shape of molecule	
2	2	0	Linear	Linear	$BeCl_2$, $HgCl_2$
3	3	0	Trigonal	Trigonal	BF_3, $AlCl_3$
3	2	1	Planar	Planar bent (V-shaped)	$SnCl_2$
4	4	0	Tetrahedral	Tetrahedral	CH_4, BF_4
4	3	1		Triangular pyramid	NH_3, PCl_3
4	2	2		Bent (V-shaped)	H_2O, H_2S
5	5	0	Trigonal bipyramid	Trigonal bipyramid	PCl_5
6	6	0	Octahedral	Octahedral	SF_6

The presence of a multiple bond in a Lewis formula has the same effect on the molecular structure as a single bond or an electron pair. Remember too that unshared electron pairs only *contribute* to the molecular shape; they are not part of the shape.

Example 9.15 Predict the structures of SbF_2^+, $SnCl_3^-$, and $TlBr_4^-$.

Answer SbF_2^+, V-shaped; $SnCl_3^-$, trigonal pyramid; $TlBr_4^-$, tetrahedral

Solution Sb (Group 5A) has five valence electrons minus one because of the 1+ charge on the ion. With 2 F atoms, each with a covalency requirement of 1, the Sb covalency is 2 or $-\ddot{Sb}-$.

The Lewis formula is $[:\ddot{F}-\ddot{Sb}-\ddot{F}:]^+$. With 2 bonding and 1 unshared pair of electrons, the structure is V-shaped.

Sn (Group 4A) has four valence electrons plus one because of the ion's 1− charge. With the 3 Cl atoms, each with a covalency requirement of 1, the covalency of Sn must be 3 or $-\ddot{Sn}-$ (with a third bond below). The Lewis formula is

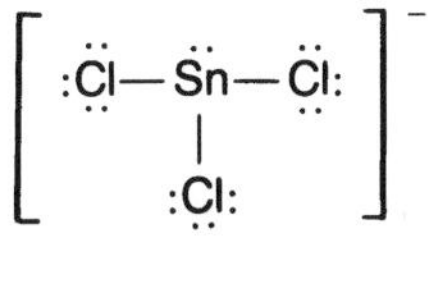

With 3 bonding pairs and 1 unshared pair, the structure is the shape of a trigonal pyramid.

Tl (Group 3A) has 3 valence electrons plus one because of the 1− charge of the ion. A covalency requirement of 4 is needed for the 4 Br atoms, each with a covalency of 1. The Lewis formula is

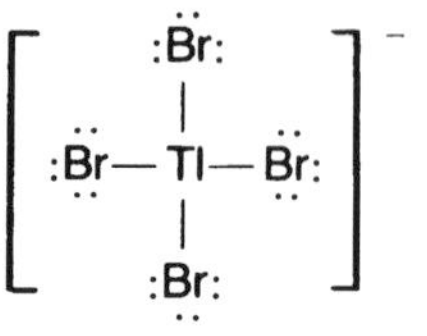

With four bonding pairs of electrons, the structure of the ion is tetrahedral.

Example 9.16 Predict the structure of SO_2 and $ClNO_2$.

Answer SO_2, bent (V-shaped); $ClNO_2$, triangular planar

Solution Always write the Lewis formula for the compound. For SO_2, see Study Guide, Example 9.14. The orientation of three electron pairs (plus a fourth in the double bond of the Lewis structure), one of which is an unshared pair, produces a bent or V-shaped structure.

For $ClNO_2$, the central atom is N because it is the least electronegative of the lone atoms, Cl and N, and has the highest covalency requirement of 3.

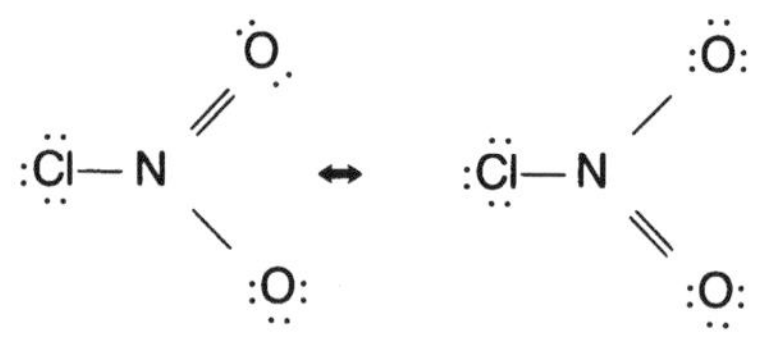

Considering the multiple bond as a pair of electrons, we have a trigonal planar structure.

SECTION 9.14 DIPOLE MOMENTS

Objectives

- To define a bond moment and a dipole moment.
- To explain the presence of a dipole moment in a covalently bonded molecule.

Focus

A **bond moment** results from a difference in electronegativity between two covalently bonded atoms. The vector sum of all the bond moments in a molecule produces the **dipole moment** of a molecule.

If the vector sum of the bond moment is equal to zero, the molecule is **nonpolar**, and its dipole moment, μ, is zero, $\mu = 0$. If the vector sum is nonzero, the molecule is **polar**, and μ is nonzero.

Example 9.17 Is the $SnCl_2$ molecule polar or nonpolar?

Answer Polar

Solution Write the Lewis formula and then determine its shape. The Lewis formula is

$$:\ddot{\underset{..}{Cl}} - \ddot{Sn} - \ddot{\underset{..}{Cl}}:$$

With 2 bonding and 1 nonbonding pairs of electrons the shape of the molecule is bent.

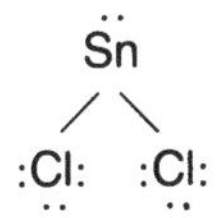

Each Sn — Cl bond is polar with the bond moment directed toward the Cl; Cl is more electronegative than Sn (from Table 9.3, Sn = 1.7, Cl = 2.8).

$$\delta^+ \mapsto \delta^-$$

$$-\ddot{Sn} - \ddot{\underset{..}{Cl}}:$$

Adding the two vectors and maintaining the bond angle

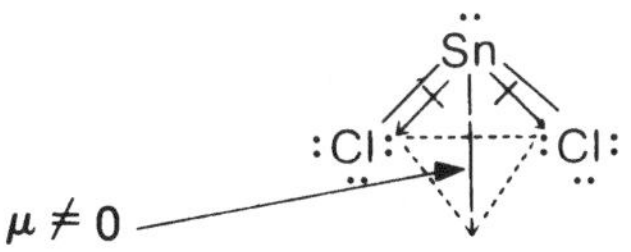

Therefore, the molecule is polar.

QUESTIONS

Section 9.1

9.1 Identify the type of bonding in the following substances.

(a) Naphthalene (mothballs) vaporizes (sublimes) at room temperature, but neither the solid nor the liquid conducts current.

(b) Molten KI conducts current.

(c) Boron has a melting point of 2300°C and does not conduct current.

9.2 Explain why $BeCl_2$ has a lower melting point than $CaCl_2$.

Section 9.2

9.3 Write the Lewis symbols for I and I^-; Sr and Sr^{2+}; Te; Sn and Sn^{4+}; Li and Li^+; Ge and Ge^{3+}; and Ar.

9.4 Determine the number of valence electrons on the Ca, Cs, Tl, Si, and Te atoms.

Section 9.3

9.5 Write the Lewis formulas for the ionic compounds (a) lithium hydride; (b) potassium sulfide; (c) calcium carbide; and (d) germanium bromide.

9.6 Write the Lewis formula for the ionic compound formed between the following elements; assume each atom is isoelectronic with a noble gas: (a) K and N; (b) Li and S; and (c) Tl and F.

Section 9.4

9.7 Circle the one with the

(a) most positive ΔH_{EA}: O, F.

(b) least positive ΔH_{IE}: Mg, Ca.

(c) most negative ΔH_{LE}: NaCl, $MgCl_2$.

(d) most negative ΔH_{EA}: I, Br.

(e) most positive ΔH_{LE}: Mg_3N_2, MgO.

9.8 From the following data, calculate ΔH_f for

$Li(c) + \frac{1}{2}F_2(g) \rightarrow LiF(c)$:

$Li(c) \rightarrow Li(g)$	ΔH_{subl}	$= +155.2$ kJ/mol
$\frac{1}{2}F_2(g) \rightarrow F(g)$	$\frac{1}{2}\Delta H_{BE}$	$= +77.2$ kJ/mol
$Li(g) \rightarrow Li^+(g)$	ΔH_{IE}	$= +520.0$ kJ/mol
$F(g) \rightarrow F^-(g)$	ΔH_{EA}	$= -333.0$ kJ/mol
$Li^+(g) + F^-(g) \rightarrow LiF(c)$	ΔH_{LE}	$= -1005.1$ kJ/mol

Section 9.5

9.9 What is the covalency of N, S, Br, and Ar?

9.10 What is the charge on the NH_2 and the PH_4 polyatomic groups, assuming that each atom is isoelectronic with a noble gas?

9.11 Write the Lewis formula consistent with the octet rule and covalency requirement for SF_2, P_2I_4, SiF_4, NH_3O, PCl_2Br and CH_5As.

Section 9.6

9.12 Arrange the bonds in order of increasing polarity: C — F; O — H; P — O; Si — I; and Ca — Br.

Section 9.7

9.13 Write the Lewis formula consistent with the octet rule and covalency requirements for C_2H_4O, O_3, and C_3H_4 (two possibilities).

Section 9.8

9.14 Write the Lewis formula for CH_3, InCl, IF_5, $AsCl_5$, and XeF_2, none of which obeys the octet rule or adheres to the predicted covalency requirement.

Section 9.9

9.15 Determine the oxidation number of the underlined element in the following compounds: $K_2\underline{Mn}O_4$, $K\underline{Mn}O_4$, $\underline{Mn}O_2$, $\underline{Mn}Cl_2$, $H_2\underline{S}O_3$, $\underline{S}O_3$, $H_2\underline{S}$, $\underline{S}F_2$, $\underline{Fe}O$, $\underline{Fe}_2O_3$, and $\underline{Xe}F_6$.

9.16 In the following pairs of transformations, determine whether oxidation or reduction of the underlined element has occurred: $\underline{Mn}O_4^- \rightarrow \underline{Mn}O_2$; $\underline{N}H_3 \rightarrow \underline{N}O_2$; $\underline{Cl}^- \rightarrow \underline{Cl}O_3^-$; $H_2\underline{O}_2 \rightarrow H_2\underline{O}$; $\underline{Cu}_2S \rightarrow \underline{Cu}O$; $\underline{P}_4O_{10} \rightarrow \underline{P}H_3$.

Section 9.10

9.17 Write the Lewis formula for NH_2^-, PO_4^{3-}, BH_4^-, ICl_2^+, PF_4^-, and XeF_5^+.

Section 9.11

9.18 Arrange the following sets of bonds in order of increasing bond strength:

(a) I — F, I — Br, I — Cl, I — I

(b) C ≡ N, C ≡ C, N ≡ N (see text, Table 9.6)

(c) Si — F, Si — Cl, Si — Si, F — F

Section 9.12

9.19 Write the acceptable resonance structures for CN_2^{2-} (C is the central atom), O_3, and N_2O_3 (atom arrangement of $\mathrm{O\,N\,N}\genfrac{}{}{0pt}{}{\mathrm{O}}{\mathrm{O}}$). Follow the three rules from the Study Guide.

Section 9.13

9.20 Predict the shapes of the following molecules and ions: AsH_3, GeF_2, BiI_4^+, $SbOCl_3$ (Sb is central atom), $InCl_2^+$, $OPCl_3$ (P is central atom), and $OSCl_2$ (S is central atom).

Section 9.14

9.21 Indicate which of the following covalent molecules have an appreciable dipole moment: PF_3, PF_5, SCl_2, AsH_3, GeF_2, $OPCl_3$, and $OSCl_2$.

ANSWERS

9.1 (a) covalent; (b) ionic; (c)covalent

9.2 $BeCl_2$ is a covalent substance; $CaCl_2$ is ionic.

9.3 :I· and [:I:]$^-$; Sr: and Sr^{2+}; :Te: ; ·Sn: and Sn^{4+}; Li· and Li^+; Ge: and Ge^{3+}; :Ar:

9.4

Ca: 2
Cs: 1
Tl: 3
Si: 4
Te: 6

9.5 (a) $Li^+ + :H^-$ (LiH)
(b) $2K^+ + :\ddot{S}:^{2-}$ (K_2S)
(c) $2Ca^{2+} + :\ddot{C}:^{4-}$ (Ca_2C)
(d) $Ge^{3+} + 3\,:\ddot{Br}:^-$ ($GeBr_3$)

9.6 (a) $3K^+ + :\ddot{N}:^{3-}$
(b) $2Li^+ + :\ddot{S}:^{2-}$
(c) $Pb^{4+} + 4:\ddot{F}:^-$

9.7 (a) O; (b) Ca; (c) $MgCl_2$; (d) I; (e) Mg_3N_2

9.8 −585.6 kJ/mol

9.9 N, 3; S, 2; Br, 1; Ar, 0

9.10 NH_2^-; PH_4^+

9.11

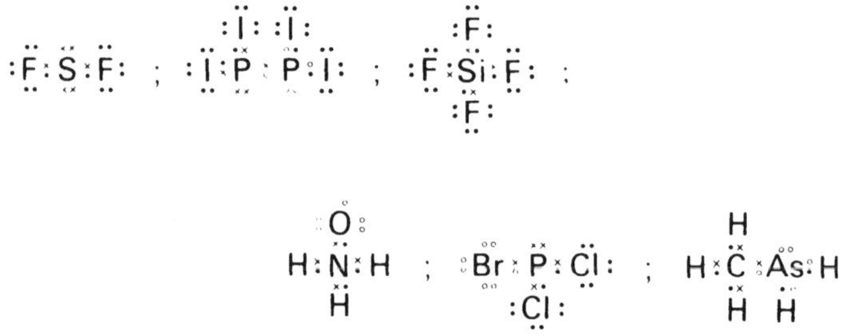

9.12 SiI < OH < PO < CF < CaBr

9.13

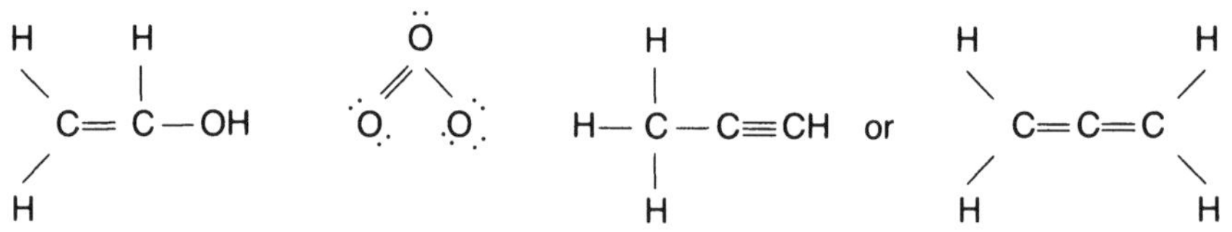

9.14

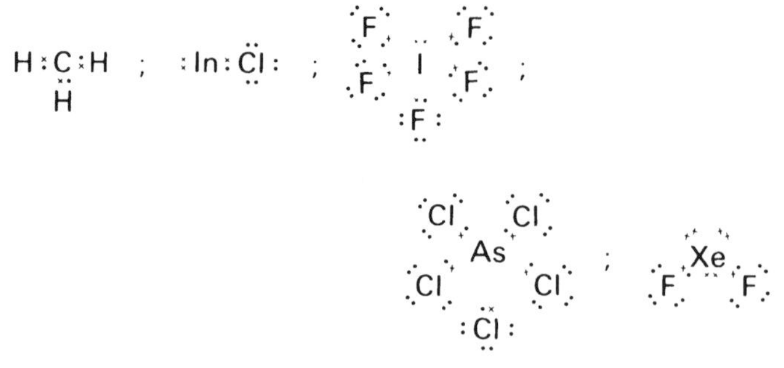

9.15 Mn = 6+; Mn = 7+; Mn = 4+; Mn = 2+; S = 4+; S = 6+; S = 2−; S = 2+; Fe = 2+; Fe = 3+; Xe = 6+

9.16 Mn—reduction; N—oxidation; Cl—oxidation; O—reduction; Cu—oxidation; P—reduction

9.17

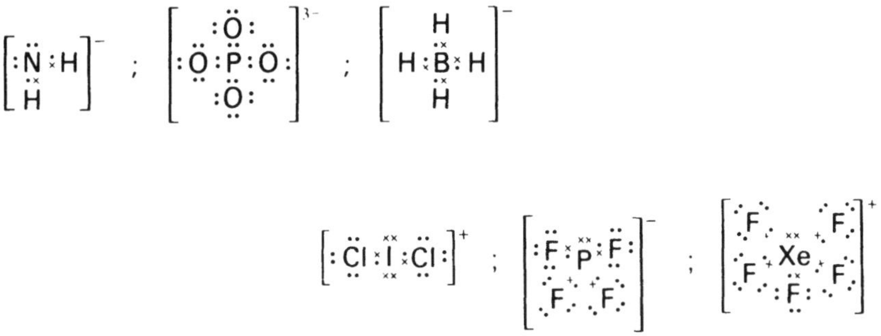

9.18 (a) II < IBr < ICl < IF

(b) C ≡ C < N ≡ N < C ≡ N

(c) Si — Si < F — F < SiCl < SiF

9.19 CN_2^{2-}

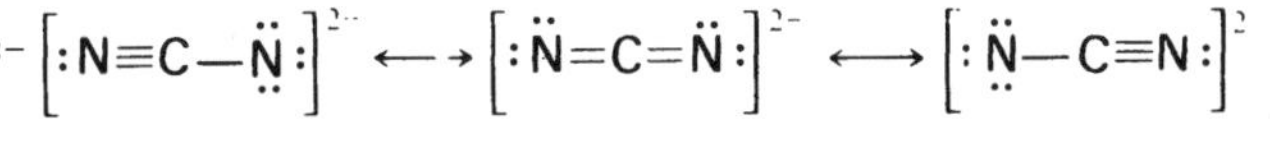

O_3

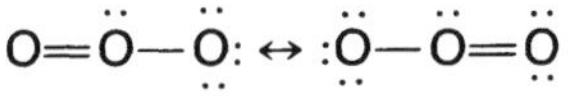

N_2O_3

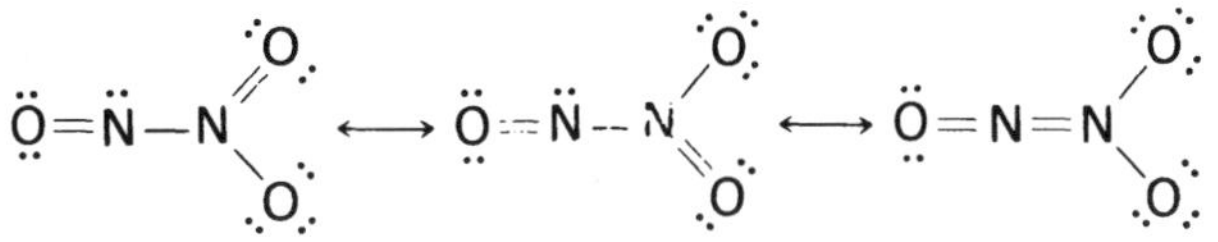

9.20 AsH_3, trigonal pyramid
GeF_2, bent
BiI_4^+, tetrahedral
$SbOCl_3$, tetrahedral
$InCl_2^+$, linear
$OPCl_3$, tetrahedral
$OSCl_2$, trigonal pyramid

9.21 PF_3, SCl_2, AsH_3, GeF_2, $OPCl_3$, $OSCl_2$

Chapter 10

INTERMOLECULAR FORCES AND CONDENSED STATES OF MATTER

SECTION 10.1 THE STATES OF MATTER

Objectives

- To identify various physical states of homogeneous matter.
- To define terms associated with changes of state.

Focus

Table 10.1 in the text lists most physical states of homogeneous matter. Study this table. The two broad classifications are based upon the arrangement of molecules in matter. Each classification is subdivided by the strength with which the molecules are held together.

Salt (NaCl) crystals have an orderly arrangement of Na^+ and Cl^- ions with a high degree of cohesiveness, state 6; in molten NaCl the arrangement of the Na^+ and Cl^- ions is disorderly, state 3.

Example 10.1 List each of the following according to states 1 through 6 from Table 10.1.

(a) tar	(d) quartz
(b) gasoline	(e) air
(c) copper metal	(f) tea (no ice)

Answer

(a) state 3 (noncrystalline solid)
(b) state 2 (liquid)
(c) state 6 (orderly solid)
(d) state 6 (orderly solid)
(e) state 1 (gas)
(f) state 2 (liquid)

Various terms are associated with the changes of state in homogeneous systems. Define those listed in the Study Guide, Question 10.1.

SECTION 10.2 WHAT CAN WE LEARN FROM LIQUIDS?

Objective

- To describe the transition of a solid to a liquid phase and a liquid to a gas phase in terms of intermolecular (I.M.) forces.

Focus

Why don't all solids melt at the same temperature or liquids boil at the same temperature? The magnitude of the I.M. forces (forces between molecules) determines the ease with which molecules become more fluid: The stronger the I.M. forces, the higher the melting point and the higher the boiling point; as a result more energy is required to reduce the effect of the I.M. forces.

SECTION 10.3 INTERMOLECULAR FORCES

Objective

- To define the various types of I.M. forces.

Focus

The three categories of I.M. forces are the following.

1. **Dipole-dipole forces:** an electrostatic attraction between molecules with a dipole moment greater than zero.
2. **H bonding:** an extremely strong electrostatic attraction between an H atom on one molecule and a small, very electronegative atom, particularly F, O, or N, having at least one pair of lone electrons, on another molecule. Thus, NH_3, H_2O, and HF all have extremely high melting points when compared to the other hydrides of Groups 5A, 6A, and 7A, respectively.
3. **London forces:** a weak electrostatic attraction that is dominant between nonpolar molecules. The attraction arises from an instantaneous self-induced dipole of the molecule. Its magnitude is proportional to the size of the "jellylike" spherical electron cloud surrounding the molecule: the larger the molecule, or the larger its electron cloud, the more easily it is polarized and therefore the stronger the London forces. Because of London forces, all substances can be liquefied.

Molecular shape and molecular weight are two more factors that affect the boiling point of a liquid: Spherical molecules vaporize easier than linear molecules, and heavier molecules vaporize with more difficulty.

Example 10.2 Indicate the I.M. force that dominates in the following substances.

(a) CO
(b) H_2O
(c) NF_3 (trigonal pyramid)
(d) CH_4 (tetrahedral)
(e) CH_3Cl (tetrahedral)
(f) Xe

Answer

(a) dipole	(d) London
(b) H-bonding	(e) dipole
(c) dipole	(f) London

Solution

(a) $\overset{\delta^+}{C} \equiv \overset{\delta^-}{O}$, a polar molecule.

(b) H atom of one water molecule is attracted to the strongly electronegative O atom of another.

(c) NF_3 is a trigonal pyramid. The vector sum of the bond moments is nonzero with the N, being δ^+, attracted to the δ^- F atom on another molecule.

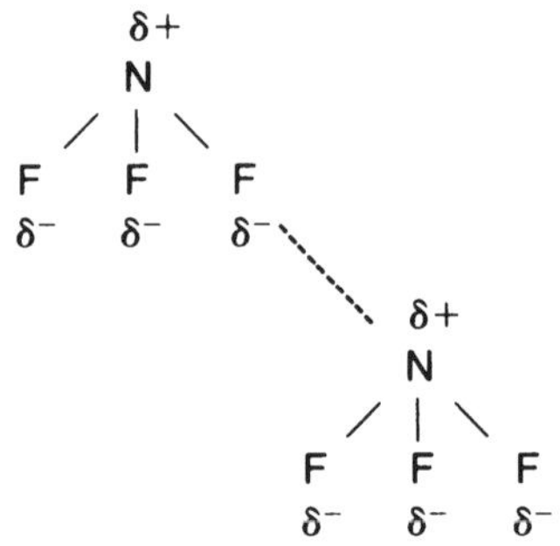

(d) CH_4 is a tetrahedral molecule. The vector sum of the bond moments is equal to zero.

(e) CH_3Cl is also a tetrahedral molecule. Because the Cl atom is more electronegative than the H atom, the vector sum of the bond moments is nonzero. The δ^- Cl atom attracts the H atom of another molecule.

(f) The Xe atom is nonpolar; only London forces can exist.

Review the effect H-bonding has on the density of ice and water at temperatures below 4°C. Also explain why H_2O has a higher boiling point than HF, even though the O atom is less electronegative than an F atom. The answer is that a three-dimensional network of hydrogen bonding exists between H_2O molecules, whereas only a linear network exists for HF.

SECTION 10.4 LIQUIDS AND GLASSES

Objective

- To describe the flow properties of liquids and glasses.

Focus

Molecules in liquids are in constant motion and travel only short distances before collisions occur. In addition, the I.M. forces between molecules are quite strong, but not as strong as those in solids. Increased temperatures produce a weakening of I.M. forces, thus causing molecules in the liquid state to flow from one "hole" in the structure to another. The holes, being scattered randomly throughout the liquid, allow molecular flow in any direction with little change in energy.

A liquid cooled to the point where the rate of molecular flow almost stops is called a **glass**. It has all the characteristics of a liquid and the appearance of a solid. Generally, glasses are composed of very bulky molecules making crystallization difficult.

Glasses do not have sharp melting points, for they are not really solids but rather cooled liquids. Substances like tar, paraffin, and butter (especially when cold) are glasses. Of course, so is ordinary window glass.

SECTION 10.5 CRYSTALLINE SOLIDS

Objective

- To explain how the crystalline structures of solids can be studied.

Focus

The internal structure of a crystalline solid, particularly the distance between atoms or molecules, is determined by studying the diffraction patterns of X rays.

SECTION 10.6 THE CRYSTAL LATTICE AND THE UNIT CELL

Objectives

- To define a unit cell.
- To identify the four different locations for an atom in a unit cell and to determine the contribution of each location of an atom to the total number of atoms per unit cell.
- To recognize the three basic cubic crystal systems.

Focus

The **unit cell** is a three-dimensional self-repeating representation of the formula and structure of a crystalline solid.

The number of atoms per unit cell is the sum of the portions of atoms located at the corners (1/8 atom per corner), at the edges (1/4 atom per edge), at the faces (1/2 atom per face), or at the center (1 atom per cell) of the unit cell. The three basic cubic crystal systems are:

1. **Simple cubic:** 1 atom/unit cell
2. **Body-centered cubic:** 2 atoms/unit cell
3. **Face-centered cubic:** 4 atoms/unit cell

Example 10.3 Nickel metal crystallizes in a face-centered lattice. How many nickel atoms are there per unit cell?

Answer Four Ni atoms per unit cell

Solution A face-centered cubic unit cell has 1 Ni atom at each of the 8 corners:

$$\frac{1/8 \text{ atom}}{\text{corner}} \times 8 \text{ corners} = 1 \text{ atom}$$

and 1 Ni atom at each of the 6 faces

$$\frac{1/2 \text{ atom}}{\text{face}} \times 6 \text{ faces} = 3 \text{ atoms}$$

for a total of 4 Ni atoms per unit cell.

SECTION 10.7 COORDINATION NUMBERS

Objectives

- To identify the coordination number for an atom in each of the three cubic crystal systems.
- To calculate the density, edge distances, and the atomic radii of atoms in a cubic crystal.

Focus

The **coordination number** is the number of nearest neighbors to an atom, ion, or molecule. The coordination numbers for the three basic cubic crystal structures are

1. Simple cubic 6
2. Body-centered cubic 8
3. Face-centered cubic 12

Review these facts by studying the text, Figure 10.22.
From experiments, we can determine the type of cubic crystal.

Example 10.4 Sodium metal has a density of 0.97 g/cm³ and crystallizes in a body-centered lattice. Calculate the length of the edge of the unit cell (u.c.)

Answer 4.29 Å per edge

Solution For a body-centered lattice, there are 2 Na atoms/unit cell. Since we are working with atoms, we need to think small. If we calculate the volume of a unit cell, then the cubed root is the length of its edge.

Given quantity and units: $\frac{2 \text{ atoms Na}}{\text{unit cell}}$ (body-centered cubic)

Desired quantity and units: volume, cm^3/u.c.
Conversion factors: density of Na: 0.97 g/cm^3

Avogadro's number: $\frac{6.02 \times 10^{23} \text{ atoms Na}}{\text{mol Na}}$

$$\text{atomic weight of Na:} \quad \frac{23.0\ \text{g Na}}{\text{mol Na}}$$

$$\text{volume, } \frac{\text{cm}^3}{\text{u.c.}} = \frac{2\ \cancel{\text{atoms Na}}}{\text{u.c.}} \times \frac{\cancel{\text{mol Na}}}{6.02 \times 10^{23}\ \cancel{\text{atoms Na}}} \times \frac{23.0\ \cancel{\text{g Na}}}{\cancel{\text{mol Na}}}$$

$$\times \frac{\text{cm}^3}{0.97\ \cancel{\text{g Na}}} = 7.87 \times 10^{-23}\ \frac{\text{cm}^3}{\text{u.c.}}$$

The length of the edge is

$$\frac{\sqrt[3]{\text{cm}^3}}{\text{u.c.}} = \frac{\text{cm}}{\text{u.c.}} = 4.29 \times 10^{-8}\ \text{cm} = 4.29\ \text{Å}$$

Example 10.5 Polonium metal crystallizes in a simple cubic unit cell whose edge measures 3.36 Å. What is the radius of the Po atom?

Answer 1.68 Å

Solution A simple cubic unit cell has a Po atom at each corner. As can be seen in Figure 10.1, the radius is one-half the length of an edge for any simple cubic system. For Po,

$$r = \frac{1}{2} \times 3.36\ \text{Å} = 1.68\ \text{Å}$$

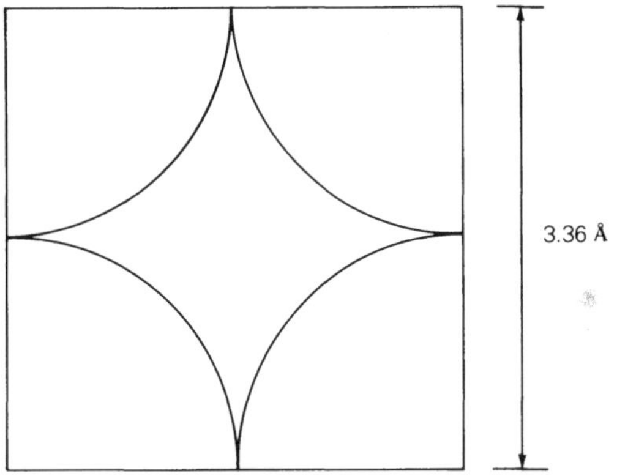

Figure 10.1 Polonium simple cubic unit cell.

Example 10.6 Copper metal has a density of 8.92 g/cm³ and crystallizes forming a cubic crystal structure with an edge length of 3.61 Å. Calculate the number of atoms per unit cell and determine the cubic crystal system. What is the coordination number of Cu in the unit cell?

Answer 4 atoms/unit cell; face-centered cubic; coordination number is 12.

Solution To calculate the number of atoms per unit cell, we need to use Avogadro's number, 6.02×10^{23} atoms/mol, the atomic weight of Cu, 63.54 g/mol, and the volume of the cube = $(3.61\text{Å})^3$/u.c.

$$\frac{\text{atoms}}{\text{u.c.}} = \frac{6.02 \times 10^{23}\ \text{atoms Cu}}{\cancel{\text{mol Cu}}} \times \frac{\cancel{\text{mol Cu}}}{63.54\ \cancel{\text{g Cu}}} \times \frac{8.92\ \cancel{\text{g Cu}}}{\cancel{\text{cm}^3}}$$

$$\times \frac{(3.61 \times 10^{-8}\ \cancel{\text{cm}})^{\cancel{3}}}{\text{u.c.}} = \frac{4.0\ \text{atoms Cu}}{\text{u.c.}}$$

This is a face-centered cubic with a coordination number of 12.

In calculations using unit cell data be careful to use units throughout the calculations and know the units of the unknown. Review the text, Examples 1, 2, and 3 in Section 10.6 by using units in the calculations.

SECTION 10.8 COMPOUND LATTICES

Objective

- To recognize the fact that the basic unit cells of ionic compounds can be more complex than those of the basic cubic crystal systems.

Focus

The crystal structure of an ionic compound, one consisting of more than one kind of atom or ion, can be a complex intertwining of different unit cells. For example, in ZnS, the Zn^{2+} and S^{2-} each has a coordination number of 4, but for CaF_2 the coordination number for Ca^{2+} is 8 and for F^- it is 4. In TiO_2, Ti^{4+} has a coordination number of 6 and O^{2-} a coordination number of 3.

SECTION 10.9 REAL CRYSTALS

Objective

- To indicate the types of defects that occur in a real crystal versus that of an ideal crystal.

Focus

Real crystals generally have defects: a missing cation, anion, or metal atom at points in the lattice structure; slippage of the ion or atom in between the lattice points; empty lattice sites; or foreign atoms or ions (an impurity) at a lattice site.

Crystal defects can result in a change of properties in pure crystals, including hardness, conductivity, and cleavage.

SECTION 10.10 COLLOIDS AND ADSORBENTS

Objectives

- To characterize a colloid in terms of its dispersed phase and the dispersion medium.
- To distinguish between hydrophilic and hydrophobic colloids.
- To explain the properties of adsorbents.

Focus

Colloids are particles with a diameter range of 1 to 300 nm that remain suspended in a dispersion medium. These homogeneous mixtures are called *collodial dispersions*. The particles may be an aggregate of hundreds or thousands of molecules or a single, huge molecule.

Tyndall light scattering effects reveal the presence of these suspended particles, as when sunlight shines through fog or smoke. Table 10.2 in the text reviews some various types of collodial dispersions.

Example 10.7 Classify each of the following collodial dispersions according to Table 10.2.

(a) paint (c) agricultural sprays
(b) mayonnaise (d) beer froth

Answer

(a) a sol: solids suspended in a liquid
(b) emulsion: olive oil, vinegar, and egg
(c) emulsion: organic compound in water
(d) foam: gas in a liquid

Even though collodial particles are heavier than the medium in which they are dispersed, they remain suspended because of constant, random molecular bombardment and because they have the same charge. This charge is acquired from the ions in the dispersion medium that are adsorbed to the high energy surface of the colloid. The colloidal particles adsorb the same charged ions on their surface and therefore they repel each other. **Hydrophilic** colloids are strongly attracted to water molecules; **hydrophobic** ones are only weakly attracted. Coagulation and/or flocculation of negatively charged colloidal clay platelets with lime, alum, or ferric chloride is a common procedure in the treatment of sewage.

Adsorbents are solid particles with a large surface area. As a result, an atom in a crystal structure satisfies its coordination number by adsorbing other atoms or molecules. Since the exposed atoms are at a lower energy state, energy is released and the adsorbent becomes warmer.

QUESTIONS

Section 10.1

10.1 Define each of the following terms.

(a) condensation (liquefaction) ______
(b) vaporization ______
(c) freezing ______
(d) fusion ______
(e) sublimation ______
(f) crystallization ______

10.2 What term in Question 10.1 defines the change that occurs when molten NaCl solidifies?

Section 10.3

10.3 Based upon the I.M. forces and the structure of the following compounds predict the one with the higher boiling point.

(a) $CH_3\ddot{O}H$ or $CH_3\ddot{S}H$

(b) Cl_2 or I_2

(c) CO_2 or CS_2 (both linear)

(d) CH_3F or CH_3Cl (both tetrahedral)

(e) $:NH_3$ or $(CH_3)_3N:$ (both trigonal pyramid)

Section 10.4

10.4 Does window glass really flow? Explain.

Section 10.6

10.5 Lead metal has a density of 11.34 g/cm^3 and crystallizes in a face-centered lattice. If the length of an edge of the unit cell is 4.95Å, calculate Avogadro's number.

10.6 Chromium metal has a density of 7.15 g/cm^3 and crystallizes in a body-centered lattice.

(a) Determine the length of the edge of the unit cell.

(b) Calculate the radius of the chromium atom.

(*Hint:* Assume the chromium atoms to touch along the *diagonal* of the cube.)

Section 10.10

10.7 Yellow light is used instead of white light in heavy fog. Based upon the Tyndall effect, can you explain the choice?

ANSWERS

10.1 (a) Condensation is the transformation of a vapor to a liquid.

(b) Vaporization is the transformation of a liquid to a vapor.

(c) Freezing is the transformation of a liquid to a solid.

(d) Fusion is the transformation of a solid to a liquid.

(e) Sublimation is the transformation of a solid to a vapor without the appearance of the liquid state.

(f) Crystallization is the transformation of a liquid to a crystalline solid.

10.2 crystallization

10.3 (a) CH_3OH, H-bonding with the O atom.

(b) I_2, London forces, larger molecule than Cl_2

(c) CS_2, London forces, larger molecule than CO_2

(d) CH_3F, H-bonding with the F atom

(e) NH_3, H-bonding with the N atom

10.4 Yes, window glass does flow. The molecules in the glass are large, but do flow over one another with time. Old window glass is thicker at the bottom than at the top.

10.5 1.81×10^{24} atoms/mol

10.6 (a) 2.89 Å

(b) 1.25 Å

10.7 The wavelength of yellow light is longer than the average diameter of the colloidal particles that constitute fog. Therefore, there is less reflection of the yellow light than for the shorter wavelengths of white light.

Chapter 11

CHANGES OF STATE, ENTROPY, AND FREE ENERGY

SECTION 11.1 STATES AND PROCESSES

Objective

• To define a property of a chemical system in terms of initial and final states of a process.

Focus

A system has many properties that change in ways that depend only upon the initial and final states of the process and *not* upon the pathway by which it occurs. For example, in heating a pot of water from room temperature to boiling, few care how it is done, who did it, or how long it took. The temperature difference between the initial state and the final state is going to be the same. We will discuss the most common chemical changes in properties of a chemical system that depend only on the initial and final states. They are ΔE, ΔH, ΔT, ΔP, ΔV, and ΔS.

From studying this chapter, you will gain a better appreciation of these properties.

SECTION 11.2 CHANGES OF STATE

Objectives

• To explain the relative energies of substances in various physical states and the degree of orderliness in their molecular arrangements.

• To calculate the heat required for a change in physical state and/or for a change in temperature.

Focus

When a solid melts, heat is absorbed from the surroundings. The heat energy weakens the intermolecular (I.M.) forces to make the molecules move more easily. If we then want to vaporize the liquid, additional heat is needed to weaken the remaining I.M. forces. The same quantity of heat is evolved in the reverse processes when the vapor condenses to liquid and the liquid freezes to solid.

Consider the following sequence in which molecular arrangement becomes more disordered: a solid is heated to its melting point; a solid melts at its melting point; a liquid is heated to its boiling point; a liquid boils at its boiling point and finally, a vapor is heated.

The energy needed for the phase change of a solid to a liquid and a liquid to a gas is absorbed without a change in temperature; only a change in the magnitude of the I.M. forces occurs. The terms associated with the energy required for a phase change are

- ΔH_{fus} (J/g): a solid melts at its normal melting point (1 atm pressure).
- ΔH_{vap} (J/g): a liquid boils at its normal boiling point (1 atm pressure).
- ΔH_{subl} (J/g): a solid sublimes (phase change from solid to gas) at its sublimation temperature, (1 atm pressure).

$$\Delta H_{sub} = \Delta H_{fus} + \Delta H_{vap}$$

when ΔH_{fus} and ΔH_{vap} are known at the sublimation temperature, *not* at the normal freezing and boiling points.

These quantities can be measured at other temperatures, but the temperatures mentioned are assumed unless otherwise specified.

Example 11.1 Calculate the energy released when 200 g Hg condenses at 357°C.

Answer -5.90×10^4 J

Solution Table 11.1 lists ΔH_{vap} for Hg = +295 J/g. This means that when 1 g Hg goes from the gaseous to the liquid state, 295 J/g is released or $\Delta H = -295$ J/g. (This ΔH is its heat of condensation.)

$$\text{heat released} = 200\ \cancel{\text{g Hg}} \times -295\ \text{J}/\cancel{\text{g Hg}} = -5.90 \times 10^4\ \text{J}$$

Example 11.2 If the same Hg in Example 11.1 is cooled from 357°C to room temperature, 22.0°C, how much additional heat is released? The specific heat of Hg is 0.137 J/(g°C). What is the total quantity of heat released?

Answer -9.18×10^3 J; -6.82×10^4 J

Solution From the specific heat of Hg a total of 0.137 J of heat is released for *each* gram of the 200 g of Hg and for *each* °C of the (357 − 22)°C temperature change.

We use the equation

$$q = m(g) \times \text{sp ht}\left(\frac{\text{J}}{\text{g °C}}\right) \times \Delta t$$

$$\Delta H \text{ released from cooling} = \frac{0.137 \text{ J}}{\text{g°C}} \times 200 \text{ g} \times 335\text{°C} = -9180 \text{ J}$$

total heat released = heat from condensation + heat from cooling = -5.90×10^4 J + $(-9.18 \times 10^3$ J$) = -6.82 \times 10^4$ J

SECTION 11.3 VAPOR PRESSURE

Objectives

- To define vapor pressure.
- To account for various factors that affect a liquid's vapor pressure.

Focus

The **vapor pressure** of a liquid is the pressure exerted by the molecules of the vapor against the walls of a closed container when the vapor and liquid coexist in an equilibrium state, or when the rate of evaporation equals the rate of condensation.

Its magnitude, proportional to the number of molecules in the vapor state, depends upon the temperature and strength of the I.M. forces. At higher temperatures more molecules are in the vapor state at equilibrium; thus the vapor pressure is higher (see text Figure 11.2). If the I.M. forces are weaker, the molecules are more likely to separate and pass into the vapor state, thus a higher vapor pressure.

Generally, the volume of the enclosed space above the liquid has no effect on vapor pressure. Although a larger volume can hold more molecules, pressure (proportional to the number of molecules hitting the wall per unit area) remains constant.

Example 11.3 The vapor pressure of methanol in a 200 mL cylinder is 580 torr. What is the vapor pressure when the volume of the cylinder is reduced to 100 mL? (See Figure 11.1.)

Answer 580 torr

Solution The vapor pressure is independent of the enclosed volume above the liquid. When the volume decreases, the pressure of the vapor momentarily increases because equilibrium between the vapor and liquid is disturbed and the rate of condensation is greater than the rate of evaporation. When the rates become equal, equilibrium is reestablished and the pressure is again 580 torr, assuming no temperature change.

SECTION 11.4 BOILING AND CONDENSATION

Objectives

- To define the boiling point of a liquid.
- To explain the processes of boiling and condensation.
- To explain the phenomena of superheating and undercooling.

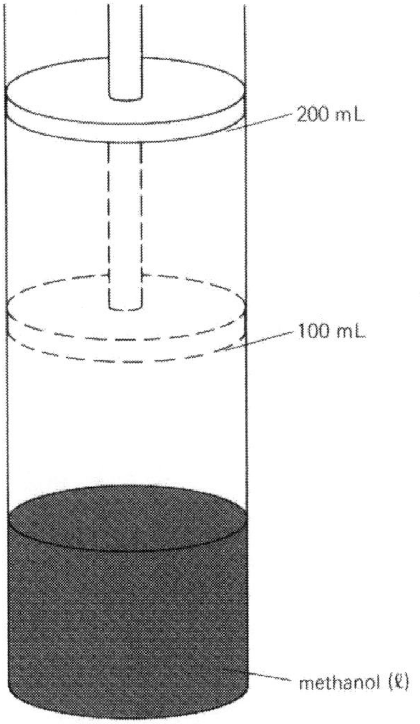

Figure 11.1 An apparatus to measure change of vapor pressure with change of volume.

Focus

A liquid boils at a temperature where its vapor pressure equals that of the surrounding atmosphere. The **normal boiling point** is the temperature at which its vapor pressure equals one atmosphere.

When water is heated, the first bubbles that form are from the escape of the dissolved gases. At the boiling point the molecules throughout the liquid have enough energy to overcome the I.M. forces and form a bubble inside the liquid. The vapor molecules exert a pressure (force per unit area) against the walls of the bubble.

Below the boiling point, the pressure inside the bubble is less than the outside pressure; if a bubble manages to form, it would be crushed out of existence.

When a phase change occurs, liquids may **superheat** (the temperature is above the boiling point, but the liquid has not started to boil) and gases may **undercool** (the temperature is below the condensation point, but the gas has not started to liquefy). These phenomena are often troublesome and may be avoided by using a substance or nucleus upon which a bubble can form or vapor can condense. A boiling chip is used in the laboratory to avoid the superheating of a liquid. Silver iodide crystals are used by rainmakers to seed the clouds and avoid undercooling. This causes precipitation.

Superheating a liquid can cause "bumping," that is, the sudden formation of a bubble consisting of a vapor with a pressure greater than the surrounding pressure. A rapid expansion of the bubble causes splattering of the superheated liquid.

SECTION 11.5 MELTING AND FREEZING

Objectives

- To define the melting point and freezing point of a substance.
- To describe a cooling curve.

Focus

The normal **melting point** or **freezing point** is the temperature at which the solid and liquid phases coexist at one atmosphere pressure. When a liquid is cooled very quickly to the freezing point, undercooling often occurs because molecules do not have time to reach the special alignment needed to form a crystalline solid. A "seed crystal" can be used as a site (nucleus) on which other molecules can quickly orient and attach. This is analagous to using a nucleus to avoid superheating of a liquid or undercooling of a vapor.

SECTION 11.6 THE PHASE DIAGRAM

Objectives

- To interpret the effect of pressure and temperature changes on a three-phase system represented by a phase diagram.
- To define the critical density, critical temperature, and critical pressure of a liquid.
- To explain the states of the liquid and vapor at the critical temperature.

Focus

A **phase diagram** indicates the temperature and pressure conditions that a substance needs to exist as a solid, liquid, or gas and the *T-P* conditions necessary for sublimation, vaporization, or fusion. You should be able to identify the following on a phase diagram: the three physical states; the solid-liquid, solid-gas, and liquid-gas equilibrium lines; and the triple point. The solid-liquid line for water (text, Figure 11.5) is unusual in that it slopes to the left. (See the text for an explanation.)

Example 11.4 Construct a phase diagram in which the normal boiling and freezing points occur at 138°C and 16°C, respectively. The triple point is at 0.33 atm and at 15°C. Remember that the liquid-gas and solid-gas equilibrium lines are *not* straight lines. Label the location of the three phases.

Answer See Figure 11.2.

Example 11.5 Referring to the phase diagram in Example 11.4, construct a warming curve for a solid that is warmed from 0°C to 120°C at 0.66 atm. Label the temperature at each phase change.

Answer See Figure 11.3.

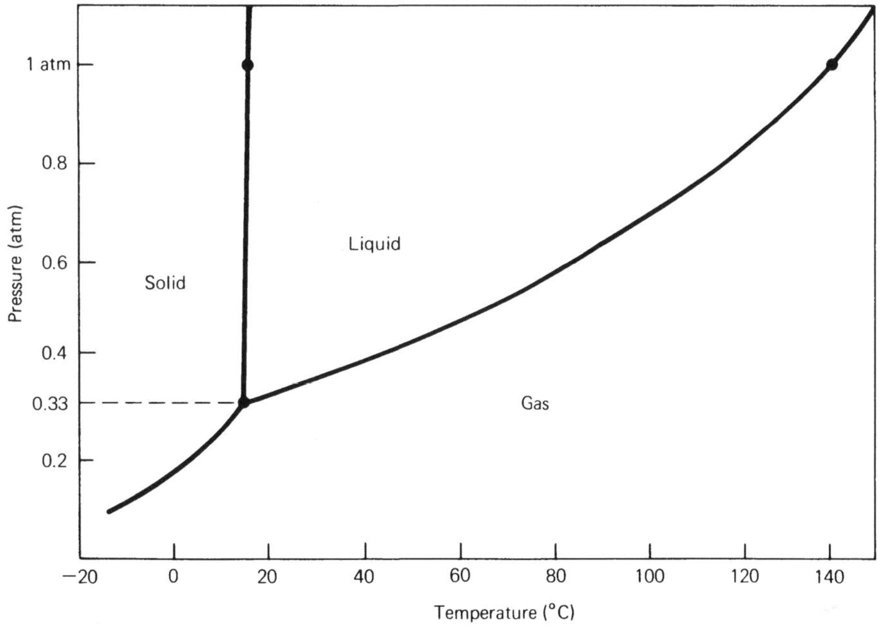

Figure 11.2 Phase diagram; answer to Example 11.4.

The **critical temperature** is the temperature above which a liquid can no longer exist regardless of the external pressure applied. Just below the critical temperature, the pressure needed for liquefaction is the **critical pressure**. Above the critical temperature, the molecules are moving so rapidly that the I.M. forces are not strong enough to hold them together in the liquid phase. Therefore, molecules with strong I.M. forces have high critical temperatures; higher temperatures make it impossible for even these molecules to condense to a liquid.

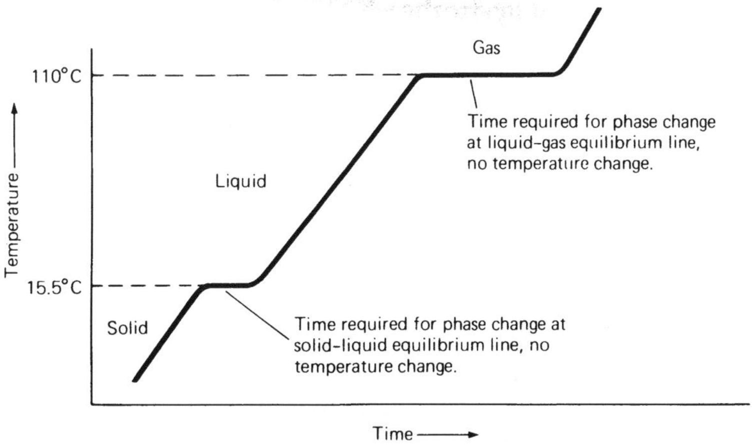

Figure 11.3 Warming curve for a solid.

Example 11.6 Select the substance which has the higher critical temperature (See text, Figure 11.2.)

(a) H_2O or Br_2
(b) Cl_2 or naphthalene
(c) mercury or ethanol

Answer (a) H_2O; (b) naphthalene; (c) mercury

Solution (a) Br_2 has a higher vapor pressure than H_2O at any given temperature. Therefore, Br_2 has weaker I.M. forces and a lower critical temperature.

(b) Naphthalene is a solid at room temperature, Cl_2 is a gas. Naphthalene has the greater I.M. forces and therefore has a higher critical temperature.

(c) Hg has a lower vapor pressure than ethanol at a given temperature; therefore fewer Hg atoms tend to go into the vapor state, implying greater I.M. forces and a higher critical temperature.

SECTION 11.7 ENTROPY

Objective

- To explain the concept of entropy.

Focus

The word **entropy** means disorder. The greater the disorder in an arrangement of particles, molecules, or ions, the greater the entropy. Nature tends toward disorder, or, to increase in entropy. For example, if a gardener rakes leaves into a nice neat pile (a greater degree of order, or a decrease in entropy) nature tends to disorder them by children, wind, or other natural phenomena. Work is required to restore order or to decrease the entropy of a system. Occurrences with a decrease in entropy also happen naturally; after all, snowflakes form, water freezes, salt crystallizes, tomatoes grow, and birds hatch.

Example 11.7 Determine whether the entropy increases or decreases in each of the following processes:

(a) Ice melts
(b) $BaSO_4$ precipitates
(c) Rocks weather
(d) Snowflakes form
(e) The chemical reactants are gathered from the stockroom and placed into a reaction vessel at the desired temperature and pressure.
(f) Raindrops form
(g) Salt dissolves in water.

Answer

(a) Entropy increases; more disorder in the molecules.
(b) Entropy decreases, Ba^{2+} and SO_4^{2-} combine to form the precipitate.
(c) Entropy increases; larger rocks become smaller flakes which scatter.
(d) Entropy decreases; a crystalline, ordered system forms.

(e) Entropy decreases; the reactants are collected and set up for the reaction.
(f) Entropy decreases; H_2O vapor condenses to form a more ordered raindrop.
(g) Entropy increases; the Na^+, Cl^- sequence in the crystal becomes random in the solution.

SECTION 11.8 ENTROPY AND HEAT

Objectives

• To differentiate between thermodynamically reversible and irreversible processes.

• To explain the relationship between the entropy change of a system and the amount of heat absorbed or evolved in reversible and irreversible processes.

Focus

A **thermodynamically reversible process** is an ideal process with no departure from equilibrium. It would have to occur at an infinitesimally slow rate. Therefore all laboratory processes are irreversible processes.

Entropy is a property of a system. The change in entropy, ΔS, is the difference between the final and initial states of entropy and is independent of the process by which the change is made:

$$\Delta S = S_{\text{final}} - S_{\text{initial}}$$

Study the text, Section 11.8, very closely.

The entropy increase of a *reversible* system is directly proportional to the maximum quantity of heat a system can absorb from its surroundings, or $\Delta S \simeq q_{\text{rev}}$; however, it is inversely proportional to the absolute temperature of the system, or $\Delta S \propto 1/T(\text{K})$. A greater increase in disorder occurs when the temperature of a substance increases from 1 K to 2 K than from 300 K to 301 K. Therefore for a reversible process,

$$\Delta S \propto \frac{1}{T(\text{K})} \simeq q_{\text{rev}} = \frac{q_{\text{rev}}}{T(\text{K})}$$

In an *irreversible* (real) process less heat is absorbed by the system than in a reversible process. The reason is that some heat is used by the molecules in the form of work (to overcome I.M. forces). This work has the same effect on the system as the heat it replaces. q_{actual} is therefore less than q_{rev}. However, since ΔS is defined in terms of the maximum heat (q_{rev}), ΔS is greater than the value expected from the $q_{\text{actual}}/T(\text{K})$ measurement.

$$\Delta S = \frac{q_{\text{rev}}}{T(\text{K})} > \frac{q_{\text{actual}}}{T(\text{K})}$$

An example of a reversible process is a phase change. At the boiling and melting points of a substance, an equilibrium exists between two phases. An infinitesimal change in temperature causes a phase change to occur. Therefore, the process is reversible and

$$q_{\text{actual}} = q_{\text{rev}} = \Delta H$$

at the temperature of the phase change. Of course, this does not apply when the phase change occurs for a superheated or undercooled system.

The maximum amount of heat absorbed by a gaseous system (which is that for an ideal gas) is calculated from

$$q_{\text{rev}} = 2.303nRT \log \frac{P_1}{P_2} = 2.303nRT \log \frac{V_2}{V_1}$$

Example 11.8 Calculate ΔS for the vaporization of 20.0 g Hg at its boiling point and ΔS for the fusion of 20.0 g Hg at its melting point.

Answer 9.36 J/K; 1.03 J/K

Solution Since this occurs at the temperature of a phase change, the process is reversible and $\Delta H = q_{\text{rev}}$. From Table 11.1, ΔH_{vap} for Hg is 295 J/g, and the boiling point of Hg is 357°C.

$$T(\text{K}) = 357°\text{C} + 273 = 630\ \text{K}$$

$$q_{\text{rev}} = 295\ \text{J/g} \times 20.0\ \text{g} = 5900\ \text{J}$$

$$\Delta S = \frac{q_{\text{rev}}}{T(\text{K})} = \frac{5900\ \text{J}}{630\ \text{K}} = 9.36\ \frac{\text{J}}{\text{K}}$$

At the melting point of Hg, −39°C,

$$\Delta H_{\text{fus}} = 12.0\ \text{J/K}$$

$$T(\text{K}) = -39°\text{C} + 273 = 234\ \text{K}$$

$$q_{\text{rev}} = 12.0\ \text{J/g} \times 20.0\ \text{g} = 240\ \text{J}$$

$$\Delta S = \frac{q_{\text{rev}}}{T(\text{K})} = \frac{240\ \text{J}}{234\ \text{K}} = 1.03\ \frac{\text{J}}{\text{K}}$$

Can you explain why $\Delta S_{\text{vap}} > \Delta S_{\text{fus}}$?

SECTION 11.9 THE SECOND LAW OF THERMODYNAMICS

Objective

- To understand that for every spontaneous process in a system the entropy increases.

Focus

All real spontaneous processes that occur in an isolated system are irreversible and therefore ΔS for the system is always greater than zero, $\Delta S > 0$ because $\Delta S > q_{\text{actual}}/T(\text{K})$. At best $\Delta S = 0$ for a reversible process, but ΔS is never less than zero. This is the second law of thermodynamics.

A system left alone tends to drift toward a more disordered array, $\Delta S > 0$. This is the driving force for any chemical reaction: *nature tends toward disorder*.

SECTION 11.10 FREE ENERGY

Objective

- To explain how the signs and magnitudes ΔH and ΔS relate to a spontaneous process.

Focus

Two factors contribute to the *spontaneity of a reaction* (decreased stability of a system).

1. The reaction should be exothermic, ΔH is negative.
2. The molecular arrangement of atoms or ions should proceed toward a disordered system, ΔS is positive.

Considering that for any isolated system

$$\Delta S_{\text{total}} = \Delta S_{\text{system}} + \Delta S_{\text{surroundings}}$$

and that if

$$\Delta S_{\text{surroundings}} = -\frac{\Delta H\text{ system}}{T}$$

then

$$\Delta S_{\text{total}} = \Delta S - \frac{\Delta H}{T}$$

or, rearranging the equation,

$$-T\ \Delta S_{\text{total}} = \Delta H - T\ \Delta S$$

A new term, ΔG, the change in free energy of the system, is defined as $\Delta G = -T\ \Delta S_{\text{total}}$. This equation states that since $\Delta S_{\text{total}} > 0$ for an irreversible process, then for any spontaneous, irreversible process, ΔG must be negative, $\Delta G < 0$.

Learn the equation

$$\Delta G = \Delta H - T\Delta S$$

and all the changes in ΔH, T, and ΔS which make ΔG negative—the criterion for a spontaneous process.

Example 11.9 Using Example 11.8, calculate ΔG for the vaporization of 20.0 g Hg at 340°C. Explain the meaning of the answer.

Answer $\Delta G = +160$ J. This shows that below the boiling point the vaporization of Hg is not spontaneous.

Solution From Table 11.1, $\Delta H_{vap} = +295$ J/g.

$$T(\text{K}) = 340°\text{C} + 273 = 613\,\text{K}$$

$$\Delta G = \Delta H - T\Delta S$$

$$= 295\ \frac{\text{J}}{\cancel{\text{g}}}\ (20.0\cancel{\text{g}}) - 613\cancel{\text{K}}(9.36\ \frac{\text{J}}{\cancel{\text{K}}})$$

$$= 5900\,\text{J} - 5740\,\text{J} = +160\,\text{J}$$

Example 11.10 For the phase transition $M(s_1) \rightarrow M(s_2)$, $\Delta H = -14.07$ J/mol and $\Delta S = -0.0480$ J/(mol K) at -84°C. (a) What is ΔG? (b) At what temperature does $\Delta G = 0$? (c) Which solid state is favored at 0°C?

Answer

(a) $\Delta G = -5.00$ J/mol

(b) $\Delta G = 0$ at 293 K

(c) $M(s_2)$ is favored at 0°C because $\Delta G < 0$ and therefore the reaction is spontaneous in the direction it is written.

Solution

(a) $T(\text{K}) = -84°\text{C} + 273 = 189\,\text{K}$

$$\Delta G = \Delta H - T\Delta S$$

$$\Delta G = -14.07\ \frac{\text{J}}{\text{mol}} - 189\cancel{\text{K}}\left(\frac{-0.0480\,\text{J}}{\text{mol}\cancel{\text{K}}}\right)$$

$$= -5.00\ \frac{\text{J}}{\text{mol}}$$

(b) At equilibrium, $\Delta G = 0$

$$T = \frac{\Delta H - \Delta G}{\Delta S} = \frac{-14.07\,\text{J/mol} - 0}{-0.0480\,\text{J/(mol K)}} = 293\,\text{K}$$

At 293 K, equilibrium between the phases exists.

$$M(s_1) \rightleftharpoons M(s_2)$$

(c) $\Delta G = \Delta H - T\Delta S$

$$= -14.07\ \frac{\text{J}}{\text{mol}} - 273\cancel{\text{K}}\left(\frac{-0.0480\,\text{J}}{\text{mol}\cancel{\text{K}}}\right) = -0.966\ \frac{\text{J}}{\text{mol}}$$

The stable form at 0°C is $M(s_2)$ since ΔG is negative. Above 293 K, the reverse reaction is spontaneous; the stable form is $M(s_1)$.

The equation

$$\Delta G = \Delta H - T\Delta S$$

summarizes many concepts which govern the spontaneity of a chemical reaction. The change in enthalpy of a system at constant pressure, ΔH, and the change in entropy of the system, ΔS, are the most prominent.

Let's review a few relationships that are a part of the equation.

1. If ΔG is positive, the reaction is not spontaneous because from $\Delta G = -T\Delta S_{total}$, the total entropy change of the universe has to decrease (ΔS would be negative). We have already seen (Section 11.9) that ΔS for any irreversible process is greater than zero. If ΔG is positive, the *reverse* reaction is negative (and is spontaneous).
2. If ΔG is zero, the reaction is at equilibrium because $\Delta S_{total} = 0$, one in which $\Delta S_{system} = -\Delta S_{surrounding}$.
3. If ΔG is negative, the reaction is spontaneous because ΔS_{total} is positive, pointing to a total entropy increase, a factor necessary for an irreversible process.

What factors contribute to:

1. A positive ΔG? The reaction may be endothermic (ΔH positive) and/or be accompanied by a decrease in entropy (ΔS negative). If ΔH is positive and ΔS is positive, only a temperature increase, an increase of $-T\Delta S$, can override the $+\Delta H$.
2. A negative ΔG? The reaction may be exothermic (ΔH negative) and/or be accompanied by an increase in entropy (ΔS positive). If ΔH is positive and ΔS is positive, then a temperature increase makes ΔG more negative. If ΔH is negative and ΔS is also negative, then a lower temperature lessens the entropy decrease within the system and makes ΔG more negative. See text, Table 11.3, for a summary of the relations between ΔH, ΔS, T, and ΔG.

QUESTIONS

Section 11.1

11.1 Which of the following quantities depend only on the initial and final states of the process and *not* upon pathway?

(a) The heat produced from a 2.0 kg chunk of coal.
(b) The time required to inflate a tire.
(c) The energy used in moving a 145 g paper weight 3 inches.
(d) The work used in hoeing a garden.
(e) The weeding (or deweeding) of a garden.
(f) The potential energy difference between the 16th floor and ground level.
(g) The energy consumed in driving from Dallas to Omaha.

Section 11.2

11.2 Calculate the quantity of heat (kJ) released in the condensation of 20.0 g H_2O at 700 K to H_2O (ℓ) at the boiling point. The specific heat of $H_2O(g)$ is 2.01 J/(g °C).

Section 11.3

11.3 From the text, Figure 11.2, which substance—Br_2, C_2H_5OH, or water—has the highest vapor pressure at 50°C?

11.4 Arrange the following in order of increasing strength of I.M. forces: Br_2, He, Cl_2, H_2O, C_2H_5OH.

11.5 One liter of air at 1 atm pressure is saturated with H_2O vapor at 50°C. What is the total pressure of the combined gases in the 1 liter volume? What is the partial pressure of H_2O? (See text, Figure 11.2.)

11.6 If H_2O at 50°C is added to column b in the text, Figure 11.3, what is the reading of the Hg level when $H_2O(\ell) \rightleftharpoons H_2O(g)$?

Section 11.4

11.7 Explain why more heat is needed to vaporize 1 mole of a substance than to melt it.

11.8 What is a boiling chip? Describe its function.

11.9 If the pressure exerted by the atmosphere is less than 1 atm, do bubbles in the liquid form at a temperature above or below the normal boiling point? Explain.

Section 11.5

11.10 When a solid melts, heat energy is absorbed. How is this energy used? Why doesn't the temperature of the system increase before all the solid melts?

Section 11.6

11.11 Does CO_2 have a normal boiling point and freezing point? Explain. (See text, Figure 11.6.)

11.12 Complete the following statements about the effect of I.M. forces on the physical state of a substance.

(a) The higher the boiling point of a liquid the (stronger, weaker) the I.M. forces.

(b) Substances with strong I.M. forces show a (high, low) critical temperature.

(c) O_2 has (strong, weak) I.M. forces and therefore has a (high, low) critical temperature.

(d) H_2O has a high critical temperature and therefore has a (high, low) vapor pressure.

(e) Substances with low vapor pressures have (strong, weak) I.M. forces.

(f) The weak I.M. forces in Cl_2 indicate that Cl_2 also has a (high, low) vapor pressure and a (high, low) critical temperature.

(g) NH_3 has a critical temperature of 405.6 K, whereas H_2O has one of 647.2 K. NH_3 has the (stronger, weaker) I.M. forces. (NH_3, H_2O) has the lower vapor pressure. (NH_3, H_2O) has the lower melting point. (NH_3, H_2O) has the higher boiling point.

Section 11.7

11.13 Which has a higher entropy, a mole of $H_2O(\ell)$ or 1 mole of steam, $H_2O(g)$?

Section 11.8

11.14 When does $\Delta S = q/T$? Which is greater for an irreversible process, q or $T\Delta S$?

11.15 Calculate the entropy change for 10.0 g CCl_4 (ℓ) at 76.8°C to CCl_4 (g) at 76.8°C. The ΔH_{vap} for CCl_4 (Table 11.1) is 194 J/g. Is this a reversible or an irreversible process? How does this value of ΔS compare with that for the same process at 80°C?

Section 11.10

11.16 For the reaction

$$2CHCl_3(\ell) + O_2(g) \rightarrow 2COCl_2(g) + 2HCl(g)$$

ΔG is −467 kJ and ΔS is +340 J/K.

(a) Calculate ΔH at 25°C.

(b) Can ΔG ever equal zero for this reaction? Explain.

11.17 For the reaction

$$2Ag(c) + Br_2(\ell) \rightarrow 2AgBr(c)$$

$\Delta S = -25.5$ J/(mol K) and $\Delta H = -199$ kJ/mol.

(a) What does the fact that ΔS is negative imply?

(b) Calculate ΔG at 298 K. What significance does this value have?

(c) At what temperature does $\Delta G = 0$? Is this temperature experimentally feasible?

ANSWERS

11.1 (a), (c), (d), (f)

11.2 58.3 kJ

11.3 Br_2

11.4 He, Cl_2, Br_2, C_2H_5OH, H_2O

11.5 760 + 93 = 853 torr; 93 torr

11.6 667 mm Hg

11.7 When a substance melts, the I.M. forces are weakened such that molecules slide by one another; but when a substance vaporizes, the I.M. forces are eliminated entirely. This requires more energy.

11.8 A boiling chip serves as a site upon which a bubble can form when a substance reaches its boiling point. This avoids superheating of the liquid.

11.9 Bubbles in the liquid form below the normal boiling point because the pressure exerted by the molecules in the bubble push against a lower external pressure than normal.

11.10 Heat is used to reduce the strength of the I.M. forces. The temperature does not increase because molecules in the liquid state are constantly moving and transferring kinetic energy to other molecules in the system until all the I.M. forces are weakened and the substance becomes a liquid.

11.11 No, CO_2 has a normal sublimation point only. Remember normal refers to 760 torr. CO_2 has a boiling point and a freezing point only when the pressure on the system exceeds 5.2 atm.

11.12 (a) stronger

(b) high

(c) weak, low

(d) low

(e) strong

(f) high, low

(g) NH_3, weaker; H_2O; NH_3; H_2O

11.13 A mole of steam has a higher entropy because the $H_2O(g)$ molecules are randomly distributed over the entire volume.

11.14 $\Delta S = q/T$ for a thermodynamically reversible process. q for an irreversible process is less than $T\ \Delta S$.

11.15 5.54 J/K; reversible; 5.49 J/K

11.16 (a) $\Delta H = -366$ kJ

(b) No, when ΔH is negative and ΔS is positive, $\Delta G = -367 - T\Delta S$. ΔG cannot equal zero.

11.17 (a) When ΔS is negative, there is a decrease in entropy, or the system is becoming more ordered.

(b) $\Delta G = -191$ kJ/mol; the reaction is spontaneous.

(c) $\Delta G = 0$ at 7800 K; yes, but at a very high temperature.

Chapter 12

SOLUTIONS

SECTION 12.1 WHAT IS A SOLUTION?

Objectives

- To differentiate between a solute and a solvent in a solution.
- To distinguish between a dilute and a concentrated solution.

Focus

A **solution** is a homogeneous mixture of two or more pure substances that has a uniform and constant composition. In a solution, the **solvent** is the dissolving substance and the **solute** is the substance being dissolved.

A dilute solution has "a little bit" of solute dissolved in the solvent and a concentrated solution has "a lot." These two terms are relative and do not refer to any exact amount of solute dissolved.

SECTION 12.2 MEASURES OF COMPOSITION FOR SOLUTIONS

Objectives

- To calculate the molarity, mass percentage, and volume percentage of solute in an aqueous (water) solution.
- To prepare a dilute solution from a more concentrated one.

Focus

Two common methods that express the amount of solute dissolved in a solution are

$$\text{molarity, M} = \frac{\text{moles solute}}{\text{liters of solution}}$$

$$\text{mass percentage of solute, } w\% = \frac{w_{\text{solute}}}{w_{\text{solute}} + w_{\text{solvent}}} \times 100\%$$

Know these and work as many problems as time allows.

Example 12.1 Calculate the mass in grams of $FeCl_2$ in 1.00 liter of 0.670 M $FeCl_2$.

Answer 85.1 $FeCl_2$/L solution

Solution 0.670 M $FeCl_2$ means $\dfrac{0.670 \text{ mol } FeCl_2}{\text{L soln}}$.

Convert mol $FeCl_2$ to g $FeCl_2$:

$$1 \text{ mol } FeCl_2 = 127 \text{ g } FeCl_2$$

Therefore, using dimensional analysis,

$$\frac{\text{g } FeCl_2}{\text{L soln}} = \frac{0.670 \cancel{\text{mol } FeCl_2}}{\text{L soln}} \times \frac{127 \text{ g } FeCl_2}{\cancel{\text{mol } FeCl_2}} = \frac{85.1 \text{ g } FeCl_2}{\text{L soln}}$$

Example 12.2 A mass of 48.7 g $FeCl_2(H_2O)_6$ dissolves in 600 mL of solution. Calculate the molarity of the solution.

Answer 0.345 mol/L

Solution Ferrous chloride can be purchased only as $FeCl_2(H_2O)_6$. In weighing the salt, the six H_2O's are also weighed; therefore the molecular weight (also called formula weight) of the entire compound, $FeCl_2(H_2O)_6$, must be considered.

$$\text{molecular wt. of } FeCl_2(H_2O)_6 = 235 \text{ g/mol}$$

Analyze the problem:

Given quantity: $\dfrac{48.7\text{g } FeCl_2(H_2O)_6}{0.600 \text{ L}}$

Desired quantity: $\dfrac{\text{mol } FeCl_2}{\text{L}}$

Conversion factor:

$$1 \text{ mol } FeCl_2(H_2O)_6 = 235 \text{ g}$$

$$1 \text{ mol } FeCl_2 = 1 \text{ mol } FeCl_2(H_2O)_6$$

$$\frac{\text{mol } FeCl_2}{\text{L}} = \frac{48.7 \cancel{\text{g } FeCl_2(H_2O)_6}}{0.600 \text{ L soln}} \times \frac{\cancel{1 \text{ mol } FeCl_2(H_2O)_6}}{235 \cancel{\text{ g } FeCl_2(H_2O)_6}} \times \frac{1 \text{ mol } FeCl_2}{\cancel{1 \text{ mol } FeCl_2(H_2O)_6}}$$

$$= 0.345 \frac{\text{mol } FeCl_2}{\text{L soln}}$$

$$= 0.345 \text{ M } FeCl_2$$

The mass percent of a solute in solution is merely its mass fraction × 100.

Example 12.3 Calculate the mass of $KMnO_4$ in 80.0 g of a 2.00% $KMnO_4$ solution.

Answer 1.60g $KMnO_4$

Solution What does 2.00% $KMnO_4$ mean?

Answer: $\dfrac{2.00 \text{ g } KMnO_4}{100 \text{ g soln}}$

To calculate the number of g $KMnO_4$ in 80.0 g solution, then

$$\text{mass } KMnO_4 = 80.0 \cancel{\text{g soln}} \times \frac{2.00 \text{ g } KMnO_4}{100 \cancel{\text{g soln}}} = 1.60 \text{ g } KMnO_4$$

Example 12.4 (a) Calculate the volume (mL) of 37.23% HCl solution (conc HCl) that provides 20.0 g HCl. The density of conc. HCl is 1.19 g/mL. (b) If the conc HCl in part (a) is diluted to 250 mL, what is the molarity of the solution?

Answer (a) 45.1 mL solution; (b) 2.19 M HCl

Solution (a) What does 37.23% HCl mean?

$$\frac{37.23 \text{ g HCl}}{100 \text{ g soln}}$$

Given quantity: 20.0 g HCl

Desired quantity: mL soln

Conversion factors:

$$\frac{100 \text{ g soln}}{37.23 \text{ g HCl}} ; \frac{1 \text{ mL soln}}{1.19 \text{ g soln}}$$

$$\text{mL soln} = 20.0 \cancel{\text{g HCl}} \times \frac{100 \cancel{\text{g soln}}}{37.23 \cancel{\text{g HCl}}} \times \frac{1 \text{ mL soln}}{1.19 \cancel{\text{g soln}}}$$

$$= 45.1 \text{ mL soln}$$

(b) Given quantity: 20.0 g HCl/250 mL soln

Desired quantity: $\frac{\text{mol HCl}}{\text{L soln}}$

$$\frac{\text{mol HCl}}{\text{L soln}} = \frac{20.0 \cancel{\text{g HCl}}}{250 \cancel{\text{mL soln}}} \times \frac{1 \text{ mol HCl}}{36.46 \cancel{\text{g HCl}}} \times \frac{1000 \cancel{\text{mL soln}}}{\text{L soln}} = 2.19 \frac{\text{mol HCl}}{\text{L soln}}$$

$$= 2.19 \text{ M HCl}$$

If 45.1 mL of conc HCl is diluted to 250 mL, the molarity is 2.19 M HCl.

To prepare a dilute solution from a more concentrated one, just add water. The question is how much? The equation for dilution is

$$V_c M_c = V_d M_d$$

V_c = volume of concentrated solution
M_c = molarity of concentrated solution
V_d = volume of diluted solution
M_d = molarity of diluted solution

The units on both sides of the equation are moles of solute, the only quantity that is constant in a dilution.

$$\cancel{\text{L soln}_c} \times \frac{\text{mol solute}}{\cancel{\text{L soln}_c}} = \cancel{\text{L soln}_d} \times \frac{\text{mol solute}}{\cancel{\text{L soln}_d}}$$

Example 12.5 Describe the preparation of 600 mL of 0.371 M Na_2SO_4, starting with 4.60 M Na_2SO_4.

Answer Dilute 48.4 mL of 4.60 M Na_2SO_4 to 600 mL with water.

Solution (a) Convert 0.371 M Na_2SO_4 and 4.60 M Na_2SO_4 to

$$\frac{0.371 \text{ mol } Na_2SO_4}{\text{L}} \quad \text{and} \quad \frac{4.60 \text{ mol } Na_2SO_4}{\text{L}}$$

(b) Remember the mol Na_2SO_4 is constant; therefore the mol Na_2SO_4 in 0.600 L of solution is

$$0.600\cancel{\text{L}} \times \frac{0.371 \text{ mol } Na_2SO_4}{\cancel{\text{L}}} = 0.223 \text{ mol } Na_2SO_4$$

(c) The volume of 4.60 M Na_2SO_4 which provides 0.223 mol Na_2SO_4 in

$$0.223 \cancel{\text{mol } Na_2SO_4} \times \frac{\text{L soln}}{4.60 \cancel{\text{mol } Na_2SO_4}} = 0.0484 \text{ L} = 48.4 \text{ mL}$$

or

$$V_c = \frac{V_d M_d}{M_c} = \frac{0.600 \text{ L} \times \frac{0.371 \cancel{\text{mol } Na_2SO_4}}{\cancel{\text{L}}}}{\frac{4.60 \cancel{\text{mol } Na_2SO_4}}{\cancel{\text{L}}}} = 0.0484\text{L} = 48.4 \text{ mL}$$

To make 600 mL of 0.371 M Na_2SO_4, dilute the 48.4 mL of 4.60 M Na_2SO_4 with water to 600 mL. Notice that we have *not* added (600 − 48.4 mL) = 551.6 mL of H_2O. Rather we add enough water to reach a final volume of 600 mL. This is the correct procedure because a combination of the volumes of two solutions is not always equal to their sum. However, it is very close to 551.6 mL.

When mixing two liquids, volume percent is commonly used.

$$\% \text{ volume} = \frac{\text{volume of solute}}{\text{volume of solution}} \times 100\%$$

SECTION 12.3 SOLUBILITY

Objectives

- To define solubility and saturated solution.
- To convert solubility of a salt into percent composition.

Focus

In a saturated solution, the rate at which a solid solute precipitates equals the rate at which the solid solute dissolves so that there is dynamic equilibrium between the dissolved and the undissolved solute. A saturated solution of NaCl is represented as

$$NaCl(c) \rightleftharpoons Na^+(aq) + Cl^-(aq)$$

The solubility of NaCl is the maximum quantity of NaCl dissolved in a certain amount of solvent or solution.

Example 12.6 The solubility of sodium arsenate-water (1/12), $Na_3AsO_4(H_2O)_{12}$, is 38.9 g per 100 g H_2O. Calculate the mass percentage of Na_3AsO_4 in the saturated solution.

Answer 13.8% Na_3AsO_4

Solution $$w\%_{\text{salt in hydrate}} = \frac{w_{Na_3AsO_4}}{w_{Na_3AsO_4(H_2O)_{12}}} \times 100\%$$

We need to calculate the total mass of the solution:

$$38.9\text{ g } Na_3AsO_4(H_2O)_{12} + 100\text{ g } H_2O = 138.9\text{ g solution}$$

What is the mass of Na_3AsO_4 in 38.9 g $Na_3AsO_4(H_2O)_{12}$?

1 mol Na_3AsO_4 has a mass of 207.9 g (formula weight). 1 mol $Na_3AsO_4(H_2O)_{12}$ has a mass of 424.1 g (formula weight).

Therefore the mass ratio of Na_3AsO_4 to $Na_3AsO_4(H_2O)_{12}$ is

$$\frac{207.9\text{ g } Na_3AsO_4}{424.1\text{ g } Na_3AsO_4(H_2O)_{12}}$$

$$\text{mass } Na_3AsO_4 = 38.9\,\cancel{\text{g } Na_3AsO_4(H_2O)_{12}} \times \frac{207.9\text{ g } Na_3AsO_4}{424.1\,\cancel{\text{g } Na_3AsO_4(H_2O)_{12}}}$$

$$= 19.1\text{ g } Na_3AsO_4$$

$$w\%_{Na_3AsO_4} = \frac{19.1\text{ g } Na_3AsO_4}{138.9\text{ g soln}} \times 100\% = 13.8\%\ Na_3AsO_4$$

Remember that the 12 H_2O molecules in $Na_3AsO_4(H_2O)_{12}$ become indistinguishable from the solvent water molecules in solution.

The term **miscible** defines the solubility of one liquid in another. If two liquids are completely miscible, regardless of their relative amounts, they are infinitely soluble in each other and a saturated solution is never realized. If a saturated solution is realized between two liquids, they are **partially miscible**. Each liquid layer is saturated with the other liquid.

SECTION 12.4 SOLUBILITY AND PRESSURE

Objective

- To explain the effect of pressure on the solubility of a gas in a liquid.

Focus

Henry's law states that the solubility of a gas in a liquid is directly proportional to the pressure of the gas above the liquid. This is true for gases which do not react with the liquid solvent.

This solubility law may be stated in several ways as long as the units are consistent with the constant in Henry's law.

The solubility of a gas in a liquid is measured when the rate at which the gas molecules enter the liquid equals the rate at which they leave it, or when the process

$$\text{gas (in solution)} \rightleftharpoons \text{gas (undissolved)}$$

has come to equilibrium.

Remember that pressure changes have essentially no effect on the solubility of liquids and solids in liquid solvents.

Example 12.7 Calculate the solubility of N_2 in water (g N_2/100 g H_2O) at 675 torr and at 25°C. Use the text, Table 12.2.

Answer 0.00160 g N_2/100 g H_2O

Solution From Table 12.2, the solubility of N_2 is 0.00175 g/100 g H_2O at a pressure of 1 atm. The total pressure

$$P_{\text{total}} = 760 \text{ torr}$$

equals $p_{N_2} + p_{H_2O}$, the partial pressure of N_2 plus the vapor pressure of H_2O at 25°C (24 torr).

$$p_{N_2} = P_{\text{total}} - p_{H_2O} = 760 \text{ torr} - 24 \text{ torr}$$
$$= 736 \text{ torr}$$

Henry's law says that the solubility of a gas is directly proportional to its partial pressure above the solution.

Since 0.00175 g N_2 dissolves in 100 H_2O at 736 torr, then the g N_2 that dissolve in 100 g H_2O at 675 torr is

$$\frac{\text{g } N_2}{100 \text{ g } H_2O} = \frac{0.00175 \text{ g } N_2}{100 \text{ g } H_2O} \times \frac{675 \cancel{\text{torr}}}{736 \cancel{\text{torr}}}$$
$$= \frac{0.00160 \text{ g } N_2}{100 \text{ g } H_2O}$$

SECTION 12.5 SOLUBILITY AND TEMPERATURE

Objectives

- To predict the effect of temperature changes on the solubility of gases and solids in a liquid.
- To explain the existence of a supersaturated solution.

Focus

Heat may be evolved (ΔH is negative, exothermic) or absorbed (ΔH is positive, endothermic) when a solid dissolves in a liquid. Two theoretical steps for the dissolving of a solid are:

Step 1. Solid particles (ions or molecules) are separated—energy absorbed.

Step 2. Separated solid particles are attracted to the liquid solvent—energy evolved.

If the energy absorbed (step 1) > energy evolved (step 2), then the dissolving process is endothermic: ΔH is positive. The solubility increases with a rise in temperature (supplying the heat absorbed).

If the energy absorbed (step 1) < energy evolved (step 2), then the dissolving process is exothermic: ΔH is negative. The solubility increases with a decrease in the temperature (removing the heat evolved).

> An endothermic (ΔH is positive) process for dissolving a salt is spontaneous. Recall from Chapter 11 that ΔG is negative for a spontaneous process. For ΔG to be negative and ΔH positive, ΔS must also be positive.
>
> $$\Delta G = \Delta H - T\,\Delta S$$
>
> The entropy increase, from the ordered array in the solid salt to a very random distribution of the salt in solution, causes the negative $T\,\Delta S$ to be greater than the positive ΔH, making ΔG negative.

For gases, there is no step 1; therefore, nearly all dissolving processes involving gases are exothermic: solubility decreases with a rise in temperature.

Example 12.8 Given the following data for LiBr:

Step 1. $LiBr(c) \rightarrow Li^+(g) + Br^-(g)$
$\Delta H_1 = +787$ kJ

Step 2. $Li^+(g) + Br^-(g) \rightarrow Li^+(aq) + Br^-(aq)$
$\Delta H_2 = -854$ kJ

determine whether the solubility of LiBr increases or decreases with an increase in temperature.

Answer Decreases

Solution The ΔH for the solution process is

$$\begin{aligned}\Delta H &= \Delta H_1 + \Delta H_2 \\ &= 787\text{ kJ} + (-854\text{ kJ}) = -67\text{ kJ}\end{aligned}$$

The dissolving process is exothermic. Therefore, adding heat disfavors the exothermic process and decreases the solubility of LiBr.

A **supersaturated solution** is an unstable system in which more than the equilibrium amount of solute is dissolved in a liquid. The solution returns to a saturated system (the equilibrium condition) when we add a seed crystal, stir rapidly, or agitate the solution.

The free energy for dissolving more solute in solutions at various states of saturation is:

Saturated solution:	$\Delta G = 0$	(at equilibrium)
Supersaturated solution:	$\Delta G > 0$	(positive value)
Unsaturated solution:	$\Delta G < 0$	(negative value and therefore the dissolving of more solute is spontaneous)

SECTION 12.6 SOLUBILITY AND MOLECULAR STRUCTURE

Objective

• To explain why some solute-solvent pairs form solutions more easily than others, or why "like dissolves like."

Focus

Three classifications of I.M. forces are:

1. **London forces**—dominant between nonpolar molecules.
2. **Dipole-dipole forces**—dominant between polar molecules.
3. **Hydrogen bonding**—an exceptionally strong dipole-dipole force in which an H atom on one molecule is strongly attracted to a more electronegative atom such as N, O, or F on another molecule.

A common interpretation for "like dissolves like" is that if the polarity of the *solute* molecules is more like the polarity of the *solvent* molecules, dissolving of the solute occurs; if the polarity of *solute* molecules is unlike the polarity of *solvent* molecules, then the solute does *not* dissolve in the solvent.

Nonpolar carbon tetrachloride, CCl_4, molecules do *not* dissolve in water because the H_2O-H_2O forces of attraction are greater than those of H_2O-CCl_4. H_2O stays with H_2O molecules and therefore CCl_4 remains with CCl_4 molecules.

Carbon tetrachloride is soluble in benzene because both substances are nonpolar and have similar London forces.

SECTION 12.7 SOLUBILITY OF IONIC CRYSTALS

Objectives

• To explain why salts dissolve in water.

• To determine the effect of temperature on the solubility of a salt in water given its heat of solution.

• To recall from memory whether a salt may be labeled as soluble or insoluble.

Focus

When a salt dissolves in water, positive ions are attracted to the negative end of the water molecules and negative ions to the positive end. The ions are hydrated. Because of this sphere of water molecules around the ions, the attraction between oppositely charged ions is reduced, and they freely move about in solution.

Nonpolar solvent molecules have no negative or positive end and therefore have little or no attraction for positive and negative solute ions.

Dissolving a salt in water may be an endothermic or exothermic process depending upon the two factors discussed in Section 12.5 of this Study Guide. For a salt, the energy for step 1 is the lattice energy, ΔH_ℓ, the energy needed to separate the solid salt into its gaseous ions:

Step 1. $AB(c) \rightarrow A^+(g) + B^-(g)$ $\quad \Delta H_\ell$ is positive

The energy for step 2 is the hydration energy, ΔH_h, the energy released from the natural attraction between gaseous ions and polar water molecules:

Step 2. $A^+(g) + B^-(g) \rightarrow A^+(aq) + B^-(aq)$ $\quad \Delta H_h$ is negative

Often salts are labeled soluble or insoluble. Insoluble salts are actually slightly soluble in water. Memorize a major part of Table 12.3 in the text.

SECTION 12.8 NET IONIC EQUATIONS

Objective

- To write the net ionic equation for a reaction involving an insoluble salt.

Focus

A **net ionic equation** *includes only the ions that undergo a chemical change;* those remaining are spectator ions. Spectator ions maintain the neutrality of the salt solution but are not involved in any chemical reaction.

Soluble salts are written in ionic form and insoluble salts in molecular form. Review Table 12.3, the solubility rules for salts.

Example 12.9 Write the net ionic equation for each of the following salt mixtures in aqueous solution.

(a) $BaCl_2 + Hg_2(NO_3)_2 \rightarrow$
(b) $FeSO_4 + NaOH \rightarrow$
(c) $(NH_4)_3PO_4 + Ca(CO_3)_2 \rightarrow$

Answer

(a) $2Cl^- + Hg_2^{2+} \rightarrow Hg_2Cl_2(c)$
(b) $Fe^{2+} + 2OH^- \rightarrow Fe(OH)_2(c)$
(c) $2PO_4^{3-} + 3Ca^{2+} \rightarrow Ca_3(PO_4)_2(c)$

Solution Attack these reactions step-by-step:

1. Write the products of the reaction in molecular form by exchanging cation-anion partners.
2. Rewrite the equation, putting the soluble salts in ionic form and the insoluble salts in molecular form. Use Table 12.3.
3. Omit the spectator ions common to both sides of the equation. We will assume for now that if a reactant salt is insoluble, no reaction occurs.
4. Write the net ionic equation.

(a) 1. $BaCl_2 + Hg_2(NO_3)_2 \rightarrow Hg_2Cl_2 + Ba(NO_3)_2$

2. $BaCl_2$, $Hg_2(NO_3)_2$, $Ba(NO_3)_2$—soluble; Hg_2Cl_2—insoluble. Rewrite according to step 2.

$$Ba^{2+} + 2Cl^- + Hg_2^{2+} + 2NO_3^- \rightarrow Hg_2Cl_2(c) + Ba^{2+} + 2NO_3^-$$

3. Omit spectator ions common to both sides:

$$\cancel{Ba^{2+}} + 2Cl^- + Hg_2^{2+} + \cancel{2NO_3^-} \rightarrow Hg_2Cl_2(c) + \cancel{Ba^{2+}} + \cancel{2NO_3^-}$$

4. Write the net ionic form of the equation:

$$2Cl^- + Hg_2^{2+} \rightarrow Hg_2Cl_2(c)$$

(b) 1. $FeSO_4 + 2NaOH \rightarrow Fe(OH)_2 + Na_2SO_4$

2. $FeSO_4$, $NaOH$, Na_2SO_4—soluble; $Fe(OH)_2$—insoluble. Rewrite

$$Fe^{2+} + SO_4^{2-} + 2Na^+ + 2OH^- \rightarrow Fe(OH)_2(c) + 2Na^+ + SO_4^{2-}$$

3. The SO_4^{2-} and Na^+ ions are spectator ions.

4. $Fe^{2+} + 2OH^- \rightarrow Fe(OH)_2(c)$

(c) 1. $2(NH_4)_3PO_4 + 3Ca(ClO_3)_2 \rightarrow Ca_3(PO_4)_2 + 6NH_4ClO_3$

2. $(NH_4)_3PO_4$, $Ca(ClO_3)_2$, NH_4ClO_3—soluble; $Ca_3(PO_4)_2$—insoluble. Rewrite

$$6NH_4^+ + 2PO_4^{3-} + 3Ca^{2+} + 6ClO_3^- \rightarrow Ca_3(PO_4)_2(c) + 6NH_4^+ + 6ClO_3^-$$

3. NH_4^+ and ClO_3^- are spectator ions.

4. $2PO_4^{3-} + 3Ca^{2+} \rightarrow Ca_3PO_4(c)$

SECTION 12.9 VAPOR PRESSURES OF SOLUTIONS

Objectives

- To state the method by which a solution's vapor pressure is determined.
- To explain that the vapor pressure of a solution containing a nonvolatile solute is always less than that of a pure solvent.

Focus

Solvent molecules that escape into the gaseous state exert a pressure on a closed system. If a nonvolatile solute is added to a pure liquid solvent, the ability of solvent molecules to escape into a vapor decreases. This happens because the solute particles displace some solvent molecules from the liquid's surface, thereby decreasing the number of solvent molecules that can escape.

When a solute is volatile, both the solute and solvent contribute to the total vapor pressure of the solution.

$$P_{soln} = p_{solvent} + p_{solute}$$

SECTION 12.10 VAPOR PRESSURE AND COMPOSITION

Objectives

- To calculate the mole fractions of a solvent and solute in solution.
- To explain how the vapor pressure of a volatile liquid depends upon its mole fraction in solution.
- To explain the deviations of an experimental vapor pressure from that which is calculated.

Focus

The **mole fraction** of substance, A, in solution is

$$x_A = \frac{n_A}{n_A + n_B + \cdots}$$

where n represents numbers of moles. Given the mole fraction of substance, B, we can calculate the mole fraction of substance A kg using the following equation:

$$x_A = 1 - x_B$$

Example 12.10 Calculate the mole fraction and mole percent of CH_3OH for a solution prepared by mixing 50.0 g CH_3OH and 50.0 g H_2O.

Answer $x_{CH_3OH} = 0.359$; 35.9 mol % CH_3OH

Solution Find the number of moles of each component:

$$n_{CH_3OH} = 50.0\ \text{g}\ CH_3OH \times \frac{\text{mol}\ CH_3OH}{32.0\ \text{g}\ CH_3OH} = 1.56\ \text{mol}\ CH_3OH$$

$$n_{H_2O} = 50.0\ \text{g}\ H_2O \times \frac{\text{mol}\ H_2O}{18.0\ \text{g}\ H_2O} = 2.78\ \text{mol}\ H_2O$$

$$x_{CH_3OH} = \frac{1.56}{1.56 + 2.78} = 0.359$$

$$\text{mol}\ \%\ CH_3OH = x_{CH_3OH} \times 100\% = 35.9\ \text{mol}\ \%$$

According to Raoult's law, the vapor pressure of a volatile liquid equals the product of its mole fraction in solution and its vapor pressure as a pure liquid.

$$p_A = x_A p_A{}^0$$

If several volatile liquids are in a solution, its vapor pressure equals the sum of each volatile component's partial vapor pressure.

$$P_{soln} = p_{solvent} + p_{solute\ A} + p_{solute\ B} + \cdots$$

Substituting in the Raoult's law expression,

$$\begin{aligned} P_{soln} &= x_{solvent}\, p^0_{solvent} + x_{solute\ A}\, p^0_{solute\ A} \\ &= x_s p^0_s + x_A p^0_A + x_B p^0_B + \cdots \end{aligned}$$

For ideal solutions Raoult's law is valid: the I.M. forces between solute-solute, solute-solvent, and solvent-solvent molecules are equal.

If the I.M. forces between solute and solvent are *less than* those between solute and solute or solvent and solvent, then molecules of the solute and solvent escape more easily, producing a combined vapor pressure that is greater than that calculated from Raoult's law, a *positive* deviation. Figures 12.10 and 12.12 in the text show solutions which have positive deviations from ideal behavior.

If the I.M. forces between solute and solvent are *greater* than those between solute and solute or solvent and solvent, then solute and solvent molecules are less likely to escape into the vapor, thus producing a vapor pressure lower than that calculated from Raoult's law, a *negative* deviation.

Example 12.11 Calculate the (ideal) vapor pressure for the solution in Example 12.10 at 30°C. The vapor pressure of pure CH_3OH is 160 torr at 30°C and that of pure H_2O is 31.8 torr at 30°C.

Answer 77.8 torr

Solution To find the mole fraction of H_2O, use the equation

$$x_A = 1 - x_B$$

$$x_{H_2O} = 1 - x_{CH_3OH}$$

$$x_{H_2O} = 1 - 0.359 = 0.641$$

Using Raoult's expanded equation,

$$\begin{aligned} P_{soln} &= x_{CH_3OH}P^0_{CH_3OH} + x_{H_2O}P^0_{H_2O} \\ &= 0.359(160 \text{ torr}) + 0.641(31.8 \text{ torr}) \\ &= 77.8 \text{ torr} \end{aligned}$$

Remember that we assume these two liquids form an ideal solution. The total vapor pressure for the actual solution has to be measured in the laboratory.

SECTION 12.11 COLLIGATIVE PROPERTIES OF SOLUTIONS

Objective

- To identify the four colligative properties of solutions.

Focus

When a solute is added to a solvent, the solution has

1. A lower vapor pressure.
2. A higher boiling point.
3. A lower freezing point than the pure solvent.

In addition, the solution has

4. An osmotic pressure relative to the pure solvent.

The magnitude of these four colligative properties depend only upon the *number* of solute particles (molecules or ions) in solution, not their type. As a result, a sugar molecule, a Na^+ ion, or a urea molecule has the same effect on these four colligative properties. This is strictly true only for very dilute solutions.

SECTION 12.12 VAPOR PRESSURE DEPRESSION

Objective

- To calculate the vapor pressure depression caused by the addition of a nonvolatile solute to a solvent.

Focus

Learn the equation for the **vapor pressure depression**. (See the text.)

$$p_A{}^0 - p_A = x_B p_A^0$$

or

$$\Delta p_A = x_B p_A^0$$

SECTION 12.13 FREEZING-POINT DEPRESSION AND BOILING-POINT ELEVATION

Objectives

- To calculate the molality of a solute in solution.
- To calculate the freezing-point depression and boiling-point elevation caused by the addition of a solute to a solvent.
- To explain the effect of a nonvolatile solute on the phase diagram of a pure solvent.

Focus

The **molality** of a solute in solution is

$$\text{molality, } m = \frac{\text{mol solute}}{\text{kg solvent}}$$

Notice that the definition is "per kg *solvent*," *not* solution.

The **freezing-point depression**, Δt_f, is calculated from

$$\Delta t_f = K_f m$$

and the **boiling-point elevation**, Δt_b, is calculated from

$$\Delta t_b = K_b m$$

where K_f and K_b are constants for each particular solvent. Table 12.5 in the text gives the values of these constants for various solvents.

Example 12.12 A mass of 20.0 g urea, $CO(NH_2)_2$, is dissolved in 200 g H_2O. Determine the solution's freezing and boiling points.

Answer $t_f = -3.11°C$; $t_b = 100.855°C$

Solution To use the Δt equations, we first need to calculate the molality of the solution. The formula weight of $CO(NH_2)_2$ is 60.0 g/mol

$$m_{CO(NH_2)_2} = \frac{20.0\ \cancel{g\ CO(NH_2)_2}}{200\ \cancel{g\ H_2O}} \times \frac{1\text{ mol }CO(NH_2)_2}{60.0\ \cancel{g\ CO(NH_2)_2}} \times \frac{1000\ \cancel{g\ H_2O}}{1\text{ kg }H_2O} = \frac{1.67\text{ mol }CO(NH_2)_2}{\text{kg }H_2O}$$

From Table 12.5, $K_f(H_2O) = 1.86\ \frac{°C\ kg}{mol}$ and $K_b(H_2O) = 0.512\ \frac{°C\ kg}{mol}$

$$\Delta t_f = K_f m = 1.86\ \frac{°C\ \cancel{kg}}{\cancel{mol}} \times 1.67\ \frac{\cancel{mol}}{\cancel{kg}} = 3.11°C$$

$$\Delta t_b = K_b m = 0.512\ \frac{°C\ \cancel{kg}}{\cancel{mol}} \times 1.67\ \frac{\cancel{mol}}{\cancel{kg}} = 0.855°C$$

The freezing point is

$$t_f = 0.00°C - 3.11°C = -3.11°C$$

The boiling point is

$$t_b = 100.000°C + 0.855°C = 100.855°C$$

Figure 12.13 shows the relationships between the three colligative properties. A pure solvent boils at t_b when the vapor pressure equals atmospheric pressure (1 atm for normal boiling points). Since the vapor pressure is reduced to less than 1 atm when a solute is added to the solvent, the temperature of the solution must be increased in order for boiling to occur at t_b'. Similarly, the freezing point of the solution, t_f', is less than the solvent because solute particles are in the way, thereby reducing the tendency of solvent molecules to form a crystal at t_f. To solidify, the solution needs a lower temperature. Remember that the three equations for colligative properties are valid only under the conditions described in the text. (See Study Guide, Question 12.22.)

SECTION 12.14 COLLIGATIVE PROPERTIES OF IONIC SOLUTIONS

Objectives

- To explain the effect of ionic solutes on the colligative properties of a solution.
- To calculate the van't Hoff factor for ionic solutions.

Focus

Remember that colligative properties depend only upon the number of solute particles (ions or molecules) dissolved in a solvent, not the kind of solute. (See Section 12.11.)

Therefore, we expect 1 mol NaCl ($\rightarrow Na^+ + Cl^-$) to have twice the effect on a colligative property of an aqueous solution as 1 mol sugar, which does not dissociate in water. The number of solute particles produced in solution for some salts are:

	Moles of ions per mole of salt
$KCl \rightarrow K^+ + Cl^-$	2
$K_2SO_4 \rightarrow 2K^+ + SO_4^{2-}$	3
$CuSO_4 \rightarrow Cu^{2+} + SO_4^{2-}$	2
$Na_3PO_4 \rightarrow 3Na^+ + PO_4^{3-}$	4

Generally, the effect of an ion on the colligative properties of a solution is less than that for a neutral molecule. The natural electrostatic attraction between oppositely charged ions causes some association between ions producing effectively fewer particles (ions) than if the solute were neutral molecules. The ions in solution are not independent of other ions, whereas molecules are.

The **van't Hoff factor**, i, is the ratio of the experimental value for the colligative property to that value when no dissociation occurs.

$$i = \frac{\text{experimental value}}{\text{calculated value (assuming no dissociation)}}$$

Example 12.13 The freezing point of a 0.0100 m HNO_2 solution is −0.0222°C. Calculate the van't Hoff factor.

Answer 1.19

Solution For a 0.0100 m HNO_2 solution

$$\Delta t_f = K_f m = \frac{1.86°\text{C}\ \cancel{\text{kg}}}{\cancel{\text{mol}}} \times \frac{0.0100\ \cancel{\text{mol}}}{\cancel{\text{kg}}} = 0.0186°\text{C}$$

or

$$t_f = -0.0186°\text{C}$$

$$i = \frac{\text{actual freezing point}}{\text{calculated freezing point}} = \frac{-0.0222\cancel{°\text{C}}}{-0.0186\cancel{°\text{C}}} = 1.19$$

For 100% ionization of HNO_3 ($\rightarrow H^+ + NO_3^-$), we expect the freezing-point depression to be two times the calculated freezing-point depression. The ratio, however, is only 1.19. This means that 0.0100 m HNO_2 does not dissociate 100% to form two ions:

$$HNO_2 \rightleftharpoons H^+ + NO_2^-$$

Large deviations in colligative properties occur from of the presence of ions in these examples:

1. Concentrated solutions, where the association between oppositely charged ions is greater than for dilute solutions

2. For ions with a large charge, where the electrostatic attraction between ions is great.

SECTION 12.15 OSMOTIC PRESSURE

Objective

- To explain the process of osmosis and the significance of osmotic pressure.

Focus

Osmosis is the spontaneous flow of *solvent* molecules through a semipermeable membrane from a solution with a low solute concentration to one with a high solute concentration. In the text, Figure 12.14, the flow of solvent molecules from left to right is greater than that from right to left.

The flow of solvent molecules across a semipermeable membrane can be made equal by applying an external pressure to the side containing the solution with a *high* solute concentration. The pressure is called the osmotic pressure:

$$\pi = [\text{B}]RT$$

According to this equation, the higher the concentration of solute [B] (the lower the concentration of solvent), the greater is the osmotic pressure of the solution. A greater external pressure must be applied to maintain an equilibrium flow of solvent molecules across the membrane.

Example 12.14 A 0.16 M NaCl solution is isotonic with blood. What is the osmotic pressure of blood at 37°C? For a 0.16 M NaCl solution, the van't Hoff factor is1.96, that is, the osmotic pressure is 1.96 times greater than when there is no dissociation.

Answer π = 7.98 atm

Solution Use the equation

$$\pi = [\text{B}]RT$$

$$T(\text{K}) = 273 + 37°\text{C} = 310\ \text{K}$$

$$\pi = 0.16\ \frac{\cancel{\text{mol}}}{\cancel{\text{L}}}\left(0.08206\frac{\cancel{\text{L}}\ \text{atm}}{\cancel{\text{mol}\,\text{K}}}\right)(310\ \cancel{\text{K}})$$

$$= 4.07\ \text{atm}$$

Adjusting for the dissociation of NaCl into particles,

$$\pi = 1.96 \times 4.07\ \text{atm}$$

$$= 7.98\ \text{atm}$$

SECTION 12.16 DETERMINATION OF MOLECULAR WEIGHTS

Objective

- To calculate the molecular weight of a solute from a freezing point change or an osmotic pressure determination.

Focus

In the chemistry laboratory, freezing and boiling point changes are commonly used to determine the molecular weight of low molecular weight solutes. An osmometer, an instrument that measures a solution's osmotic pressure, can also determine the molecular weight of high molecular weight solutes.

Example 12.15 A solution of 2.55 grams of unknown solute, Z, and 150 g benzene freezes at 4.99°C. What is the molecular weight of the solute?

Answer 160 g Z/mol Z

Solution From Table 12.5, $K_f(\text{benzene}) = 4.90 \dfrac{°\text{C kg}}{\text{mol}}$

and $t_f(\text{benzene}) = 5.51°\text{C}$. Calculate the molality of the solution:

$$\Delta t_f = 5.51°\text{C} - 4.99°\text{C} = 0.52°\text{C}$$

$$m = \frac{\Delta t_f}{K_f} = \frac{0.52 \cancel{°\text{C}}}{4.90 \dfrac{\cancel{°\text{C}}\ \text{kg benzene}}{\text{mol Z}}} = 0.106 \frac{\text{mol Z}}{\text{kg benzene}}$$

Calculate the moles of solute in the solution:

$$n_Z = 150 \cancel{\text{g benzene}} \times \frac{0.106 \text{ mol Z}}{1 \cancel{\text{kg benzene}}} \times \frac{1 \cancel{\text{kg benzene}}}{1000 \cancel{\text{g benzene}}}$$

$$= 0.0159 \text{ mol Z}$$

Calculate the molecular weight (g/mol) of the solute:

$$\frac{2.55 \text{ g Z}}{0.0159 \text{ mol Z}} = 160 \frac{\text{g Z}}{\text{mol Z}}$$

QUESTIONS

Section 12.1

12.1 Which of the following are solutions?

(a) gasoline (d) raw milk

(b) distilled water (e) oil and vinegar

(c) sea water (f) latex paint

12.2 Underline the solutes in the following solutions.

(a) NaCl and water

(b) mothballs and gasoline

(c) air: O_2 and N_2

(d) instant tea and water

12.3 Determine whether each is a dilute or a concentrated solution.

(a) syrupy sugar water ________

(b) household ammonia ________

(c) drinking water ________

(d) hair spray ________

(e) oven cleaner (in a jar) ________

Section 12.2

12.4 Calculate the mass in grams of KNO_3 in 25.0 grams of a 3.11% KNO_3 solution.

12.5 How many grams of $CuSO_4(H_2O)_5$ per 100 g solution are required to prepare a 4.50% $CuSO_4$ solution? Remember, the mass of $CuSO_4$ in the solution is *not* what is weighed in the laboratory. $CuSO_4(H_2O)_5$ is weighed in the laboratory.

12.6 A mass of 14.5 g KOH dissolves in 200 mL of solution. What is the molarity of the KOH solution?

12.7 Determine the number of grams of $CuSO_4(H_2O)_5$ needed to prepare 500 mL of a 0.440 M $CuSO_4$ solution. See the reminder in Question 12.5.

12.8 In order to prepare a 0.777 M K_2CrO_4 solution, 20.0 mL of 4.80 M K_2CrO_4 is diluted. What volume of 0.777 M K_2CrO_4 is prepared?

Section 12.3

12.9 Determine the mass percentage of a saturated $MgCO_3$ solution. $MgCO_3$ has a solubility of 0.0110 g $MgCO_3$ per 100 g H_2O.

12.10 Calculate the mass percentage of $FeNH_4(SO_4)_2$ in a saturated iron(III) (ferric) ammonium sulfate-water (1/12) solution. $FeNH_4(SO_4)_2(H_2O)_{12}$ has a solubility of 124 g per 100 g at 25°C.

Section 12.4

12.11 The partial pressure of ethane, C_2H_6, over a saturated solution containing 6.56×10^{-6} g C_2H_6 is 751 torr. What is the partial pressure of C_2H_6 when the solubility is 9.23×10^{-6} g C_2H_6, assuming no temperature change?

Section 12.7

12.12 The heat of solution, ΔH_s, for KI is +20.3 kJ/mol and the lattice energy, ΔH_ℓ, is +632.0 kJ/mol.

(a) Calculate the hydration energy for KI.

(b) Does the solubility of KI increase or decrease with a temperature rise? Explain.

12.13 Using Table 12.3, label these salts as soluble or insoluble.

(a) K_2CO_3 ________	(d) $FePO_4$ ________	(g) CsOH ________
(b) NaCl ________	(e) $Cu(NO_3)_2$ ________	(h) As_2SO_3 ________
(c) $BaSO_4$ ________	(f) CuS ________	(i) $Pb(C_2H_3O_2)_2$ ________

Section 12.8

12.14 Write the net ionic equation for each of the following salt mixtures in aqueous solution.

(a) $CuSO_4 + Pb(C_2H_3O_2)_2 \rightarrow$

(b) $Na_2CO_3 + AgNO_3 \rightarrow$

(c) $Zn(NO_3)_2 + K_2SO_3 \rightarrow$

Section 12.10

12.15 Calculate the mole fraction and the mole percent of ethanol in a 6.00% ethanol, C_2H_5OH, solution in water. (Remember % is by weight, or 6.00% means 6.00 g C_2H_5OH per 100 g *solution*.)

12.16 Calculate the ideal vapor pressure at 35°C for a solution containing 20.0 g CS_2 and 80.0 g $CHCl_3$. The vapor pressures of CS_2 and $CHCl_3$ at 35°C are 512 torr and 273 torr, respectively.

Section 12.11

12.17 Define a colligative property.

12.18 Identify the four colligative properties of a solution.

Section 12.12

12.19 Calculate the vapor pressure of a solution made by dissolving 26.0 g glucose, $C_6H_{12}O_6$, in 100 g water at 30°C. The vapor pressure of water at 30°C is 31.8 torr.

Section 12.13

12.20 Calculate the boiling and freezing point of an aqueous solution containing 26.0 g glucose, $C_6H_{12}O_6$, per 100 g H_2O.

12.21 The freezing point of an acetic acid solution (nonaqueous) containing a nonvolatile solute is 15.4°C. What is the boiling point of the solution? Acetic acid is the solvent.

12.22 List three conditions in which the equations for Δt_f, Δt_b, and Δp_A are valid.

(a) ________

(b) ________

(c) ________

Section 12.14

12.23 Calculate the freezing point of a 0.500 *m* NaCl solution. The van't Hoff factor for 0.500 *m* NaCl is 1.82.

12.24 Select the solution which has the larger van't Hoff factor.

(a) 0.60 *m* $CaCl_2$ or 0.47 *m* $CaCl_2$

(b) 0.22 *m* $FeSO_4$ or 0.22 *m* NaI

(c) 0.33 *m* $NiSO_4$ or 0.33 *m* Na_2SO_4

Section 12.15

12.25 A 0.030 M $CO(NH_2)_2$ solution is separated from a 0.070 M $C_{12}H_{22}O_{11}$ solution by a semipermeable membrane at 25°C. Both solutes are nonelectrolytes.

(a) Which solution has the higher osmotic pressure?

(b) Which solution becomes more dilute with the passage of solvent molecules?

(c) To which solution should an external pressure be applied in order to produce an equilibrium flow of H_2O molecules across the membrane?

(d) What pressure should be applied for part (c)?

Section 12.16

12.26 A mass of 15.933 grams of a nonvolatile solute dissolves in 200.0 g ethanol, C_2H_5OH. The boiling point of the solution is 79.7°C.

(a) Calculate the molecular weight of the solute.

(b) Calculate the freezing point of the solution.

(c) Calculate the vapor pressure depression at 78.5°C caused by the addition of the solute.

The vapor pressure of pure C_2H_5OH at 78.5°C is 760 torr.

12.27 A 2.10 gram sample of a naturally occurring medicine-man cure, dissolved in 200 mL of solution, has an osmotic pressure of 4.22 torr at 40°C. Calculate the molecular weight of the "cure" molecule.

ANSWERS

12.1 gasoline and sea water

12.2 (a) NaCl
(b) mothballs
(c) O_2
(d) tea

12.3 (a) concentrated
(b) dilute
(c) dilute
(d) dilute
(e) concentrated

12.4 0.778 g KNO_3

12.5 7.04 g $CuSO_4(H_2O)_5$

12.6 1.29 M KOH

12.7 54.9 g $CuSO_4(H_2O)_5$

12.8 124 mL

12.9 1.10×10^{-2} % $MgCO_3$

12.10 30.4% $FeNH_4(SO_4)_2$

12.11 1060 torr

12.12 (a) −611.7 kJ

(b) increase. Heat is absorbed in the process. Therefore, more heat favors the dissolving of KI.

12.12 (a), (b), (e), (g), and (i) are soluble; (c), (d), (f), and (h) are insoluble.

12.14 (a) $SO_4^{2-} + Pb^{2+} \rightarrow PbSO_4(c)$

(b) $CO_3^{2-} + 2Ag^+ \rightarrow Ag_2CO_3(c)$

(c) $Zn^{2+} + SO_3^{2-} \rightarrow ZnSO_3(c)$

12.15 $x_{C_2H_5OH} = 0.0243$

2.43 mol %

12.16 340 torr

12.17 A colligative property of a solution depends only on the number of solute particles and not on the nature of the solute.

12.18 vapor pressure lowering; boiling-point elevation; freezing-point depression; osmotic pressure

12.19 30.9 torr

12.20 b.p. = 100.737°C; f.p. = −2.69°C

12.21 b.p. = 119.0°C

12.22 (a) dilute solution

(b) nonelectrolyte solutes

(c) nonvolatile solutes

12.23 −1.69°C

12.24 (a) 0.47 *m* $CaCl_2$ (solution is more dilute)

(b) 0.22 *m* NaI (the ions have charges of only +1 and −1)

(c) 0.33 *m* Na_2SO_4 (more moles solute particles/mol solute)

12.25 (a) 0.070 M $C_{12}H_{22}O_{11}$

(b) 0.070 M $C_{12}H_{22}O_{11}$

(c) 0.070 M $C_{12}H_{22}O_{11}$

(d) 0.98 atm

12.26 (a) 80.9 g/mol

(b) −119.3°C

(c) 33.0 torr

12.27 4.86×10^4 g/mol

Chapter 13

CHEMICAL EQUILIBRIUM

SECTION 13.1 INTRODUCTION

Objective

- To describe a chemical system at equilibrium conditions.

Focus

A *chemical system at equilibrium* means that the rate at which reactants form products (forward reaction) is equal to the rate at which products form reactants (reverse reaction). The equilibrium is a dynamic one in which the forward and the reverse reactions occur continuously with no net change in the concentrations of the reactants or products.

For the reaction

$$N_2(g) + 3H_2(g) \rightleftharpoons 2NH_3(g)$$

at equilibrium, the rate at which N_2 reacts with H_2 to produce NH_3 is equal to the rate at which NH_3 reacts with itself to produce N_2 and H_2. No change in the pressures or concentrations of the three gases is detected *at equilibrium*. Note, however, that the reaction has *not* stopped. In addition, equilibrium does not mean that equal numbers of moles of reactants and products are present at equilibrium; it refers only to the equal rates of forward and reverse reactions.

SECTION 13.2 THE LAW OF CHEMICAL EQUILIBRIUM

Objectives

- To establish the mathematical expression for the reaction quotient, Q, and to explain how Q is related to the equilibrium constant, K.
- To determine whether a chemical system is at equilibrium by calculating the reaction quotient. When the system is not at equilibrium, we should be able to

predict the direction in which the reaction goes in order to reach equilibrium.

- To differentiate between the thermodynamic and the kinetic stability.

Focus

The **reaction quotient**, Q, is equal to the pressures of the products divided by the pressures of the reactants; each pressure is raised to the same power as its coefficient in the balanced equation. If more than one reactant or product exists in the reaction, the respective pressures are multiplied together, never added.

Example 13.1 Express the reaction quotient for $4NH_3(g) + 3O_2(g) \rightleftharpoons 2N_2(g) + 6H_2O(g)$ in terms of partial pressures.

Answer $Q = \dfrac{p^2_{N_2} p^6_{H_2O}}{p^4_{NH_3} p^3_{O_2}}$

Solution The reaction quotient is the pressures of the products multiplied together divided by the pressures of the reactants multiplied together.

$$Q = \frac{p_{N_2} p_{H_2O}}{p_{NH_3} p_{O_2}} \qquad \text{(We are not finished here.)}$$

Each pressure in the reaction quotient is raised to the power of its coefficient in the balanced equation:

$$Q = \frac{p^2_{N_2} p^6_{H_2O}}{p^4_{NH_3} p^3_{N_2}}$$

The value of the reaction quotient varies except at equilibrium. When the system reaches equilibrium, the reaction quotient equals the **equilibrium constant**, K, for the reaction. From then on the pressures of all the gases remain unchanged. The magnitude of K indicates the "position of equilibrium" for the reaction. A large K means that the numerator is large and the denominator small; thus the equilibrium favors the product. A small value for K means that the equilibrium favors the reactants.

A comparison of Q with K tells us in which direction a system must shift to reach equilibrium. If Q is greater than K, the partial pressures of the products are too large and that of the reactants too small. A shift to the left reduces the value of Q. If Q is less than K, then a shift to the right increases the partial pressures of the products and decreases those of the reactants. The value of Q then increases.

Example 13.2 An analysis of the following 1 liter system at 25°C

$$H_2(g) + I_2(g) \rightleftharpoons 2HI(g)$$

shows the presence of HI at 19.0 atm, H_2 at 0.662 atm, and I_2 at 4.20 atm. $K = 8.7 \times 10^2$ at 25°C. Calculate Q. If Q is not equal to K, in which direction does the chemical system shift to reach equilibrium?

Answer $Q = 130$. A shift to the right favors the formation of products.

Solution $Q = \dfrac{p^2_{HI}}{p_{H_2}p_{I_2}}$

Substituting the pressure for each gas, we have

$$Q = \frac{(19.0\ \text{atm})^2}{(0.662\ \text{atm})\ (4.20\ \text{atm})} = 130$$

Since $130 < 870$, the system shifts to the right to increase the pressure of HI (numerator) and to decrease that of H_2 and I_2 (denominator).

Example 13.3 An equilibrium mixture has the following analysis: $p_{NOCl} = 7.32$ atm; $p_{NO} = 3.11 \times 10^{-4}$ atm; $p_{Cl_2} = 36.0$ atm. Calculate K for

$$NOCl(g) \rightleftharpoons NO(g) + \tfrac{1}{2}Cl_2(g)$$

Answer $K = 2.55 \times 10^{-4}$

Solution The equilibrium condition for the reaction is

$$K = \frac{p_{NO}p^{1/2}_{Cl_2}}{p_{NOCl}}$$

Substituting equilibrium partial pressures, we have

$$K = \frac{(3.11 \times 10^{-4}\ \text{atm})\ (36.0\ \text{atm})^{1/2}}{7.32\ \text{atm}}$$

$$= 2.55 \times 10^{-4}$$

Chemists express the stability of a reaction in two ways:

1. Generally, kinetic stability refers to a preequilibrium condition. For example, the equilibrium constant may be very large, meaning the formation of products is favorable, but conditions may be such that the reaction does not go (or goes extremely slowly).
2. Thermodynamic stability refers to a system in which product formation is not favorable.

A reaction can be thermodynamically unstable but kinetically stable because the conditions are such that the rate of the reaction is practically zero. For example, a mixture of coal and oxygen is kinetically stable because the conditions for combustion are not obtained until the coal is heated to its ignition temperature. The system is thermodynamically unstable because at equilibrium the coal-oxygen reaction favors the formation of CO_2.

SECTION 13.3 FREE ENERGY AND EQUILIBRIUM

Objectives

- To identify the standard state of a substance.
- To distinguish between ΔG^0 and ΔG for a system at equilibrium.
- To summarize the relationship between K, Q, ΔG, and ΔG^0.

Focus

The **standard state** for a gas is the state in which the gas exerts a pressure of one atmosphere at some specified temperature.

The **standard free energy change**, ΔG^0, for a reaction is the sum of the standard free energies of the products minus the sum of the standard free energies of the reactants:

$$\Delta G^0 = G^0_{\text{products}} - G^0_{\text{reactants}}$$

ΔG^0 is related to K by the equation

$$\Delta G^0 = -RT \ln K$$

where $R = 8.314 \times 10^{-3}$ kJ/(mol K).

> You can convert natural logarithms to common logarithms with the following conversion: $\ln x = 2.303 \log x$, or $-RT \ln K = -2.303\ RT \log K$

The free energy changes continuously for a reaction that is approaching equilibrium. This is because the difference between the free energy of the products and that of the reactants changes. As a result, the reactants and products are probably not in their standard states during the course of the reaction. Therefore, the change in free energy of the system at any point in time during the reaction equals ΔG rather than ΔG^0. Thus ΔG is related to the reaction quotient, Q, and not the equilibrium constant, K.

The relationship between ΔG, ΔG^0, and Q is

$$\Delta G = \Delta G^0 + RT \ln Q$$

ΔG calculated from this equation tells us the direction of the spontaneous reaction that occurs in reaching equilibrium. It also corrects for the fact that the reactants and products are not in their standard states.

At equilibrium, the free energy difference between the products and reactants is zero, $\Delta G = 0$, and Q equals K.

$$0 = \Delta G^0 + RT \ln K$$

$$\Delta G^0 = -RT \ln K$$

If the system is not at equilibrium, Q does not equal K and ΔG does not equal zero but instead

$$\Delta G = RT \ln \left(\frac{Q}{K}\right)$$

If Q is greater than K, ΔG is positive and the forward reaction is not spontaneous; the reaction shifts to the left to favor the formation of reactants. If Q is less than K, ΔG is negative and the reaction shifts to the right.

Example 13.4 For the reaction

$$N_2(g) + 3H_2(g) \rightleftharpoons 2NH_3(g)$$

$\Delta G^0 = -33.0$ kJ/mol at 25°C. Calculate the equilibrium constant, K_p.

Answer 6.09×10^5

Solution At equilibrium,

$$\Delta G = 0 \quad \text{or} \quad 0 = \Delta G^0 + RT \ln K$$

$$\ln K = \frac{-\Delta G^0}{RT} = \frac{-(-33.0\ \text{kJ/mol})}{\left(\dfrac{8.314 \times 10^{-3}\ \text{kJ}}{\text{mol K}}\right)(298\ \text{K})} = 1.33 \times 10^1$$

$$K = 6.09 \times 10^5$$

Example 13.5 The partial pressures of the three gases in Example 13.4 at 25°C are $p_{NH_3} = 9.22$ atm; $p_{N_2} = 0.0311$ atm; and $p_{H_2} = 0.00844$ atm.

(a) In which direction does the reaction go to reach equilibrium?

(b) What is the free energy change in forming 2 mol NH_3 in the initial mixture? Assume the system is large enough so that the partial pressures do not change when 2 mol NH_3 form.

Answer (a) To the left; (b) 22.1 kJ

Solution (a) To determine which direction the reaction goes to establish equilibrium, compare the value of the reaction quotient, Q, to the equilibrium constant, K.

$$Q = \frac{p^2_{NH_3}}{p_{N_2} p^3_{H_2}} = \frac{(9.22\ \cancel{\text{atm}})^2}{(0.0311\ \cancel{\text{atm}})(0.00844\ \cancel{\text{atm}})^3} = 4.55 \times 10^9$$

$Q > K$; therefore, the reaction goes to the left. This reduces the value of the numerator and increases the denominator. Because $Q > K$ we know that ΔG is *positive* for the left to right reaction.

(b) The free energy change, ΔG, is calculated from

$$\Delta G = \Delta G^0 + RT \ln Q$$

From Example 13.4, $\Delta G^0 = -33.0$ kJ/mol.

$$\begin{aligned}
\Delta G &= -33.0\ \text{kJ}/\cancel{\text{mol}} + 8.314 \times 10^{-3}\ \text{kJ}/(\cancel{\text{mol}}\ \text{K}) \times 298\ \text{K} \times \ln 4.55 \times 10^9 \\
&= -33.0\ \text{kJ} + 8.314 \times 10^{-3}\ \text{kJ} \times 298 \times 22.2 \\
&= -33.0\ \text{kJ} + 55.1\ \text{kJ} \\
&= 22.1\ \text{kJ}
\end{aligned}$$

SECTION 13.4 FREE ENERGY OF FORMATION

Objectives

- To evaluate ΔG^0, ΔH^0, and ΔS^0 for a reaction from given data.
- To calculate K_p, given the ΔG^0_f for each substance in the reaction system.

Focus

Remember for gases, $K = K_p$: the equilibrium constant in terms of the partial pressures of the gases within the system. From Section 13.3

$$\Delta G^0 = -RT \ln K_p \text{ for gases.}$$

To calculate ΔG^0 for a chemical reaction from the standard free energies of formation, ΔG_f^0, remember

1. ΔG_f^0 for the elements in their standard state at one atmosphere pressure and 25°C equals zero.
2. ΔG_f^0 values (see text, Appendix D) are the standard free energies of formation *per mole*.
3. ΔG^0 for a reaction equals the sum of the standard free energies of formation of the products minus the sum of the standard free energies of formation of the reactants, or

$$\Delta G^0 = \Sigma\Delta G_f^0\,(\text{products}) - \Sigma\Delta G_f^0\,(\text{reactants})$$

Similarly for ΔH^0,

$$\Delta H^0 = \Sigma\Delta H_f^0\,(\text{products}) - \Sigma\Delta H_f^0\,(\text{reactants})$$

Knowing $\Delta G^0 = \Delta H^0 - T\Delta S^0$, we can calculate the standard entropy change:

$$\Delta S^0 = \frac{\Delta H^0 - \Delta G^0}{T}$$

Example 13.6 Calculate ΔG^0, ΔH^0, and ΔS^0 for the combustion of methane, CH_4, at 25°C:

$$CH_4(g) + 2O_2(g) \rightleftharpoons CO_2(g) + 2H_2O(g)$$

Answer $\Delta G^0 = -584.45$ kJ; $\Delta H^0 = -694.35$ kJ;
$\Delta S^0 = -0.0369$ kJ/K

Solution From the text, Appendix D:

	ΔH_f^0(kJ/mol)	ΔG_f^0(kJ/mol)
$CH_4(g)$	−74.73	−50.71
$CO_2(g)$	−393.48	−394.36
$H_2O(g)$	−187.8	−120.4
$O_2(g)$	0	0

$$\Delta G^0 = \Sigma\,\Delta G_f^0\,(\text{products}) - \Sigma\,\Delta G_f^0\,(\text{reactants})$$

$$= \left[1\,\cancel{\text{mol CO}_2} \times \left(\frac{-394.36\text{ kJ}}{\cancel{\text{mol CO}_2}}\right) + 2\,\cancel{\text{mol H}_2\text{O}} \times \left(\frac{-120.4\text{ kJ}}{\cancel{\text{mol H}_2\text{O}}}\right)\right]$$

$$- \left[1\,\cancel{\text{mol CH}_4} \times \left(\frac{-50.71\text{ kJ}}{\cancel{\text{mol CH}_4}}\right) + 2\,\cancel{\text{mol O}_2} \times \left(\frac{0\text{ kJ}}{\cancel{\text{mol O}_2}}\right)\right]$$

$$= -635.16\text{ kJ} - [-50.71\text{ kJ}]$$

$$= -584.45\text{ kJ}$$

$$\Delta H^0 = \Sigma\,\Delta H_f^0\,(\text{products}) - \Sigma\,\Delta H_f^0\,(\text{reactants})$$

$$= \left[1\,\cancel{\text{mol }CO_2} \times \left(\frac{-393.48\text{ kJ}}{\cancel{\text{mol }CO_2}}\right) + 2\,\cancel{\text{mol }H_2O} \times \left(\frac{-187.8\text{ kJ}}{\cancel{\text{mol }H_2O}}\right)\right]$$

$$\left[1\,\cancel{\text{mol }CH_4} \times \frac{-74.73\text{ kJ}}{\cancel{\text{mol }CH_4}} + 2\,\cancel{\text{mol }O_2} \times \left(\frac{0\text{ kJ}}{\cancel{\text{mol }O_2}}\right)\right]$$

$$= -769.09\text{ kJ} - [-74.73\text{ kJ}]$$

$$= -694.35\text{ kJ}$$

$$\Delta S^0 = \frac{\Delta H^0 - \Delta G^0}{T} = \frac{-694.35\text{ kJ} - (-584.45\text{ kJ})}{298\text{ K}}$$

$$= -0.0369\text{ kJ/K}$$

Thus there is a decrease in entropy in the reaction, or we can say that the products are in a more ordered state than the reactants.

SECTION 13.5 EQUILIBRIUM CONDITIONS IN TERMS OF CONCENTRATIONS

Objective

- To convert between K_p and K_c.

Focus

K_p is the equilibrium constant when partial pressures are used to express the composition of the equilibrium mixture of gases.

K_c is the equilibrium constant when molar concentrations are used to express the relative amount of each gas at equilibrium.

To convert:

$$K_p = K_c\,(RT)^{\Delta n}$$

$$R = 0.08206\text{ L atm/(mol K)}$$

when pressure is expressed in atm and concentration in mol/L.

$$\Delta n = \Sigma\text{mol (gaseous products)} - \Sigma\text{mol (gaseous reactants)}$$

Example 13.7 For the reaction

$$H_2O(g) \rightleftharpoons H_2(g) + \tfrac{1}{2}O_2(g)$$

$K_c = 1.32 \times 10^{-6}$ at 1500°C. Calculate K_p.

Answer 1.59×10^{-5}

Solution $K_p = K_c\,(RT)^{\Delta n}$

$$\Delta n = \Sigma\text{mol (gaseous products)} - \Sigma\text{mol (gaseous reactants)}$$
$$= 1 + \tfrac{1}{2} - 1$$
$$= \tfrac{1}{2}$$

$$T = 273 + 1500°\text{C} = 1773\text{ K}$$
$$K_p = 1.32 \times 10^{-6}\,(RT)^{1/2}$$
$$= 1.32 \times 10^{-6}\left(0.08206\,\frac{\text{L atm}}{\text{mol K}} \times 1773\text{ K}\right)^{1/2}$$
$$= 1.59 \times 10^{-5}$$

Remember Δn can have a positive, zero, or negative value; Δn is the sum of the moles of gaseous product minus the sum of the moles of gaseous reactant.

SECTION 13.6 CHANGE OF *K* WITH FORM OF EQUATION

Objective

- To determine the value of an equilibrium constant according to how the chemical equation is written.

Focus

Remember that the reaction quotient, Q, is always written as products divided by reactants, each raised to the power of its coefficient that appears in the balanced equation. Because a balanced equation can be written with various sets of coefficients, the corresponding equilibrium equation should also be given when K is given.

Example 13.8 K_c for the reaction

$$H_2O(g) \rightleftharpoons H_2(g) + \tfrac{1}{2}O_2(g)$$

is 1.32×10^{-6} at 1500°C. Determine the K_c for each of the following:

(a) $2H_2O(g) \rightleftharpoons 2H_2(g) + O_2(g)$

(b) $2H_2(g) + O_2(g) \rightleftharpoons 2H_2O(g)$

Answer (a) 1.74×10^{-12}
(b) 5.75×10^{11}

Solution For $H_2O(g) \rightleftharpoons H_2(g) + \tfrac{1}{2}O_2(g)$

$$K_c = \frac{[H_2][O_2]^{1/2}}{[H_2O]} = 1.32 \times 10^{-6}$$

For (a) $$K_c = \frac{[H_2]^2[O_2]}{[H_2O]^2} = \left(\frac{[H_2][O_2]^{1/2}}{[H_2O]}\right)^2$$
$$= (1.32 \times 10^{-6})^2 = 1.74 \times 10^{-12}$$

For (b) $K_c = \dfrac{[H_2O]^2}{[H_2]^2[O_2]} = \left(\dfrac{[H_2]^2[O_2]}{[H_2O]^2}\right)^{-1}$

K_c in (b) is the reciprocal of K_c in (a).

$K_c = (1.74 \times 10^{-12})^{-1} = 5.75 \times 10^{11}$

SECTION 13.7 COMBINATION OF EQUILIBRIA

Objective

- To calculate an equilibrium constant for a chemical reaction for which no K_p or K_c is given.

Focus

An equilibrium constant for any reaction can be calculated by multiplying the K's for those equations which add up to the desired equation.

Example 13.9 Calculate K_p for the reaction

$$ClO_2(g) + 2F_2(g) \rightleftharpoons \tfrac{1}{2}Cl_2(g) + 2OF_2(g)$$

given that at 25°C

$$ClO_2(g) \rightleftharpoons \tfrac{1}{2}Cl_2(g) + O_2(g) \qquad K_p = 2.74 \times 10^{21}$$

$$O_2(g) + 2F_2(g) \rightleftharpoons 2OF_2(g) \qquad K_p = 2.43$$

Answer 6.66×10^{21}

Solution Adding the two equations gives the desired reaction

$$ClO_2(g) + 2F_2(g) \rightleftharpoons \tfrac{1}{2}Cl_2(g) + 2OF_2(g)$$

A product of the two K_p's equals the K_p for the reaction

$$K_p = \frac{p_{Cl_2}^{1/2}p_{O_2}}{p_{ClO_2}} \qquad K_p = \frac{p_{OF_2}^2}{p_{O_2}p_{F_2}^2}$$

$$K_p = \frac{p_{Cl_2}^{1/2}\cancel{p_{O_2}}}{p_{ClO_2}} \times \frac{p_{OF_2}^2}{\cancel{p_{O_2}}p_{F_2}^2} = \frac{p_{Cl_2}^{1/2}p_{OF_2}^2}{p_{ClO_2}p_{F_2}^2}$$

$$K_p = (2.74 \times 10^{21})(2.43)$$

$$= 6.66 \times 10^{21}$$

SECTION 13.8 PRINCIPLE OF LE CHATELIER

Objective

- To explain and apply Le Chatelier's principle.

Focus

Le Chatelier's principle states that if a system at equilibrium is disturbed, the equilibrium shifts so as to compensate for the effect of the disturbance. Only changes in concentration, volume (for gases), and temperature disturb a system at equilibrium.

Example 13.10

$$N_2(g) + 3H_2(g) \rightleftharpoons 2NH_3(g) \qquad \Delta H^0 = -92.2 \text{ kJ}$$

How does each of the following disturbances affect the position of equilibrium in the system?

(a) $N_2(g)$ is added to the system.
(b) The temperature is increased.
(c) The volume is decreased.

Answer (a) shift right; (b) shift left; (c) shift right

Solution (a) When N_2 is added to the system, the equilibrium shifts to the *right* in an attempt to remove the effect of the added N_2. As a result, the quantity of NH_3 increases. Since the reaction of $N_2 + 3H_2$ produces NH_3, the quantity of H_2 decreases.

(b) Because the reaction is exothermic, the addition of more heat results in a shift to the *left*. The right → left reaction is endothermic and therefore absorbs the added heat. The shift increases the amount of N_2 and H_2 but decreases the amount of NH_3.

(c) If the volume decreases (or pressure increases), the system shifts to decrease the amount of gas in the system. With 1 + 3 moles of gaseous reactants and 2 moles of gaseous products, the equilibrium shifts to the *right*, causing an increase in the quantity of NH_3 and a decrease in the quantity of N_2 and H_2.

The addition of a chemically inert gas or a catalyst has no effect on shifting the equilibrium. The catalyst only affects the rate at which the system reaches equilibrium.

SECTION 13.9 HETEROGENEOUS EQUILIBRIUM

Objectives

- To understand equilibria in which pure solids or pure liquids as well as gases are present.
- To calculate ΔG^0 and K_p for a heterogeneous equilibrium system.

Focus

The concentration of a pure solid or a pure liquid is expressed as moles per cubic centimeter or as mass per unit volume (density). Since their density is constant (within a narrow temperature range), the addition of more solid or liquid does not disturb the equilibrium.

Example 13.11 Write the equilibrium conditions using K_p and K_c for

$$NH_4Cl(c) \rightleftharpoons NH_3(g) + HCl(g)$$

Answer $K_p = p_{NH_3}p_{HCl}$

$K_c = [NH_3][HCl]$

Solution Since NH_4Cl is a pure solid, it has a constant vapor pressure and therefore appears as a constant in the equilibrium condition for K_p:

$$K = \frac{p_{NH_3}p_{HCl}}{p_{NH_4Cl}} = \frac{p_{NH_3}p_{HCl}}{\text{constant}}$$

$$K \times \text{constant} = K_p = p_{NH_3}p_{HCl}$$

$$K \times \text{constant} = K_c = [NH_3][HCl]$$

For a heterogeneous equilibrium, a change in the volume affects only the quantities of gaseous reactants and products and only a change in these quantities affects the equilibrium.

Using ΔH_f^0 and ΔG_f^0 values for pure solids and liquids, we can calculate K_p and ΔS^0 values for heterogeneous equilibria.

Example 13.12 Calculate ΔG^0 and K_p at 25°C for the reaction

$$Fe(CO)_5(\ell) \rightleftharpoons Fe(s) + 5CO(g)$$

Answer $\Delta G^0 = 19.7$ kJ; $K_p = 3.52 \times 10^{-4}$

Solution From the text, Appendix D and since $\Delta G^0 = \Sigma\,\Delta G_f^0$ (products) $- \Sigma\,\Delta G_f^0$ (reactants), then

$$\Delta G^0 = \left[0 + 5\,\cancel{\text{mol CO}} \times \left(\frac{-137.15\text{ kJ}}{\cancel{\text{mol CO}}}\right)\right] - \left[1\,\cancel{\text{mol Fe(CO)}_5} \times \left(\frac{-705.4\text{ kJ}}{\cancel{\text{mol Fe(CO)}_5}}\right)\right]$$

$$= -685.75\text{ kJ} - [-705.4\text{ kJ}] = 19.7\text{ kJ}.$$

A positive ΔG^0 shows that the reactant is thermodynamically stable.

$$\ln K_p = \frac{-\Delta G^\circ}{RT} = \frac{-(19.7\text{ kJ})}{\left(\dfrac{8.314 \times 10^{-3}\text{ kJ}}{\text{mol K}}\right)(298\text{ K})} = -7.95$$

$$K_p = 3.52 \times 10^{-4};\ K_p < 1.$$

Example 13.13 Estimate ΔG^0 and K_p for the reaction in Example 13.12 at 1000 K. Assume that ΔH^0 and ΔS^0 are the same at 1000 K and at 25°C.

Answer $\Delta G^0 = -456$ kJ; $K_p = 6.29 \times 10^{23}$

Solution To calculate ΔG^0 at 1000 K, we use the equation $\Delta G^0 = \Delta H^0 - T\Delta S^0$. Therefore, first calculate ΔH^0 and ΔS^0 at 25°C or 298 K. Use Appendix D for ΔH_f^0 values.

$$\Delta H^0 = \Sigma\Delta H_f^0\text{ (products)} - \Sigma\Delta H_f^0\text{ (reactants)}$$

$$= \left[0 + 5\,\cancel{\text{mol CO}} \times \frac{-110.52\text{ kJ}}{\cancel{\text{mol CO}}}\right] - \left[1\,\cancel{\text{mol Fe(CO)}_5} \times \frac{-774.0\text{ kJ}}{\cancel{\text{mol Fe(CO)}_5}}\right]$$

$$\Delta H^0 = -552.6\text{ kJ} - [-774.0\text{ kJ}] = +221.4\text{ kJ}$$

$$\Delta S^0 = \frac{\Delta H^0 - \Delta G^0}{T} = \frac{221.4\text{ kJ} - 19.7\text{ kJ}}{298\text{ K}}$$

$$= 0.677\text{ kJ/K}$$

At 1000 K,

$$\Delta G^0 = \Delta H^0 - T\Delta S^0$$

$$\Delta G^0 = 221.4\text{ kJ} - 1000\cancel{K} \times 0.677\ \frac{\text{kJ}}{\cancel{K}}$$

$$\Delta G^0 = 221.4\text{ kJ} - 677\text{ kJ} = -456\text{ kJ}.$$

Therefore,

$$\ln K = \frac{-\Delta G^0}{RT} = \frac{-(-456\text{ kJ})}{\left(\frac{8.314 \times 10^{-3}\text{ kJ}}{\text{mol K}}\right)(1000\text{ K})}$$

$$= 54.8$$

$$K_p = 6.29 \times 10^{23}$$

SECTION 13.10 EQUILIBRIUM IN SOLUTIONS

Objective

- To express the equilibrium condition for solutes in solution.

Focus

The concentration of solvent in a dilute solution is assumed to remain unchanged when small amounts of solute are added. For example, the difference between the concentration of pure water and that in which a few grams NaCl are dissolved is insignificant. Therefore, when writing an equilibrium condition for a solute in a solution, we assume the solvent concentration is constant and thus do not include it in the equilibrium condition—it is a part of K.

Second, solids (or precipitates) in equilibrium with their solutions also have a concentration that is assumed to be constant. For example, the concentration of limestone rock on the bottom of a beaker containing water is constant whether there are two or six rocks. Remember that a concentration for solids can be calculated from its density and we assume that the densities of solids are constant.

Example 13.14 Write the equilibrium condition for a limestone rock, $CaCO_3$, in equilibrium with its ions in aqueous solution. The equilibrium is

$$CaCO_3(c) \rightleftharpoons Ca^{2+}(aq) + CO_3^{2-}(aq)$$

Answer $K = [Ca^{2+}][CO_3^{2-}]$

Solution $CaCO_3$ does not appear in the equilibrium condition because it is a solid. Therefore $K = [Ca^{2+}][CO_3^{2-}]$.

The solvent must be identified for a solution because an equilibrium may have different K values depending upon the solvent. The concentration or partial pressure of a gas in equilibrium with a solvent must appear in the equilibrium condition since its concentration or pressure is a function of volume.

Example 13.15 Write the equilibrium condition of K_c for

$$H_2SO_3(aq) \rightleftharpoons H_2O(\ell) + SO_2(g)$$

Answer $K_c = \dfrac{[SO_2]}{[H_2SO_3]}$ (In equilibrium conditions we always express solutes in terms of molar concentrations.)

Solution $[H_2O]$ does not appear because it is the solvent. The concentrations of $H_2SO_3(aq)$ and $SO_2(g)$ depend upon the volume of the system and therefore, must appear in the equilibrium condition.

SECTION 13.11 CALCULATION OF THE EQUILIBRIUM CONSTANT

Objective

- To calculate an equilibrium constant, given data from which you can determine equilibrium concentrations (or partial pressures) for gases.

Focus

This section and Section 13.12 are very important for a quantitative understanding of chemical equilibrium. Be prepared to work a number of problems in order to gain familiarity with the calculations—so charge your calculator!

Chemical equilibrium problems, in this and in later chapters, can be made easier if we follow this step-by-step procedure.

Step 1. Write or rewrite the chemical equation and leave plenty of space between the formulas in the equation.

Step 2. Directly below the formula record the initial number of moles of each substance in the equilibrium and label as "initial moles" or "mol_i."

Step 3. Record the *change* in moles of each substance that is required to establish equilibrium and label as "change in moles" or "Δmol." Use "+" and "−" to indicate an increase or decrease in moles, respectively. This information is determined by reading the coefficients of the balanced equation in step 1.

Step 4. Add steps 2 and 3 and label as "moles at equilibrium" or "mol $\rightleftharpoons$."

Step 4′. Divide each quantity in step 4 by the volume of the system and label "concentration at equilibrium" or "[ ]$\rightleftharpoons$." (You will soon see that when the initial data are expressed in pressures or molar concentrations, then step 4′ is omitted.)

Step 5. Write the equilibrium condition for K, substitute in "[ ]$\rightleftharpoons$," and solve for the unknown.

These five steps work for nearly *all* equilibrium calculations; practice them and you will be surprised at the ease and similarity of the many types of equilibrium problems you will encounter in the next few chapters.

This systematic approach can also be used when the initial data are expressed in terms of pressures or concentration for gases or molar concentrations for solutes. The steps are labeled as follows:

Given Quantities	Moles	Partial Pressures	Molar Concentrations
Step 1	←	A balanced equation	→
Step 2	mol_i	p_i	$[\ ]i$
Step 3	Δmol	Δp	$\Delta[\ ]$
Step 4	$mol_{\rightleftharpoons}$	$p_{\rightleftharpoons}$	$[\ ]_{\rightleftharpoons}$
Step 4′	$[\ ]_{\rightleftharpoons}$	Omit	Omit
Step 5	K_c	K_p	K_c

Example 13.16 A combination of 1.47 mol NO and 0.27 mol Cl_2 is placed in a 1.0 liter reaction vessel at a given temperature. At equilibrium, 0.380 mol NOCl is present. Calculate the equilibrium constant at this temperature for

$$2NO(g) + Cl_2(g) \rightleftharpoons 2NOCl(g)$$

Answer $K_c = 1.52$

Solution

Step 1.	equation:	$2NO(g)$ +	$Cl_2(g)$	$\rightleftharpoons 2NOCl(g)$
Step 2.	mol_i:	1.47	0.270	0
Step 3.	Δmol:	−0.380	−0.190	+0.380

At equilibrium there is 0.380 mol NOCl. According to the balanced equation, for each 1 mol Cl_2 consumed in the reaction, 2 mol NO are consumed. That is, 0.380 mol NO reacts with 0.190 mol Cl_2 to yield 0.380 mol NOCl at equilibrium.

Step 4.	$mol_{\rightleftharpoons}$:	1.09	0.080	0.380
Step 4′.	$[\ ]_{\rightleftharpoons}$:	$\frac{1.09\ mol}{1.0\ L}$	$\frac{0.080\ mol}{1.0\ L}$	$\frac{0.380\ mol}{1.0\ L}$

Step 5. $$K_c = \frac{[NOCl]^2}{[NO]^2[Cl_2]} = \frac{[0.380]^2}{[1.09]^2[0.080]} = 1.52$$

Example 13.17 The initial partial pressures of N_2 and H_2 are 6.59 atm and 25.40 atm, respectively, at a given temperature. At equilibrium the partial pressure of NH_3 is 10.70 atm. Determine the value of K_p for

$$N_2(g) + 3H_2(g) \rightleftharpoons 2NH_3(g)$$

Answer $K_p = 0.113$

Solution

Step 1.	equation:	$N_2(g)$	+	$3H_2(g)$	$\rightleftharpoons$	$2NH_3(g)$
Step 2.	p_i (atm):	6.59		25.40		0

Step 3.†	Δp (atm):	$-\frac{1}{2}(10.70) = -5.35$	$-\frac{3}{2}(10.70) = -16.05$	$+10.70$
Step 4.	$p_{\rightleftharpoons}$ (atm):	1.24	9.35	10.70

Step 5. $$K_p = \frac{p^2_{NH_3}}{p_{N_2}\,p^3_{H_2}} = \frac{(10.70)^2}{(1.24)(9.35)^3} = 0.113$$

SECTION 13.12 CALCULATIONS FROM THE EQUILIBRIUM CONSTANT

Objective

- To calculate equilibrium concentrations (or partial pressures for gases), given the initial quantity of each substance and the corresponding equilibrium constant.

Focus

The five steps in Section 13.11 of this Study Guide can also be used to solve this type of problem.

Example 13.18 A mixture of 0.482 mol N_2 and 0.933 mol O_2 is placed in a 10.0 liter reaction vessel. Calculate the equilibrium concentration of N_2O at a temperature where $K_c = 2.00 \times 10^{-37}$ for the reaction

$$2N_2(g) + O_2(g) \rightleftharpoons 2N_2O(g)$$

Answer 6.58×10^{-21} mol/L

Solution

Step 1.	equation:	$2N_2(g)$	$+\ O_2(g)$	$\rightleftharpoons$	$2N_2O(g)$
Step 2.	mol_i:	0.482	0.933		0
Step 3.	Δmol:	$-2y$	$-y$		$+2y$

(Since we wish to determine how far to the right the reaction goes, we define the magnitude of the shift in terms of the number of mol O_2, defined as y. If y mol O_2 reacts, then, according to the balanced equation, $2y$ mol N_2 reacts and $2y$ mol N_2O forms to reach equilibrium conditions.)

Step 4.	$\text{mol}_{\rightleftharpoons}$:	$0.482 - 2y$	$0.933 - y$	$2y$
Step 4′	$[\ \ ]_{\rightleftharpoons}$:	$\frac{0.482 - 2y}{10.0}$	$\frac{0.933 - y}{10.0}$	$\frac{2y}{10.0}$

Step 5. $$K_c = \frac{[N_2O]^2}{[N_2]^2[O_2]} = 2.0 \times 10^{-37}$$

$$= \frac{\left(\frac{2y}{10.0}\right)^2}{\left(\frac{0.482 - 2y}{10.0}\right)^2\left(\frac{0.933 - y}{10.0}\right)} = \frac{10.0(2y)^2}{(0.482 - 2y)^2(0.933 - y)}$$

It is very difficult to solve this equation without any approximations. But if we look at the chemistry behind the mathematics we see that because K_c is very small, the

† In step 3 always determine the *change* according to the coefficients in the balanced equation.

reaction goes a very small amount to the right. Therefore let's assume for N_2 and O_2 that:

$$0.482 >> 2y \quad \text{and} \quad 0.933 >> y$$

Thus

$$2.00 \times 10^{-37} \simeq \frac{10.0(2y)^2}{(0.482)^2(0.933)}$$

$$y \simeq 3.29 \times 10^{-20}$$

Certainly the two approximations are valid. The equilibrium *concentration* of N_2O is

$$[N_2O] = \frac{2y}{10.0} = \frac{2(3.29 \times 10^{-20})\text{ mol}}{10.0\text{ L}} = 6.58 \times 10^{-21}\frac{\text{mol}}{\text{L}}$$

How do we know when to approximate? When is y negligible compared to the original concentration or pressure? Experience tells us this or we can try the following:

1. If K is very small, say 10^{-4} or smaller, *and* the initial quantities (moles, concentration, or partial pressures) are 10^{-2} or greater, we can assume that the shift (y in step 3) is insignificant compared to the initial quantities.
2. Make the approximated calculation.
3. If the calculated value of y is 5% or less of the initial quantity (moles, concentration, or partial pressure), the approximation is considered valid. If it is not, the quadratic or high order polynomial equation must be solved.

To summarize,

$$\text{if } \frac{y}{\text{original}} \times 100 \le 5\%, \text{ approximation is O.K.}$$

Example 13.19 K_p for the reaction

$$CO(g) + 3H_2(g) \rightleftharpoons CH_4(g) + H_2O(g)$$

is 2.70×10^{-7} at a given temperature. The partial pressures for two of the gases are $p_{CO} = 9.22$ atm and $p_{H_2} = 31.9$ atm. Calculate the equilibrium partial pressure of CH_4. What percentage of the H_2 is converted in the process?

Answer $p_{CH_4} = 0.284$ atm; 2.67% H_2 converted

Solution

Step 1.	equation:	$CO(g)$	$+\ 3H_2(g)$	$\rightleftharpoons CH_4(g)$	$+\ H_2O(g)$
Step 2.	p_i(atm):	9.22	31.9	0	0
Step 3.	Δp(atm):	$-y$	$-3y$	$+y$	$+y$
Step 4.	p (atm):	$9.22 - y$	$31.9 - 3y$	y	y

Step 5. $$K_p = \frac{p_{CH_4}p_{H_2O}}{p_{CO}p^3_{H_2}} = 2.70 \times 10^{-7} = \frac{(y)(y)}{(9.22 - y)(31.9 - 3y)^3}$$

Since K_p is small, assume y is negligible compared to the initial partial pressures for each gas:

$$9.22 >> y \qquad 31.9 >> 3y$$

$$2.70 \times 10^{-7} \simeq \frac{y^2}{(9.22)(31.9)^3}$$

$$y \simeq 0.284$$

Was the approximation valid? Is $\frac{y}{\text{original}} \times 100 \leq 5\%$?

$$\frac{0.284}{9.22} \times 100 = 3.08\% \quad \text{and} \quad \frac{3 \times 0.284}{31.9} \times 100 = 2.67\%$$

Yes, the approximations are valid; both are less than 5%.

$$p_{CH_4} \text{ at equilibrium} = y = 0.284 \text{ atm}$$

The percentage of H_2 converted is

$$\frac{\text{amount converted}}{\text{original}} \times 100 = \frac{3y}{31.9} \times 100$$

$$= \frac{3(0.284)}{31.9} \times 100 = 2.67\%$$

QUESTIONS

Section 13.1

13.1 Indicate which of the following systems are in a state of dynamic equilibrium.

(a) A saturated solution of sugar/water

(b) The aging process of a lake

(c) Nature's water cycle (or any of the biogeochemical cycles)

(d) The amount of $CO_2(g)$ in a sealed bottle of cola

(e) A big rock sitting at the top of a cliff

Section 13.2

13.2 Write the reaction quotient in terms of pressure for

$$PCl_3(g) + Cl_2(g) \rightleftharpoons PCl_5(g)$$

13.3 Write the reaction quotient in terms of pressure for

$$4NH_3(g) + 3O_2(g) \rightarrow 2N_2(g) + 6H_2O(g)$$

13.4 The equilibrium constant, K_p, for the reaction

$$NO(g) + O_3(g) \rightleftharpoons NO_2(g) + O_2(g)$$

is 1.32×10^{10}. If the equilibrium partial pressures are $p_{NO} = 2.17 \times 10^{-8}$ atm, $p_{NO_2} = 7.72$ atm, and $p_{O_2} = 6.29$ atm, what is the partial pressure of O_3 at equilibrium?

Section 13.3

13.5 Identify whether each of the following systems is kinetically stable and/or thermodynamically stable.

(a) A boulder which has been sitting at the top of a cliff for thousands of years

(b) A stick of dynamite

(c) $2H_2O \rightleftharpoons 2H_2 + O_2$

(d) $NaCl(c) \rightleftharpoons Na + \frac{1}{2}Cl_2$

13.6 For the reaction

$$NO_2(g) \rightleftharpoons NO(g) + \tfrac{1}{2}O_2(g)$$

at 500°C, $K_p = 0.943$. Calculate ΔG^0.

13.7 Calculate the free energy change, ΔG, and the direction in which the equilibrium shifts for the reaction in Question 13.6 when the partial pressures of the three gases are $p_{NO_2} = 0.422$ atm, $p_{NO} = 0.915$ atm, and $p_{O_2} = 0.714$ atm.

Section 13.4

13.8 Calculate ΔG^0, ΔH^0, ΔS^0, and K_p for the reaction

$$2NO(g) + 3H_2O(g) \rightleftharpoons 2NH_3(g) + \tfrac{5}{2}\,O_2(g)$$

For NO, $\Delta G_f^0 = 86.57$ kJ/mol and $\Delta H_f^0 = 90.25$ kJ/mol. (See text, Appendix D for additional values.)

Section 13.5

13.9 For the reaction

$$N_2(g) + 3H_2(g) \rightleftharpoons 2NH_3(g)$$

$K_c = 6.00 \times 10^{-2}$ at 500°C. Calculate K_p.

Section 13.7

13.10 Determine the equilibrium constant for the reaction

$$4HCl(g) + O_2(g) \rightleftharpoons 2H_2O(g) + 2Cl_2(g)$$

given the information

$$2HCl(g) \rightleftharpoons H_2(g) + Cl_2(g) \qquad K_c = 3.98 \times 10^{-34}$$

$$2H_2(g) + O_2(g) \rightleftharpoons 2H_2O(g) \qquad K_c = 1.34 \times 10^{80}$$

(The coefficients for the first equation must be doubled. How does this affect K_c?)

Section 13.8

13.11 The four gases, NH_3, O_2, NO, and H_2O, are mixed in a reaction chamber and allowed to reach equilibrium:

$$\text{heat} + 4NH_3 + 5O_2 \rightleftharpoons 4NO + 6H_2O$$

Certain changes (see left column in table below) are then made upon this mixture. Considering each separately, state the effect (increase, decrease, no change) that the change has upon the original equilibrium value of the quantity in the center column. The temperature and volume are constant unless otherwise noted.

	Change	Equilibrium Value	Effect
(a)	Adding NO	Number of moles H_2O	i d nc
(b)	Adding NO	Number of moles O_2	i d nc
(c)	Removing NO	Magnitude of K_c	i d nc
(d)	Increasing pressure	Number of moles NH_3	i d nc
(e)	Decreasing pressure	Number of moles NO	i d nc
(f)	Adding helium gas	Number of moles NH_3	i d nc
(g)	Adding heat	Number of moles NO	i d nc
(h)	Adding heat	Magnitude of K_c	i d nc
(i)	Adding a catalyst	Number of moles NO	i d nc
(j)	Removing H_2O	Number of moles H_2O	i d nc
(k)	Removing O_2	Number of moles NO	i d nc
(l)	Reducing volume	Number of moles NH_3	i d nc

Section 13.9

13.12 Write the equilibrium condition using K_p for

$$CaCO_3(c) \rightleftharpoons CaO(c) + CO_2(g).$$

13.13 For the reaction

$$ZnO(c) + H_2S(g) \rightleftharpoons ZnS(c) + H_2O(\ell)$$

$\Delta H^0 = -120.0$ kJ and $\Delta G^0 = -84.1$ kJ.

(a) Calculate the entropy change for the reaction at 25°C.

(b) Calculate the free energy change at 200°C, assuming ΔH^0 and ΔS^0 remain unchanged.

(c) Calculate K_p at 200°C.

Section 13.10

13.14 Write the equilibrium condition using K_c for

(a) $C(\textit{graphite}) + H_2O(g) \rightleftharpoons CO(g) + H_2(g)$

(b) $HCOOH(aq) + H_2O(\ell) \rightleftharpoons H_3O^+(aq) + HCOO^-(aq)$

(c) $PbSO_4(c) \rightleftharpoons Pb^{2+}(aq) + SO_4^{2-}(aq)$

Section 13.11

13.15 A mixture of 0.2440 mol CO and 0.3110 mol H_2 are placed in a 3.00 liter reaction vessel and heated to 225°C. At equilibrium 0.0602 mol $CH_3OH(g)$ is present.

(a) Calculate K_c and K_p for the reaction

$$CO(g) + 2H_2(g) \rightleftharpoons CH_3OH(g)$$

Refer to Section 13.5 for K_p, K_c conversion.

13.16 The equilibrium concentrations for a gaseous mixture of NO, O_2, and NO_2 at 500 K are 5.00×10^{-6}, 1.50, and 0.050 mol/L, respectively. Calculate K_c, K_p, and ΔG^0 for the reaction

$$2NO(g) + O_2(g) \rightleftharpoons 2NO_2(g)$$

Section 13.12

13.17 $$COCl_2(g) \rightleftharpoons CO(g) + Cl_2(g)$$

A mixture of 0.400 mol $COCl_2$ and 1.20 mol Cl_2 is heated in a 2.00 liter reaction vessel to a temperature at which $K_c = 6.20$. Calculate the equilibrium concentrations of $COCl_2$, CO, and Cl_2.

13.18 If 0.620 atm N_2 and 0.330 atm O_2 are heated in a flask to a temperature at which $K_p = 10.0$ for the reaction

$$N_2 + O_2 \rightleftharpoons 2NO$$

calculate the equilibrium partial pressures of N_2, O_2, and NO.

13.19 The following system is analyzed at a temperature where $K_c = 3.00$:

$$NO_2(g) + SO_2(g) \rightleftharpoons SO_3(g) + NO(g)$$

A total of 0.488 mol NO_2, 0.731 mol SO_2, 0.941 mol SO_3, and 0.294 mol NO are present in 1 liter.

(a) Is the system at equilibrium? If not, in which direction does the reaction go?

(b) Is ΔG positive or negative?

(c) Calculate ΔG^0. For this reaction note that $K_c = K_p$.

(d) Calculate ΔG for the system at 25°C knowing the values for ΔG^0 and Q.

(e) What are the equilibrium concentrations of NO_2, SO_2, SO_3, and NO?

ANSWERS

13.1 (a) dynamic equilibrium

(b) Equilibrium is never established.

(c) This is a cyclic process, not necessarily a dynamic equilibrium, although H_2O molecules continuously go from the solid to the liquid to the gas and vice versa.

(d) dynamic equilibrium

(e) static equilibrium

13.2 $Q = \dfrac{p_{PCl_5}}{p_{PCl_3}p_{Cl_2}}$

13.3 $Q = \dfrac{p_{N_2}^2 p_{H_2O}^6}{p_{NH_3}^4 p_{O_2}^3}$

13.4 0.170

13.5 (a) kinetically stable

(b) kinetically stable

(c) kinetically stable and thermodynamically stable

(d) kinetically stable and thermodynamically stable

13.6 $\Delta G^0 = +0.377$ kJ/mol

13.7 Reaction goes to the left because $Q = 1.83 > K = 0.943$; $\Delta G = +4.27$ kJ/mol

13.8 $\Delta G^0 = +479.63$ kJ

$\Delta H^0 = 452.74$ kJ

$\Delta S^0 = \dfrac{-0.0902 \text{ kJ}}{\text{K}}$

$K_p = 8.42 \times 10^{-85}$

13.9 $K_p = 1.49 \times 10^{-5}$

13.10 $K_c = 2.12 \times 10^{13}$

13.11

(a) d	(e) i	(i) nc
(b) i	(f) nc	(j) d
(c) nc	(g) i	(k) d
(d) i	(h) i	(l) i

13.12 $K_p = p_{CO_2}$

13.13 (a) $\Delta S^0 = -0.120$ kJ/K

(b) $\Delta G = -63.2$ kJ

(c) $K_p = 9.54 \times 10^6$

13.14 (a) $K_c = \dfrac{[CO][H_2]}{[H_2O]}$

(b) $K_c = \dfrac{[H_3O^+][HCOO^-]}{[HCOOH]}$

(c) $K_c = [Pb^{2+}][SO_4^{2-}]$

13.15 $K_c = 81.1$; $K_p = 0.0486$

13.16 $K_c = 6.67 \times 10^7$; $K_p = 1.63 \times 10^6$

$\Delta G^0 = -35.4$ kJ/mol

13.17 $[COCl_2] = 0.022$

$[CO] = 0.178$

$[Cl_2] = 0.778$

13.18 $p_{NO} = 0.516$ atm

$p_{O_2} = 0.072$ atm

$p_{N_2} = 0.362$ atm

13.19 (a) No, $Q = 0.776 < K_c = 3.00$. Reaction goes to the right.

(b) ΔG is negative.

(c) $\Delta G^0 = -2.72$ kJ/mol

(d) $\Delta G = -3.35$ kJ/mol

(e) $[NO_2] = 0.313$

$[SO_2] = 0.556$

$[SO_3] = 1.116$

$[NO] = 0.469$

Chapter 14

ACIDS AND BASES

SECTION 14.1 EARLY DEFINITION

Objectives

- To differentiate between the properties of acids and bases.
- To identify an Arrhenius acid and base.
- To distinguish between a strong and a weak electrolyte.

Focus

According to **Arrhenius**, an acid produces H^+ and a base produces OH^- in water. The properties of an acidic solution are due to H^+ and those of a basic solution to OH^-. Strong acids and bases are strong electrolytes which are essentially 100% ionized in water. They readily conduct electric currents. Weak acids and bases are weak electrolytes that are ionized below 100%, generally less than 5%. Weak electrolytes are poor conductors.

An acid, H^+, reacts with a base, OH^-, producing water, while the anion of H^+ and the cation of OH^- remain in solution forming a salt.

$$\text{acid} + \text{base} \rightarrow \text{salt} + \text{water}$$

$$HX + MOH \rightarrow MX + H_2O$$

SECTION 14.2 BRÖNSTED-LOWRY CONCEPT

Objectives

- To define an acid and a base according to Brönsted-Lowry.
- To identify the conjugate base of an acid and the conjugate acid of a base.
- To predict the relative strengths of conjugate acids and bases.
- To explain the leveling effect of water for acids and bases.
- To show that polyprotic acids and bases dissociate stepwise.

Focus

The four major divisions of this section require considerable study. Read closely and answer carefully the questions in each division:

- Conjugate Acid-Base Pairs
- Ionization Constants
- Relative Acidity and Basicity
- Ionization of Polyprotic Acids and Polyprotic Bases

A **Brönsted-Lowry (B/L) acid** is a proton donor. A **Brönsted-Lowry (B/L) base** is a proton acceptor that has an unshared pair of electrons. Good proton donors are strong acids; good proton acceptors are strong bases.

Conjugate Acid-Base Pairs

When a B/L acid, HA, reacts with a B/L base, B, a proton is transferred, forming the conjugate base of the acid, A^-, and the conjugate acid of the base, BH^+. The transferred proton is the difference between an acid and its conjugate base.

$$\text{acid} \underset{+H^+}{\overset{-H^+}{\rightleftharpoons}} \text{base}$$

$$HA \underset{+H^+}{\overset{-H^+}{\rightleftharpoons}} A^-$$

$$BH^+ \underset{+H^+}{\overset{-H^+}{\rightleftharpoons}} B$$

Example 14.1 List the conjugate base for each acid.

(a) HSO_4^- (d) NH_3
(b) HNO_3 (e) H_2SO_3
(c) H_2O (f) HCl

Answer

(a) SO_4^{2-} (d) NH_2^-
(b) NO_3^- (e) HSO_3^-
(c) OH^- (f) Cl^-

Solution The conjugate base of the acid has one *less* proton; *subtract* an H and a + charge.

Example 14.2 List the conjugate acid for each base.

(a) HSO_4^- (d) NH_3
(b) HNO_3 (e) CN^-
(c) H_2O (f) HPO_4^{2-}

Answer

(a) H_2SO_4 (d) NH_4^+
(b) $H_2NO_3^+$ (e) HCN
(c) H_3O^+ (f) $H_2PO_4^-$

Solution The conjugate acid has one proton *more* than the base; *add* on H and a + charge. Watch the charge on the ions.

Example 14.3 Consider H_2O a B/L base and show by an equation that each of the following is a B/L acid, a proton donor. Label the acid and the base on each side of the equation and designate the conjugate acid-base pairs.

(a) $HC_2H_3O_2$ (c) $H_2PO_4^-$
(b) $HClO_4$ (d) HNO_2

Answer

	Acid$_1$		Base$_2$		Acid$_2$		Base$_1$
(a)	$HC_2H_3O_2$	+	H_2O	$\rightleftharpoons$	H_3O^+	+	$C_2H_3O_2^-$
(b)	$HClO_4$	+	H_2O	$\rightleftharpoons$	H_3O^+	+	ClO_4^-
(c)	$H_2PO_4^-$	+	H_2O	$\rightleftharpoons$	H_3O^+	+	HPO_4^{2-}
(d)	HNO_2	+	H_2O	$\rightleftharpoons$	H_3O^+	+	NO_2^-

Solution All conform to the general equation:

$$HA + H_2O \rightleftharpoons H_3O^+ + A^-$$

$$\text{acid}_1 + \text{base}_2 \rightleftharpoons \text{acid}_2 + \text{base}_1$$

For a conjugate acid-base pair, strong acids (good proton donors) have weak conjugate bases (poor proton acceptors). In the $HClO_4/ClO_4^-$ acid-base pair (Table 14.1, text), ClO_4^- has no tendency to gain a proton, H^+; if it did, $HClO_4$ would reform. Remember $HClO_4$ wants to rid itself of a proton (a good proton donor).

Ionization Constants

The magnitude of K is an indicator of the strength of an acid and its conjugate base. HCl is a strong acid and a good proton donor;

$$HCl + H_2O \rightarrow H_3O^+ + Cl^-$$

$$\text{acid}_1 + \text{base}_2 \rightarrow \text{acid}_2 + \text{base}_2$$

but Cl^-, its conjugate base, is very weak. If Cl^- were a strong base, it would accept a proton from H_3O^+ to form HCl. Because the equilibrium is far to the right, Cl^- seems to have no real affinity for protons, hence a poor proton acceptor. Similarly, H_3O^+ is not as good a proton donor as HCl. However, H_2O eagerly accepts a proton, more so than Cl^-.

To summarize, HCl is a stronger acid than H_3O^+; H_2O is a stronger base than Cl^-. The equilibrium lies far to the right, in the direction of the weaker acid, H_3O^+, and the weaker base, Cl^-.

$HC_2H_3O_2$ is a weak acid so that the equilibrium for its dissociation lies to the left:

$$HC_2H_3O_2 + H_2O \rightleftharpoons H_3O^+ + C_2H_3O_2^-$$
$$\text{acid}_2 + \text{base}_2 \rightleftharpoons \text{acid}_2 + \text{base}_1$$

Therefore, H_3O^+ gives up its proton more easily than does $HC_2H_3O_2$. H_3O^+ is a stronger acid than $HC_2H_3O_2$. $C_2H_3O_2^-$ is a better proton acceptor than H_2O. $C_2H_3O_2^-$ is a stronger base than H_2O. Again the equilibrium lies in the direction of the weaker acid and weaker base—to the left.

A few rules of B/L acid-base equilibrium are:

1. An acid has one more proton than its conjugate base.
2. A base has one less proton than its conjugate acid.
3. Stronger acids have weaker conjugate bases.
4. Weaker acids have stronger conjugate bases.
5. The position of equilibrium lies toward the weaker acid and base.
6. In Table 14.1, the acids are listed in order of *decreasing* acidity and the bases in order of *increasing* basicity.

Example 14.4 List the conjugate acids of the following bases, and then using the text, Table 14.1, list them in order of increasing acidity.

$$H_2PO_4^-, SO_3^{2-}, NO_3^-, NO_2^-, H_2O$$

Answer HSO_3^-, HNO_2, H_3PO_4, H_3O^+, HNO_3

Relative Acidity and Basicity

Analyze Table 14.1. Notice that water appears twice in the table:

$$H_3O^+ \qquad H_2O$$

and

$$H_2O \qquad OH^-$$

All substances *above H_2O* in the acid column are acids and donate a proton to H_2O, the base, producing their conjugate base of the acid and H_3O^+. All substances *below H_2O* in the base column are bases and accept a proton from H_2O, the acid, producing their conjugate acid and OH^-.

Example 14.5 (a) HSO_3^- is listed above H_2O in the acid column of Table 14.1. Write an equation showing that HSO_3^- is an acid in H_2O. Watch the charge on the ions.

(b) CO_3^{2-} is listed below H_2O in the base column. Write an equation showing that CO_3^{2-} is a base in H_2O.

Answer (a) $HSO_3^- + H_2O \rightleftharpoons H_3O^+ + SO_3^{2-}$
(b) $CO_3^{2-} + H_2O \rightleftharpoons HCO_3^- + OH^-$

From Table 14.1, we may conclude the following:

1. Acid strengths decrease from the A Group on down; base strengths (the conjugate bases of the corresponding acids) increase from H_2O on down.

2. If H_2O functions as the only base in a water solution, its conjugate acid, H_3O^+, is the strongest acid. Therefore, all acids above H_3O^+ (they ionize completely) on the left produce H_3O^+ as the only acid in an aqueous solution so that we cannot tell the difference in acid strength between, let's say, HNO_3 and $HClO_4$. This is the **leveling effect** of H_2O functioning as a base.
3. If H_2O functions as the only acid in H_2O, the strongest bases listed below OH^- on the right produce OH^- as the only base in H_2O solutions, thereby obscuring differences in base strength. This is the leveling effect of H_2O functioning as an acid.

Example 14.6 Which of the following acids have the same strength in water (the leveling effect): HCl, HNO_2, HNO_3, HCN, HSO_4^-?

Answer HCl, HNO_3

Solution All acids listed above H_3O^+ are better proton donors than H_3O^+ in water.

4. Since an acid-base equilibrium favors the formation of a weaker acid and base, any acid in the acid column reacts with any base stronger than (below) its conjugate base, (in the base column) to form a weaker conjugate acid and base.

Example 14.7 Which reaction proceeds more toward completion?

(a) $HNO_2 + C_2H_3O_2^- \rightleftharpoons HC_2H_3O_2 + NO_2^-$
(b) $HNO_2 + NH_3 \rightleftharpoons NH_4^+ + NO_2^-$

Answer (b)

Solution NH_3 is a stronger base (lower in the base column) and a better proton acceptor than $C_2H_3O_2^-$, producing an acid, NH_4^+, that is weaker than $HC_2H_3O_2$. The NH_4^+ has less of a tendency to donate a proton to the common base, NO_2^-, than does $HC_2H_3O_2$. Therefore, equation (b) produces the weaker acid/base.

5. Bases that are poorer proton acceptors than H_2O can distinguish between the strengths of various strong acids. For example, pure acetic acid is an extremely poor proton acceptor; but it is not if an acid is strong enough to donate to its proton.

$$\underset{\text{acid}_1}{HA} + \underset{\text{base}_2}{HC_2H_3O_2} \rightleftharpoons \underset{\text{acid}_2}{H_2C_2H_3O_2^+} + \underset{\text{base}_1}{A^-}$$

Using a base that is weaker than water is one way to determine the protonating strengths of the strong acids. A base stronger than water does not distinguish between strong acids. Because a strong base demands protons, all acids above its conjugate would be strong proton donors. When liquid NH_3 is the solvent, all acids above NH_4^+ are strong acids. All bases below NH_2^- are the only strong bases.

Example 14.8 Write an equation representing HNO_3 as a strong acid in liquid NH_3.

Answer $HNO_3 + NH_3 \rightleftharpoons NH_4^+ + NO_2^-$

$\text{acid}_1 + \text{base}_2 \quad \text{acid}_2 + \text{base}_1$

Solution To generalize, for any solvent, SoH: an acid in the SoH solvent has the equilibrium

$$HA + SoH \rightleftharpoons SoH_2^+ + A^-$$

and for a base:

$$B + SoH \rightleftharpoons So^- + HB^+$$

The conjugate acid-base pairs for the solvent are

$$SoH \rightleftharpoons SoH_2^+$$

$$SoH \rightleftharpoons So^-$$

Ionization of Polyprotic Acids and Polyprotic Bases

Polyprotic acids lose protons in sequence; the first H^+ is donated more readily than the second H^+ and the second more readily than the third. Similarly, for **polyprotic bases**, the first H^+ is more readily accepted than the second H^+ and the second more readily than the third.

SECTION 14.3 AMPHOTERISM; SELF-IONIZATION

Objectives

- To explain the amphoteric properties of pure solvents.
- To write an equation showing the amphoteric behavior of metallic hydroxides in water.

Focus

An **amphoteric substance** can function as both a B/L acid *and* a B/L base. It must be able to donate a proton and have an unshared pair of electrons to accept a proton.

Amphoteric solvents self-ionize to produce their conjugate acid and base ions in solution. For the general solvent,

$$SoH + SoH \rightleftharpoons SoH_2^+ + So^-$$

$$H_2O + H_2O \rightleftharpoons H_3O^+ + OH^-$$

$$NH_3 + NH_3 \rightleftharpoons NH_4^+ + NH_2^-$$

The reaction written from right to left is, of course, the neutralization reaction for the solvent.

Some metallic hydroxides are amphoteric. There is no need to memorize those that are amphoteric, but do learn how they react with H^+ or OH^-. As a proton donor reacting with a strong base, OH^-, one additional OH^- and one less H_2O are associated with the metal ion, M:

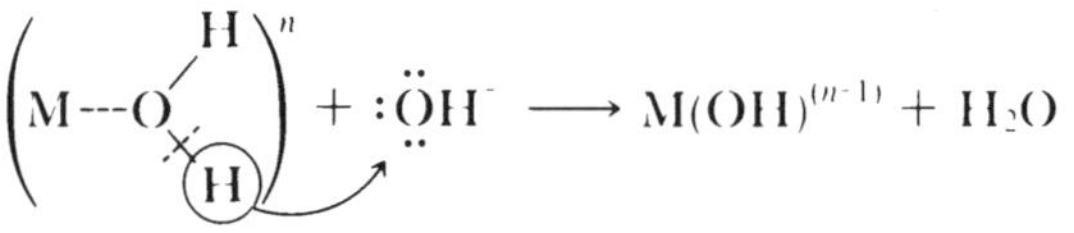

For example,

$$M(OH)_3(H_2O)_3 + OH^- \rightarrow M(OH)_4(H_2O)_2^- + H_2O$$

As a proton acceptor reacting with a strong acid, H^+, one additional H_2O and one less OH^- are associated with the metal ion:

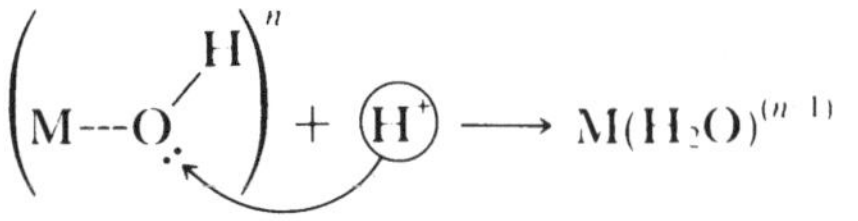

For example,

$$M(OH)_3(H_2O)_3 + H^+ \rightarrow M(OH)_2(H_2O)_4^+$$

Some amphoteric hydroxides are $Al(OH)_3(H_2O)_3$, $Zn(OH)_2(H_2O)_2$, $Cr(OH)_3(H_2O)_3$, $Be(OH)_2(H_2O)_2$, and to a lesser extent, $Cu(OH)_2(H_2O)_2$.

SECTION 14.4 TYPES OF ACID-BASE REACTIONS

Objectives

- To write the neutralization reaction in any solvent containing a proton.
- To write an equation showing the basicity of negative ions in water.
- To write an equation showing the acidity of positive ions in water.

Focus

A reaction between a solvent's conjugate acid and base constitutes a neutralization reaction in that solvent. For example, in water,

$$H_3O^+ + OH^- \rightarrow H_2O + H_2O$$

and in liquid ammonia,

$$NH_4^+ + NH_2^- \rightleftharpoons NH_3 + NH_3$$

From Table 14.1, the conjugate bases of the weak acids below H_3O^+ are stronger bases than H_2O. Therefore they accept a proton from H_2O according to the general equation

$$\begin{array}{ccccccc} A^- & + & H_2O & \rightleftharpoons & HA & + & OH^- \\ \text{base}_1 & + & \text{acid}_2 & \rightleftharpoons & \text{acid}_1 & + & \text{base}_2 \end{array}$$

producing a basic solution. Because OH^- is a stronger base than A^- (below A^-), the equilibrium lies to the left. The relative basic strengths of some negative ions are listed in Table 14.3. Negative ions that have proton donor and proton acceptor properties (amphoteric ion) are not listed as being predictable bases in water. For example, HSO_3^- may function as an acid:

$$HSO_3^- + H_2O \rightleftharpoons H_3O^+ + SO_3^{2-}$$

or as a base:

$$HSO_3^- + H_2O \rightleftharpoons H_2SO_4 + OH^-$$

Example 14.9 Write the equation for the reaction of KNO_2 with water. How does this equation differ from the effect that KH has in H_2O?

Answer $NO_2^- + H_2O \rightleftharpoons HNO_2 + OH^-$. For H^-, the equilibrium is far to the right.

Solution Ignore the presence of K^+. Later we shall see that all monovalent positive ions have no effect on the acidity or basicity of a solution. NO_2^- is a proton acceptor in the weakly basic group (Table 14.3).

$$NO_2^- + H_2O \rightleftharpoons HNO_2 + OH^-$$

H^- is a stronger base than OH^- (Tables 14.1 and 14.3). Thus, H^- reacts with H_2O to form the weaker base, OH^-, and the weaker acid, H_2.

$$H^- + H_2O \rightarrow H_2 + OH^-$$

Conjugate acids of weak bases donate a proton to water, forming an acidic solution according to the general equation

$$BH^+ + H_2O \rightleftharpoons H_3O^+ + B$$

Example 14.10 Write the equilibrium for $CH_3NH_3^+Cl^-$ in water. Cl^- is a neutral negative ion (Table 14.3) and therefore can be omitted from the equation.

Answer $CH_3NH_3^+ + H_2O \rightleftharpoons H_3O^+ + CH_3NH_2$

All trivalent, positive metal ions produce an acidic solution because of the strong affinity of the negative end of polar H_2O molecules for the positively charged ion:

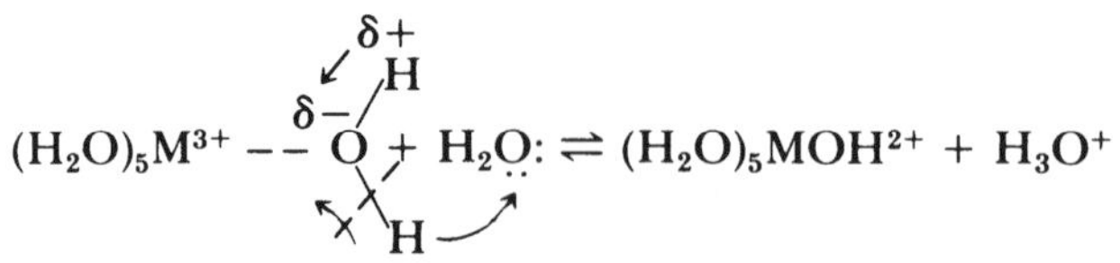

cleavage

All divalent positive ions, except those of Group 2A, produce acidic solutions. All monovalent positive ions are assumed to have no effect on the acidity of a solution. Refer to text Table 14.4 for some positive ions that produce acidic solutions.

Example 14.11 Which of the following ions change the acidity or basicity of an aqueous solution: Cl^-, Fe^{2+}, NO_2^-, Ba^{2+}, CO_3^{2-}, Br^-, Na^+, Ti^{2+}, Ag^+, Zn^{2+}, Cu^+, $C_2H_3O_2^-$, NH_2^-?

Answer Fe^{2+}, NO_2^-, CO_3^{2-}, Ti^{2+}, Zn^{2+}, $C_2H_3O_2^-$, NH_2^-

Solution Cl^- and Br^- are conjugate bases of strong acids (Table 14.3); Na^+, Ag^+, and Cu^+ are monovalent ions and Ba^{2+} is in Group 2A (Table 14.4).

SECTION 14.5 MECHANISM OF PROTON TRANSFER

Objective

- To account for the proton transfer in a B/L reaction.

Focus

The B/L acid donates its proton to a B/L base as a result of the physical closeness of the two substances in solution. This is called the **reaction mechanism** of the acid-base reaction according to B/L theory.

SECTION 14.6 LEWIS ACID-BASE CONCEPTS

Objective

- To define a Lewis acid and a Lewis base.

Focus

A **Lewis acid** is a substance that accepts a pair of electrons to form a covalent bond, an electron-pair acceptor. Lewis acids are electrophilic (electron-loving). A **Lewis base** is a substance which donates an unshared pair of electrons to form a covalent bond, an electron-pair donor. Lewis bases are nucleophilic.

SECTION 14.7 EXAMPLES OF LEWIS ACIDS

Objective

- To identify Lewis acids.

Focus

Lewis acids are identified according to four categories. Learn to recognize these.

1. Molecules with less than an octet of valence electrons on the central atom, for example, H^+, BF_3, and AlF_3.
2. Most transition metal ions and representative metal ions with a high charge density, for example, Li^+, Be^{2+}, Mg^{2+}, and Al^{3+}.
3. Molecules having a central atom that has empty available *d* orbitals for accepting an electron pair, for example, $SbCl_3$, SF_4, and $SnCl_4$.
4. Molecules with a multiple bond between atoms with different electronegativities, for example, CO_2, SO_2, and H_2CO. The acid site on the molecule is the less electronegative atom.

Example 14.12 Which of the following are Lewis acids? Identify the category which applies for each acid: H^+, Cl^-, SO_2, PCl_3, Fe^{2+}, Ca^{2+}.

Answer H^+, SO_2, PCl_3, Fe^{2+}

Solution H^+, acid (category 1) capable of accepting a pair of electrons.

$:\ddot{Cl}.^-$, base, no place to accept an additional pair of electrons.

SO_2, acid (category 4; see text, Example 13); also

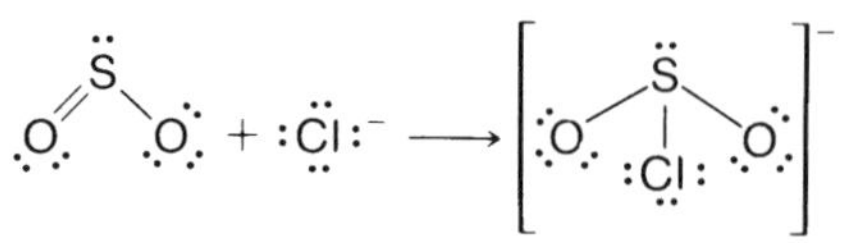

PCl_3, acid (category 3, *P* has 3*d* orbitals available for bonding, $:PCl_3 + :\ddot{Cl}:^- \rightarrow :PCl_4^-$).

Fe^{2+}, acid (category 2, a transition metal ion).

Ca^{2+}, no effect (category 2, a divalent cation).

Example 14.13 For each reaction write the structural formula for the reactants and products. Identify the Lewis acid and base.

(a) $PF_5 + F^- \rightarrow$
(b) $SO_2 + OH^- \rightarrow$
(c) $H^- + H_2C{=}O \rightarrow$
(d) $Cu^{2+} + 4NH_3 \rightarrow$

Solution

(a)

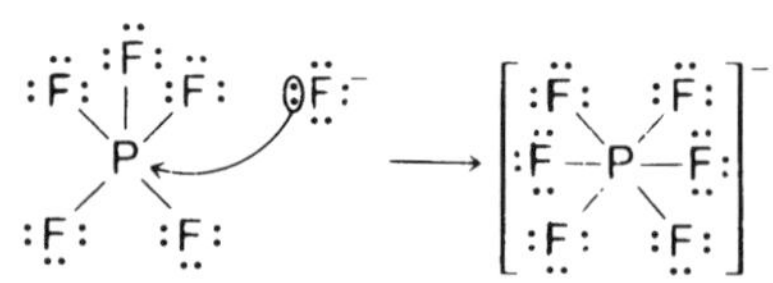

= PF_6^-

acid (category 3) base

(b)

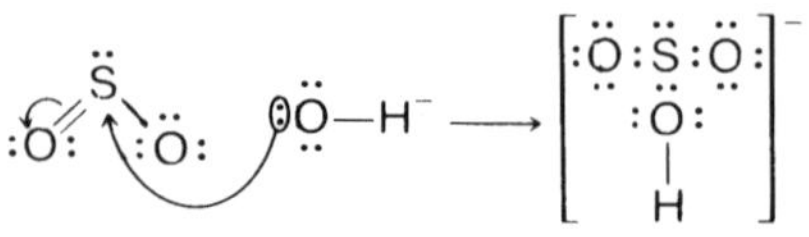

= HSO_3^-

acid (category 4) base

(c)

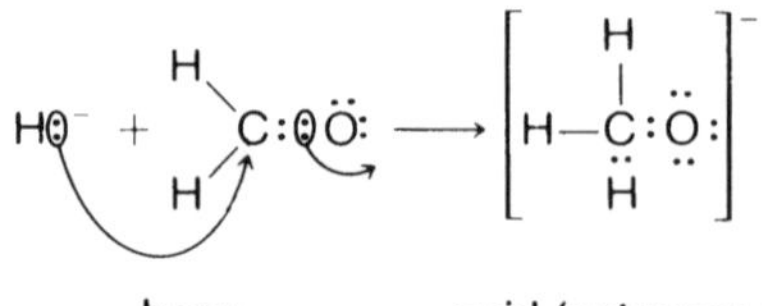

= H_3CO^-

base acid (category 4)

(d)

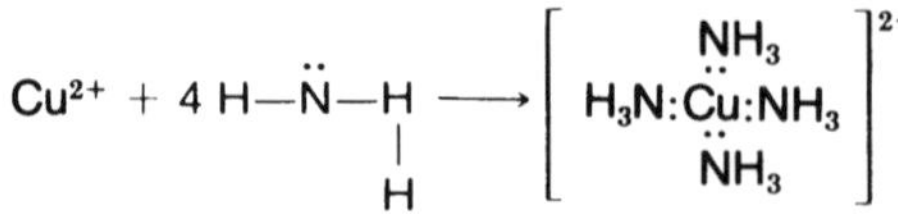

= $[Cu(NH_3)_4]^{2+}$

acid (category 4) base

SECTION 14.8 NOMENCLATURE OF ACIDS

Objective

- To identify and name binary acids, oxo acids, and their respective salts.

Focus

Binary acids are water solutions of compounds which consist of hydrogen and one other element. A binary acid is named as a hydro________ic acid; the root of the other element is inserted in the blank.

Example 14.14 Name the following binary acids: (a) HCl; (b) HI; (c) H_2S.

Answer (a) hydro*chlor*ic acid; (b) hydro*iod*ic acid; (c) hydro*sulfur*ic acid.

Oxo acids are compounds of hydrogen, oxygen, and one other element (the central element) dissolved in water having the general formula, H_mXO_n. The number of *ionizable hydrogens* are listed first in the formula. The acids are named according to the table in the text, Section 14.8. Learn the rules for the prefixes and suffixes of the oxo acids. In addition, learn the relationship between the names of the oxo acids and their anions which are listed in the third table of Section 14.8.

Another important category of oxo acids (which are discussed in Chapter 22) is the carboxylic acids (or organic acids) which have the general formula $R\text{—}\underset{\underset{\displaystyle \mathrm{O}}{\|}}{\mathrm{C}}\text{—}\mathrm{OH}$. To briefly summarize Section 14.8:

Acid	Anion	Salt
hydro________ic acid	________ide	sodium ________ide
________ic acid	________ate	sodium ________ate
________ous acid	________ite	sodium ________ite
per________ic acid	per________ate	sodium per________ate
hypo________ous acid	hypo________ite	sodium hypo________ite

For two oxidation states of "X" in the H_mXO_n acids, the higher one is termed the ________ic acid, the lower the ________ous acid. For three oxidation states, it is ________ic, ________ous, and hypo________ous acids. For four oxidation states it is per________ic, ________ic, ________ic, ________ous and hypo________ous acids, with the hypo________ous acids, in each case, having the lowest oxidation state for "X."

Example 14.15 Name the following oxo acids and their anions.

(a) HNO_3 (N has only two oxidation states)
(b) HNO_2
(c) H_2SeO_4 (Se has only two oxidation states)

(d) H_2SeO_3
(e) H_2CO_3 (C has one oxidation state)
(f) HBrO (Br has four oxidation states)

Answer

(a) nitr*ic* acid, nitr*ate* $(H\overset{+5}{N}O_3,\ \overset{+5}{N}O_3^-)$

(b) nitr*ous* acid, nitr*ite* $(H\overset{+3}{N}O_2,\ \overset{+3}{N}O_2^-)$

(c) selen*ic* acid, selen*ate* $(H_2\overset{+6}{Se}O_4,\ \overset{+6}{Se}O_4^{2-})$

(d) selen*ous* acid, selen*ite* $(H_2\overset{+4}{Se}O_3,\ \overset{+4}{Se}O_3^{2-})$

(e) carbon*ic* acid, carbon*ate* $(H_2\overset{+4}{C}O_3,\ \overset{+4}{C}O_3^{2-})$

(f) *hypo*brom*ous* acid, *hypo*brom*ite* $(H\overset{+1}{Br}O,\ \overset{+1}{Br}O^-)$

Solution For (a) and (b), (c) and (d): ________ic acid is the name for the oxo acid with the highest oxidation state, ________ous acid for the lowest. (e) ________ic acid is the name for the one oxo acid. (f) hypo________ous acid is the name for the oxo acid in which "X" is in the lowest of four oxidation states. The corresponding anions are:

________ic acid has the ________ate anion.
________ous acid has the ________ite anion.
hypo________ous acid has the hypo________ite anion.

Example 14.16 Name each salt, given the corresponding acid in Example 14.15.

(a) $Ca(NO_3)_2$ (c) Na_2SeO_4 (e) Na_2CO_3
(b) NH_4NO_2 (d) $MgSeO_3$ (f) NaBrO

Answer

(a) calcium nitr*ate*
(b) ammonium nitr*ite*
(c) sodium selen*ate*
(d) magnesium selen*ite*
(e) sodium carbon*ate*
(f) sodium *hypo*brom*ite*

SECTION 14.9 STRUCTURE AND BRÖNSTED ACIDITY

Objective

- To predict the relative acidity of acids using the periodic table.

Focus

Remember these rules for predicting the relative acidity of acids.

1. The acidity of binary acids (a) increases from left to right across a period of elements and (b) increases from top to bottom down a group of elements.
2. The acidity of oxo acids (a) increases from left to right across a period of elements but (b) decreases from top to bottom down a group of elements when the element, X, is in the same oxidation state. For example, H_2SO_3 (S = +4) is a stronger acid than H_2SeO_3 (Se = +4).
3. The acidity of an oxo acid increases with increasing oxidation state of X.

Example 14.17 Predict which is more acidic, and identify the rule used in making this selection.

(a) H_2S or H_2Se
(b) H_3AsO_4 or H_2SeO_4
(c) H_2S or HCl
(d) H_2CO_3 or H_2SiO_3
(e) $HMnO_4$ or H_2MnO_4
(f) HF or HI

Answer

(a) H_2Se, rule 1b
(b) H_2SeO_4, rule 2a
(c) HCl, rule 1a
(d) H_2CO_3, rule 2b
(e) $HMnO_4$, rule 3
(f) HI, rule 1b

SECTION 14.10 PREPARATION OF BRÖNSTED ACIDS

Objective

- To write a chemical equation for the preparation of various acids and bases.

Focus

The text lists five different preparation of acids. Become familiar with these.

1. Direct combination of H_2 and a nonmetal—for binary acids only.
2. The reaction of concentrated H_2SO_4 with a salt containing the anion of the acid—for binary and oxo acids.
3. The addition of a nonmetallic oxide (an acid anhydride) to H_2O with no change in the oxidation state of the central element—for oxo acids only. The addition of metallic oxides (basic anhydrides) to H_2O produces basic solutions. For example,

$$Na_2O + H_2O \rightarrow 2Na^+ + 2OH^-$$

$$BaO + H_2O \rightarrow Ba^{2+} + 2OH^-$$

4. The addition of a nonmetallic oxide to H_2O in which self-oxidation-reduction produces *two* oxo acids—one with an oxidation state that is greater than the oxide, and the other one less than the oxide. In the

reaction of NO_2 with H_2O, NO_2 (N = +4) forms the acids HNO_3 (N = +5) and HNO_2 (N = +3).

5. The reaction of a nonmetallic halide and H_2O with no change in oxidation state for binary and oxo acids.

Example 14.18 Write the product(s) for the following reactions and balance the equation.

(a) $KCN + H_2SO_4$ (conc.) $\rightarrow$
(b) $SO_2 + H_2O \rightarrow$
(c) $AsBr_3 + H_2O \rightarrow$
(d) $P_4O_6 + H_2O \rightarrow$
(e) $NaNO_3 + H_2SO_4$ (conc.) $\rightarrow$
(f) $IF_7 + H_2O \rightarrow$
(g) $Cl_2O_6 + H_2O \rightarrow$
(h) $H_2 + I_2 \rightarrow$

Answer

(a) $2KCN + H_2SO_4 \rightarrow 2HCN + K_2SO_4$ (type 2)

(b) $\overset{+4}{S}O_2 + H_2O \rightarrow H_2\overset{+4}{S}O_3$ (type 3)

(c) $\overset{+3}{As}Br_3 + 3H_2O \rightarrow H_3\overset{+3}{As}O_3 + 3HBr$ (type 5)

(d) $\overset{+3}{P}_4O_6 + 6H_2O \rightarrow 4H_3\overset{+3}{P}O_3$ (type 3)

(e) $2NaNO_3 + H_2SO_4 \rightarrow 2HNO_3 + Na_2SO_4$ (type 2)

(f) $\overset{+7}{I}\ \overset{-1}{F}_7 + 4H_2O \rightarrow H\overset{+7}{I}O_4 + 7H\overset{-1}{F}$ (type 5)

(g) $\overset{+6}{Cl}_2O_6 + H_2O \rightarrow H\overset{+5}{Cl}O_3 + H\overset{+7}{Cl}O_4$ (type 4)

(h) $H_2 + I_2 \rightarrow 2HI$ (type 1)

SECTION 14.11 SOME APPLICATIONS OF BRÖNSTED ACID-BASE REACTIONS

Objectives

- To explain the chemistry of carbonate and noncarbonate hardness of water.
- To explain the function of hemoglobin in maintaining the acid/base balance of the blood.

Focus

Carbonate hardness of water contains bicarbonate, HCO_3^-, which is precipitated by the addition of lime water, $Ca(OH)_2$:

$$2Ca^{2+} + 2OH^- + HCO_3^- \rightarrow 2CaCO_3(c) + 2H_2O$$

or in the home with an $NH_3/Ca(OH)_2$ solution:

$$Ca^{2+} + NH_3 + HCO_3^- \rightarrow CaCO_3(c) + NH_4^+$$

Noncarbonate hardness of water is due to the presence of Ca^{2+}, Mg^{2+}, Fe^{2+} SO_4^{2-}, or Cl^-. Ion exchange methods or washing soda, $Na_2CO_3(H_2O)_{10}$, can be used to precipitate the ion.

$$(Ca^{2+}, Mg^{2+}, \text{or } Fe^{2+}) + CO_3^{2-} \rightarrow (Ca, Mg, \text{or } Fe)CO_3(c)$$

We can understand the complexity of controlling the blood's acidity to within a very narrow range by using B/L acid-base theory. Closely review the set of equilibria in order to explain how hemoglobin maintains the blood slightly basic.

QUESTIONS

Section 14.1

14.1 Nitric acid, HNO_3, is a strong electrolyte and propionic acid, $HC_3H_5O_2$, is a weak electrolyte. Show that both are acidic but that one is strong and the other weak by means of a chemical equation.

14.2 Write a chemical equation for the reaction between nitric acid and barium hydroxide.

Section 14.2

14.3 If H_2O is a B/L acid, show with an equation that each of the following is a B/L base. Label the acid and base on each side of the equation and point out the conjugate acid-base pairs.

(a) CN^- (c) NH_3
(b) HPO_4^{2-} (d) $C_2H_3O_2^-$

14.4 In the following equilibria, identify the conjugate acid-base pairs:

(a) $H_2O_2 + CN^- \rightleftharpoons HO_2^- + HCN$

(b) $NH_4^+ + H_2O \rightleftharpoons NH_3 + H_3O^+$

(c) $H_2S + NO_2^- \rightleftharpoons HS^- + HNO_2$

(d) $HI + H_2O \rightleftharpoons H_3O^+ + I^-$

14.5 Using Table 14.1, list the conjugate bases of the following acids in order of increasing basic strength: HSO_4^-, H_2O_2, H_2S, HCl.

14.6 Using Table 14.1, answer the following:

(a) Which has the stronger conjugate base, HNO_2 or H_2SO_3?

(b) Which is the stronger acid, H_2S or H_3PO_4?

(c) Which is the stronger base, HCO_3^- or CN^-?

(d) Which has the stronger conjugate acid, H_2O or SO_4^{2-}?

(e) Which reaction

$$HNO_2 + H_2O \rightleftharpoons H_3O^+ + NO_2^-$$

or

$$NH_4^+ + H_2O \rightleftharpoons H_3O^+ + NH_3$$

lies farther to the right?

14.7 Show with an equation how each of the following produces an acidic solution.

(a) HSO_4^-
(b) HCN
(c) $H_2NNH_3^+$
(d) HNO_2
(e) $H_2PO_4^-$

14.8 Show with an equation how each of the following produces a basic solution.

(a) CN^-
(b) NH_3
(c) PO_4^{3-}
(d) H^-
(e) HCO_3^-

14.9 Would HCN or $HC_2H_3O_2$ serve as a better solvent for distinguishing the proton donor strength between HNO_3 and HBr?

14.10 (a) What is the conjugate acid of HCN? of $HC_2H_3O_2$? of HNO_2?
(b) What is the conjugate base of NH_3? of H_2SO_4? of N_2H_4?

14.11 Write an equation representing H_3PO_4 as a base in liquid HCl.

14.12 Using Table 14.1, determine which acid in each pair is the better proton donor.

(a) H_2SO_4 or HSO_4^-
(b) HPO_4^{2-} or $H_2PO_4^-$
(c) H_2SO_3 or HSO_3^-
(d) HSO_4^- or SO_4^{2-}

14.13 Using Table 14.1, determine which base in each pair is the better proton acceptor.

(a) PO_4^{3-} or $H_2PO_4^-$
(b) S^{2-} or H_2S
(c) SO_3^{2-} or HSO_3^-
(d) HSO_4^- or SO_4^{2-}

Section 14.3

14.14 (a) Write the equations for the self-ionization of $HC_2H_3O_2$ and C_2H_5OH.
(b) Write two equations showing the amphoteric behavior of hydrated chromium(III) hydroxide, $Cr(OH)_3(H_2O)_3$.

Section 14.4

14.15 Write the equation for the neutralization reaction between the conjugate acid and base of H_2SO_4.

14.16 Using Tables 14.3 and 14.4, identify whether each of the following ions produces an acidic, basic, or neutral aqueous solution.

(a) SO_4^{2-}
(b) H^-
(c) Cd^{2+}
(d) Cu^{2+}
(e) Ca^{2+}
(f) Cr^{3+}
(g) Li^+
(h) Ag^+
(i) ClO_4^-
(j) CO_3^{2-}
(k) NO_2^-
(l) Cl^-
(m) NO_3^-
(n) NH_4^+
(o) Sn^{2+}

Section 14.7

14.17 For each reaction write the structural formulas for the reactants and products. Identify the Lewis acid and base:

(a) $SF_4 + F^- \rightarrow$

(b) $GaCl_3 + Cl^- \rightarrow$

(c) $S + S^{2-} \rightarrow$

(d) $H_2O + SO_2 \rightarrow$

Section 14.8

14.18 Name the following acids.

(a) HF

(b) HI

(c) $HClO_3$ (four oxidation states)

(d) HNO_3 (two oxidation states)

(e) H_2SO_3 (two oxidation states)

(f) H_3AsO_4 (two oxidation states)

(g) H_2CrO_4 (1 oxidation state)

(h) H_2SiO_3 (1 oxidation state)

14.19 Name the following salts.

(a) KF	(e) Na_2SO_3
(b) MgI_2	(f) Li_3AsO_4
(c) $KClO_3$	(g) $CaCrO_4$
(d) $Mg(NO_3)_2$	(h) $CaSiO_3$

Section 14.10

14.20 Write the products for the following reactions and balance the equation.

(a) $N_2O_3 + H_2O \rightarrow$

(b) $Cl_2O + H_2O \rightarrow$

(c) $Cl_2O_7 + H_2O \rightarrow$

(d) $PCl_3 + H_2O \rightarrow$

(e) $P_4O_6 + H_2O \rightarrow$

(f) $H_2 + S_8 \rightarrow$

(g) $SiBr_4 + H_2O \rightarrow$

(h) $ClF_3 + H_2O \rightarrow$

14.21 ClO_2 undergoes self-oxidation-reduction in water. Write the corresponding oxo acids produced.

ANSWERS

14.1 $HNO_3 \rightarrow H^+ + NO_3^-$
$HC_3H_5O_2 \rightleftharpoons H^+ + C_3H_5O_2^-$

14.2 $2HNO_3 + Ba(OH)_2 \rightarrow 2H_2O + Ba(NO_3)_2$

14.3

	$Acid_1$		$Base_2$		$Acid_2$		$Base_1$
(a)	H_2O	+	CN^-	$\rightleftharpoons$	HCN	+	OH^-
(b)	H_2O	+	HPO_4^{2-}	$\rightleftharpoons$	$H_2PO_4^-$	+	OH^-
(c)	H_2O	+	NH_3	$\rightleftharpoons$	NH_4^+	+	OH^-
(d)	H_2O	+	$C_2H_3O_2^-$	$\rightleftharpoons$	$HC_2H_3O_2$	+	OH^-

14.4 (a) H_2O_2 and HO_2^-; CN^- and HCN

(b) NH_4^+ and NH_3; H_2O and H_3O^+

(c) H_2S and HS^-; NO_2^- and HNO_2

(d) HI and I^-; H_2O and H_3O^+

14.5 $Cl^- < SO_4^{2-} < HS^- < HO_2^-$

14.6 (a) HNO_2

(b) H_3PO_4

(c) CN^-

(d) H_2O

(e) $HNO_2 + H_2O \rightleftharpoons H_3O^+ + NO_2^-$

14.7 (a) $HSO_4^- + H_2O \rightleftharpoons H_3O^+ + SO_4^{2-}$

(b) $HCN + H_2O \rightleftharpoons H_3O^+ + CN^-$

(c) $H_2NNH_3^+ + H_2O \rightleftharpoons H_3O^+ + H_2NNH_2$

(d) $HNO_2 + H_2O \rightleftharpoons H_3O^+ + NO_2^-$

(e) $H_2PO_4^- + H_2O \rightleftharpoons H_3O^+ + HPO_4^{2-}$

14.8 (a) $CN^- + H_2O \rightleftharpoons OH^- + HCN$

(b) $NH_3 + H_2O \rightleftharpoons OH^- + NH_4^+$

(c) $PO_4^{3-} + H_2O \rightleftharpoons OH^- + HPO_4^{2-}$

(d) $H^- + H_2O \rightleftharpoons OH^- + H_2$

(e) $HCO_3^- + H_2O \rightleftharpoons OH^- + H_2CO_3$

14.9 $HC_2H_3O_2$

14.10 (a) H_2CN^+; $H_2C_2H_3O_2^+$; $H_2NO_2^+$

(b) NH_2^-; HSO_4^-; $N_2H_3^-$

14.11 $H_3PO_4 + HCl \rightleftharpoons H_4PO_4^+ + Cl^-$

14.12 (a) H_2SO_4

(b) $H_2PO_4^-$

(c) H_2SO_3

(d) HSO_4^-

14.13 (a) PO_4^{3-} (c) SO_3^{2-}

(b) S^{2-} (d) SO_4^{2-}

14.14 (a) $HC_2H_3O_2 + HC_2H_3O_2 \rightleftharpoons H_2C_2H_3O_2^+ + C_2H_3O_2^-$

$C_2H_5OH + C_2H_5OH \rightleftharpoons C_2H_5OH_2^+ + C_2H_5O^-$

(b) $\underset{\text{acid}_1}{Cr(OH)_3(H_2O)_3} + \underset{\text{base}_2}{OH^-} \rightleftharpoons \underset{\text{base}_1}{Cr(OH)_4(H_2O)_2^-} + \underset{\text{acid}_2}{H_2O}$

$\underset{\text{base}_1}{Cr(OH)_3(H_2O)_3} + \underset{\text{acid}_2}{H^+} \rightleftharpoons \underset{\text{acid}_1}{Cr(OH)_2(H_2O)_4^+} + \underset{\text{base}_2}{H_2O}$

14.15 $H_3SO_4^+ + HSO_4^- \rightarrow 2H_2SO_4$

14.16 (a) SO_4^{2-}; feebly basic
(b) H^-; strongly basic
(c) Cd^{2+}; acidic, type b
(d) Cu^{2+}; acidic, type b
(e) Ca^{2+}; neutral, type b
(f) Cr^{3+}; acidic, type c
(g) Li^+; neutral, type a
(h) Ag^+; neutral, type a
(i) ClO_4^-; neutral
(j) CO_3^{2-}; distinctly basic
(k) NO_2^-; weakly basic
(l) Cl^-; neutral
(m) NO_3^-; neutral
(n) NH_4^+; acidic, type a
(o) Sn^{2+}; acidic, type b

14.17 (a)

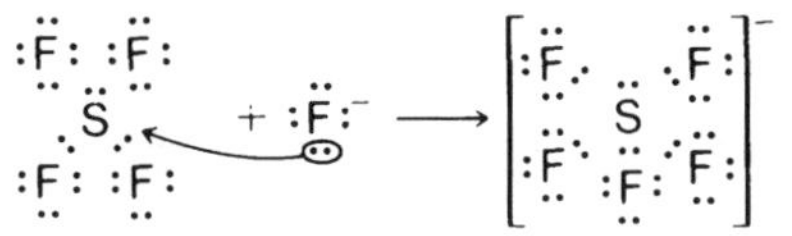

acid
(category 3)

(b)

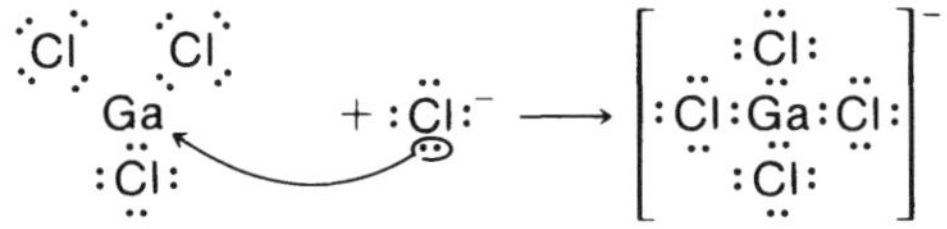

acid (category 1)

(c)

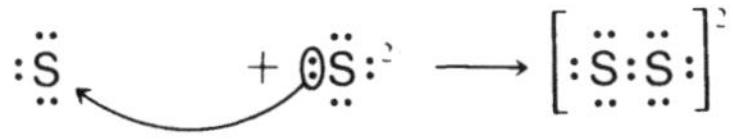

acid (category 1)

(d)

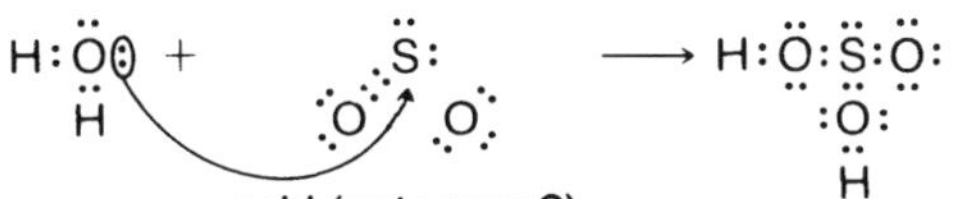

acid (category 2)

14.18 (a) hydrofluoric acid
(b) hydroiodic acid
(c) chloric acid
(d) nitric acid
(e) sulfurous acid
(f) arsenic acid
(g) chromic acid
(h) silicic acid

14.19 (a) potassium fluoride
(b) magnesium iodide
(c) potassium chlorate
(d) magnesium nitrate
(e) sodium sulfite
(f) lithium arsenate
(g) calcium chromate
(h) calcium silicate

14.20 (a) $N_2O_3 + H_2O \rightarrow 2HNO_2$ (type 3)
(b) $Cl_2O + H_2O \rightarrow 2HClO$ (type 3)
(c) $Cl_2O_7 + H_2O \rightarrow 2HClO_4$ (type 3)
(d) $PCl_3 + 3H_2O \rightarrow H_3PO_3 + 3HCl$ (type 5)
(e) $P_4O_6 + 6H_2O \rightarrow 4H_3PO_3$ (type 3)
(f) $8H_2 + S_8 \rightarrow 8H_2S$ (type 1)
(g) $SiBr_4 + 3H_2O \rightarrow H_2SiO_3 + 4HBr$ (type 3)
(h) $ClF_3 + 2H_2O \rightarrow HClO_2 + 3HF$ (type 5)

14.21 $2ClO_2 + H_2O \rightarrow HClO_3 + HClO_2$

Chapter 15

IONIC EQUILIBRIUM

SECTION 15.1 HOW GOOD IS THE LAW OF CHEMICAL EQUILIBRIUM?

Objective

• Know the limits of the law of chemical equilibrium when applied to ionic solutions.

Focus

The law of chemical equilibrium is valid for gases at low pressures, for molecules in dilute solution, and for ions in very dilute aqueous solutions. More detail on the theory of ionic solutions is covered in subsequent chemistry courses.

SECTION 15.2 THE SELF-IONIZATION OF WATER

Objective

• To determine the $[H^+]$ of an aqueous solution, given the $[OH^-]$, and vice versa.

Focus

Water self-ionizes to form an equilibrium solution containing H^+ and OH^-:

$$H_2O \rightleftharpoons H^+ + OH^-$$

The ion product for water, K_w, is

$$K_w = [H^+][OH^-] = 1.00 \times 10^{-14}, \quad \text{at } 25°\text{C}$$

The value of K_w is temperature dependent (see text, Table 15.1).

According to the equilibrium condition for K_w, the addition of an acid increases the $[H^+]$ of water and the addition of $[OH^-]$ decreases the $[H^+]$ of water.

Example 15.1 Calculate the $[OH^-]$ of a solution in which $[H^+] = 0.040$ mol/L.

Answer 2.5×10^{-13} mol/L

Solution

$$[H^+] = 0.040 \text{ mol/L}$$

$$K_w = [H^+][OH^-]$$

$$[OH^-] = \frac{K_w}{[H^+]} = \frac{1.00 \times 10^{-14}}{0.040} = 2.5 \times 10^{-13} \frac{\text{mol}}{\text{L}}$$

SECTION 15.3 pH AND p*K*

Objectives

- To calculate the pH of a solution, knowing the $[H^+]$, and vice versa.
- To express the "p" value for other small numbers.

Focus

pH is a very important measurement in environmental and biochemical systems. A scientist must be familiar with the pH scale. The definition is

$$\text{pH} = -\log [H^+]$$

$$\text{pOH} = -\log [OH^-]$$

Similarly,

$$\text{p}K_w = -\log K_w$$

$$\text{p}X = -\log X$$

Generally, X is a molar concentration or an equilibrium constant.

Example 15.2 Calculate the pH of a solution with $[H^+] = 0.00730$ mol/L

Answer pH = 2.14

Solution Change 0.00730 to scientific notation or 7.30×10^{-3}.

$$\text{pH} = -\log [H^+]$$

$$= -\log 7.30 \times 10^{-3}$$

On a calculator with scientific notation enter 7.30×10^{-3}, press the LOG button and record the *negative* of the number on the display. Do not, however, record all the digits seen on display. See text, Appendix B-1, for significant figures using logarithms and antilogarithms. If a calculator is unavailable, then

(a) Look up the log of 7.30. log 7.30 = 0.86.
(b) The log of 10^{-3} is -3.
(c) Add the values of the two log terms and record the negative value.

$$\text{pH} = -[0.86 + (-3)] = -[-2.14]$$

(d) pH = +2.14

Example 15.3 The pH of an acidic solution is 4.61. Calculate the $[H^+]$ of the solution.

Answer 2.46×10^{-5} mol/L

Solution $pH = -\log [H^+]$

$$4.61 = -\log [H^+]$$

On most calculators, two buttons are required for the calculation. First convert to

$$\log [H^+] = -4.61$$

Then in sequence use the INV (inverse) and LOG buttons. Check the instructions for operating your calculator to make certain this procedure is proper.

If no calculator is available, continue undoing the $\log [H^+] = -4.61$ by taking the antilog of both sides of the equation:

$$\text{antilog} (\log [H^+]) = \text{antilog} (-4.61)$$

Since negative numbers are not a part of the log table, make -4.61 a positive number by adding $+5$ and -5, or

$$[H^+] = \text{antilog} [(-4.61 + 5) + (-5)]$$
$$= \text{antilog} [0.39 + (-5)]$$
$$= \text{antilog}\ 0.39 \times \text{antilog} (-5)$$

From the log table, $[H^+] = 2.46 \times 10^{-5}$

Recall from this section that a solution with a $pH = 7$ is neutral at 25°C. Solutions with a pH less than 7 are acidic, and those greater than 7 are basic at 25°C. Also remember that $pH + pOH = pK_w = 14$ at 25°C.

SECTION 15.4 STRONG ACIDS AND BASES

Objective

- To calculate the pH of strong acids and strong bases.

Focus

For dilute solutions of strong acids and strong bases, we assume 100% dissociation in water, that is, the equilibrium lies all the way to the right.

At this point memorize the common strong acids—HCl, HBr, HI, $HClO_4$, HNO_3 and H_2SO_4 (see text, Table 14.1), and the common strong bases—$NaOH$, KOH, $CsOH$, and $Ba(OH)_2$.

SECTION 15.5 WEAK ACIDS AND BASES

Objective

- To calculate the pH and pOH for weak acid and weak base solutions.

Focus

The method used for calculating the pH of a weak acid or base follows the same format used in Chapter 13. Review this Study Guide, Section 13.11.

We may work with moles, or concentrations, in the equilibrium table, but if dilution occurs from the mixing of two solutions, we must carry out our calculations using moles instead of molar concentrations.

The equilibrium equation for any weak acid, whether a molecular substance, an acidic anion, or a metallic cation in water, is written as

$$HA \rightleftharpoons H^+ + A^-$$

The equilibrium equation for any weak base, whether it be a molecular substance or basic anion, is

$$B + H_2O \rightleftharpoons BH^+ + OH^-$$

Example 15.4 Calculate the pH of a 0.0940 M benzoic acid, $HC_7H_5O_2$, solution. From the text, Table 15.3, $K_a = 6.46 \times 10^{-5}$.

Answer 2.61

Solution

Step 1.	equation	$HC_7H_5O_2$	$\rightleftharpoons$	H^+	+	$C_7H_5O_2^-$
Step 2.	[ ]*i*	0.0940		0		0
Step 3.	Δ[ ]	$-y$		$+y$		$+y$
Step 4.	[ ] $\rightleftharpoons$	$0.0940 - y$		y		y

Step 5. $K = 6.46 \times 10^{-5} = \dfrac{(y)(y)}{0.0940 - y}$

Remember the 5% rule? (See Study Guide, Section 13.12.) Assume $0.0940 >> y$; therefore,

$$K_a = 6.46 \times 10^{-5} \simeq \frac{y^2}{0.0940}$$

$$y = 2.46 \times 10^{-3} = [H^+]$$

Let's now check to see if our approximation is valid:

Is $\dfrac{y}{\text{original}} \times 100 \leq 5\%$?

$$\frac{2.46 \times 10^{-3}}{0.0940} \times 100 = 2.62\%$$

Approximation is O.K.

$$[H^+] = 2.46 \times 10^{-3}$$

$$pH = -\log(2.46 \times 10^{-3})$$

$$= 2.61$$

Example 15.5 Calculate the pH of 0.0940 M sodium benzoate, $NaC_7H_5O_2$. From Table 15.3, $K_b = 1.55 \times 10^{-10}$.

Answer 8.58

Solution Na^+ is a spectator ion; since it does not affect the pH of the solution, it does *not* appear in the equilibrium equation from which pH is calculated.

Sodium benzoate is a soluble salt which produces Na^+ and $C_7H_5O_2^-$ at the same concentration as the molarity of the salt solution. According to the general equation for all weak bases in solution, B = $C_7H_5O_2^-$.

Step 1.	equation	$C_7H_5O_2^-$	+	H_2O	$\rightleftharpoons$	$HC_7H_5O_2$	+	OH^-
Step 2.	[ ]i	0.0940				0		0
Step 3.	Δ[ ]	$-y$				$+y$		$+y$
Step 4.	[ ]$\rightleftharpoons$	$0.0940 - y$				y		y

Step 5. $K_b = 1.55 \times 10^{-10} = \dfrac{(y)(y)}{(0.0940 - y)}$

Let's approximate, and assume $0.0940 >> y$:

$$K_b = 1.55 \times 10^{-10} \simeq \frac{y^2}{0.0940}$$

$$y = 3.82 \times 10^{-6} = [OH^-]$$

Approximation check:

$$\frac{3.82 \times 10^{-6}}{0.0940} \times 100 = 0.0041\% << 5\%$$

Approximation is O.K.

$$[OH^-] = 3.82 \times 10^{-6}$$

$$[H^+] = \frac{K_w}{[OH^-]} = \frac{1.00 \times 10^{-14}}{3.82 \times 10^{-6}} = 2.62 \times 10^{-9}$$

$$pH = -\log (2.62 \times 10^{-9})$$

$$pH = 8.58$$

We do expect the pH to be greater than 7.0 since we are dealing with the conjugate base of a weak acid. (*Note*: This process is referred to as hydrolysis.)

SECTION 15.6 CONJUGATE ACID-BASE PAIRS

Objective

- To derive the K_a values for an acid from the listed K_b value for its conjugate base.

Focus

$$K_a K_b = K_w$$

The product of the equilibrium constant of a weak acid, K_a, and the equilibrium constant for its conjugate base, K_b, equals K_w. Also,

$$pK_a + pK_b = 14$$

SECTION 15.7 POLYPROTIC ACIDS AND BASES

Objective

- To calculate the pH of polyprotic acids and bases.

Focus

A **polyprotic acid** is capable of producing more than 1 mole H^+ per mole of acid:

$$H_2A \rightleftharpoons H^+ + HA^- \qquad K_1 = \frac{[H^+][HA^-]}{[H_2A]}$$

$$HA^- \rightleftharpoons H^+ + A^{2-} \qquad K_2 = \frac{[H^+][A^{2-}]}{[HA^-]}$$

Generally, $K_1 > K_2$; therefore most of the H^+ in solution is a result of the first ionization of the polyprotic acid.

A **polyprotic base** is capable of accepting more than 1 mole H^+ per mole of base:

$$B^{2-} + H^+ \rightleftharpoons HB^- \qquad K_1 = \frac{[HB^-]}{[B^{2-}][H^+]}$$

$$HB^- + H^+ \rightleftharpoons H_2B \qquad K_2 = \frac{[H_2B]}{[HB^-][H^+]}$$

Example 15.6 Calculate the pH, the phthalate ion concentration, and the undissociated phthalic acid concentration in a 0.10 M phthalic acid, $H_2C_8H_4O_4$, solution. From Table 15.3,

$$K_1 = 1.26 \times 10^{-3} \qquad K_2 = 3.9 \times 10^{-6}$$

Answer pH = 1.97; $[C_8H_4O_4^{2-}] = [A^{2-}] = 3.9 \times 10^{-6}$; $[H_2C_8H_4O_4] = [H_2A] = 0.089$

Solution As the text suggests, write the two equilibria using the generalized diprotic acid, H_2A. Treat each equilibrium equation separately!

Step 1.	equation	H_2A	$\rightleftharpoons$	H^+	+	HA^-
Step 2.	$[\ ]i$	0.10		0		0
Step 3.	$\Delta[\ ]$	$-y$		$+y$		$+y$
Step 4.	$[\ ]\rightleftharpoons$	$0.10 - y$		y		y

Step 5. $K_1 = 1.26 \times 10^{-3} = \dfrac{(y)(y)}{0.10 - y}$

Let's approximate $y << 0.10$ or

$$1.26 \times 10^{-3} \simeq \frac{y^2}{0.10}$$

$$y = 1.12 \times 10^{-2}$$

Approximate check:

$$\frac{1.12 \times 10^{-2}}{0.10} \times 100 = 11.2\% >> 5\%.$$

The approximation is *not permissible*. Therefore, we need to solve the quadratic equation for y (see text, Appendix B-2).

$$1.26 \times 10^{-3} = \frac{y^2}{0.10 - y}$$

$$y^2 + 1.26 \times 10^{-3}y - 1.26 \times 10^{-4} = 0$$

$$y = \frac{-1.26 \times 10^{-3} \pm \sqrt{(1.26 \times 10^{-3})^2 - 4(-1.26 \times 10^{-4})}}{2}$$

$$= 1.06 \times 10^{-2}$$

Use the positive value of the square root term for a positive concentration.

$$y = [HA^-] = [H^+] = 1.06 \times 10^{-2}$$

$$[H_2A] = 0.10 - 1.06 \times 10^{-2} = 0.089$$

To calculate $[A^{2-}]$, the phthlate ion concentration, use K_2.

Step 1.	equation	HA^-	$\rightleftharpoons$	H^+ +	A^{2-}
Step 2.	$[\ ]i$	1.06×10^{-2}		1.06×10^{-2} from first ionization	0
Step 3.	$\Delta[\ ]$	$-z$		$+z$	$+z$
Step 4.	$[\ ]\rightleftharpoons$	$1.06 \times 10^{-2} - z$		$1.06 \times 10^{-2} + z$	z

Step 5. $K_2 = 3.9 \times 10^{-6} = \dfrac{(1.06 \times 10^{-2} + z)(z)}{(1.06 \times 10^{-2} - z)}$

Let's approximate $z << 1.06 \times 10^{-2}$.

$$3.9 \times 10^{-6} = \frac{1.06 \times 10^{-2}\,(z)}{1.06 \times 10^{-2}}$$

$$z = 3.9 \times 10^{-6}$$

Approximation check:

$$\frac{3.9 \times 10^{-6}}{1.06 \times 10^{-2}} \times 100 = 0.037\% << 5\%.$$

Our approximation is O.K.

$$z = [A^{2-}] = 3.9 \times 10^{-6}$$

$$[H^+] = 1.06 \times 10^{-2} + 3.9 \times 10^{-6} \simeq 1.06 \times 10^{-2}$$

which equals the $[H^+]$ produced in the first ionization step:

$$pH = -\log(1.06 \times 10^{-2})$$

$$pH = 1.97$$

Be very careful in selecting the K-value for the conjugate of a polyprotic acid or base appearing in Table 15.3.

Example 15.7 What is the conjugate acid of HSO_3^-? What is the K-value for the basic ionization of HSO_3^-?

Answer H_2SO_3; $K_b = 5.9 \times 10^{-13}$

Solution An easy way to start is to write the equilibrium for HSO_3^-, functioning as a base (proton acceptor), producing its conjugate acid.

$$HSO_3^- + H_2O \rightleftharpoons H_2SO_3 + OH^-$$

We see that it is necessary to look up the K-value for the conjugate acid appearing in the equation.

To determine the K_b for HSO_3^- use the K-value for the conjugate acid, H_2SO_3, appearing in the equation. The K-value for H_2SO_3 is $K_1 = 1.7 \times 10^{-2}$ (Table 15.3).

$$K_b = \frac{K_w}{K_1} = \frac{1.00 \times 10^{-14}}{1.7 \times 10^{-2}} = 5.9 \times 10^{-13} \quad \text{a very weak base.}$$

SECTION 15.8 LE CHATELIER'S PRINCIPLE FOR ACID-BASE REACTIONS

Objective

- To calculate the change in pH when an ion, common to the equilibrium in solution, is added.

Focus

When an ion, common to one of the species in an equilibrium system, is added to a solution, the equilibrium shifts according to **Le Chatelier's principle**. This is called the *common-ion effect*. The added ion usually comes from a strong acid, strong base, or soluble salt. The following example illustrates this.

Example 15.8 Calculate the pH and the benzoate ion concentration of a solution that is 0.0300 M benzoic acid, $HC_7H_5O_2$, and 0.400 M HNO_3. What is the benzoate concentration in a 0.0300 M $HC_7H_5O_2$ solution in the *absence* of 0.400 M HNO_3? Compare the two $C_7H_5O_2^-$ concentrations.

Answer pH = 0.40; $[C_7H_5O_2^-] = 4.85 \times 10^{-6}$ mol/L; $[C_7H_5O_2^-] = 1.39 \times 10^{-3}$ mol/L.

Solution We solve this problem the same way we solved previous equilibrium problems.

Step 1. When two acids (or two bases) are given, always write the equation that represents the weak equilibrium; in this problem $HC_7H_5O_2$ is the weaker acid. For simplicity, let's use HA.

	equation	HA	$\rightleftharpoons$	H^+ +	A^-

The H^+ from HNO_3 is the common ion added to the weak acid equilibrium, suppressing the ionization of HA.

Step 2. Remember that HNO_3 is a strong acid, producing a $[H^+]$ of 0.400 mol/L.

	$[\ ]i$	0.0300	0.400	0
Step 3.	$\Delta[\ ]$	$-y$	$+y$	$+y$
Step 4.	$[\ ]\rightleftharpoons$	$0.0300 - y$	$0.400 + y$	y

Step 5. From Table 15.3, K_a for the weak benzoic acid equilibrium is 6.46×10^{-5}.

$$K_a = 6.46 \times 10^{-5} = \frac{(0.400 + y)(y)}{(0.0300 - y)}$$

Let's approximate 0.0300 >> y and 0.400 >> y:

$$6.46 \times 10^{-5} \simeq \frac{(0.400)y}{0.0300}$$

$$y = 4.85 \times 10^{-6}$$

Check *both* approximations

$$\frac{4.85 \times 10^{-6}}{0.0300} \times 100 = 0.0161\% << 5\%$$

$$\frac{4.85 \times 10^{-6}}{0.400} \times 100 = 0.00121\% << 5\%$$

If either of the approximation checks had been greater than 5%, we would have had to solve the quadratic equation.

$$y = [A^-] \text{ from step 4}$$

$$[A^-] = [\text{benzoate ion}] = 4.85 \times 10^{-6}$$

$$[H^+] = 0.400 + y \simeq 0.400 = 4.00 \times 10^{-1}$$

$$\text{pH} = -\log(4.00 \times 10^{-1}) = 0.3979 \simeq 0.40$$

In the absence of 0.400 M HNO_3, the $[C_7H_5O_2^-]$ is 1.39×10^{-3} mol/L. (Solve this problem as we did Example 15.4). The concentration of benzoate ion increases in the absence of 0.400 M HNO_3 (compare with $[C_7H_5O_2^-] = 4.85 \times 10^{-6}$ in the presence of 0.400 M HNO_3). This is because removal of H^+, the common ion, shifts the weak acid equilibrium to the right, increasing production of $C_7H_5O_2^-$, as predicted by Le Chatelier's principle.

SECTION 15.9 BUFFER SOLUTIONS

Objectives

- To identify a buffered solution.
- To predict the effect of a strong acid or strong base on a buffered solution and calculate the resulting change in pH.
- To prepare a buffered solution with a desired pH.

Focus

A **buffer solution** resists large changes in pH. The addition of a strong acid or base to a buffered solution shifts the position of a weak acid/conjugate base equilibrium according to Le Chatelier's principle.

For the weak acid, HA/A^- equilibrium,

$$HA \rightleftharpoons H^+ + A^-$$

The addition of strong acid, H^+ (from HCl, HNO_3, HBr, . . .), causes the equilibrium to shift to the *left*, because H^+ reacts with A^- to produce HA. The addition of strong base, OH^- (from NaOH, KOH, $Ba(OH)_2$), shifts the equilibrium to the *right* because OH^- removes H^+ from the equilibrium to produce H_2O and A^-.

$$HA \rightleftharpoons \boxed{\begin{array}{c} H^+ \\ + \\ OH^- \end{array}} \begin{array}{l} + A^- \\ \\ \rightarrow H_2O \end{array}$$

Example 15.9 A mass of 21.7 grams of NaOH is added to 1.00 liter of a 0.840 M formic acid, $HCHO_2$, solution. Assume no volume change.

(a) Is the resulting solution a buffer solution?

(b) What is the pH of the final solution?

Answer (a) yes; (b) pH = 4.02

Solution (a) Formic acid in solution has the equilibrium

$$HA \rightleftharpoons H^+ + A^-$$

Addition of OH^- shifts the equilibrium to the *right*, decreasing [HA], increasing $[A^-]$, but neutralizing H^+. Therefore, after the OH^- addition, we have in solution HA and A^-, a conjugate weak acid/base pair. The equilibrium shifts by the amount of OH^- added or

$$\frac{21.7\text{ g}}{\text{L}} \times \frac{\text{mol}}{40.0\text{ g}} = 0.543\text{ mol OH}^-/\text{L}$$

(b) From Table 15.3, K_a ($HCHO_2$) = 1.76×10^{-4}

Step 1.	equation	HA	$\rightleftharpoons$	H^+	+ A^-
Step 2.	The original [HA] decreases by 0.543 mol/L as a result of the OH^- addition and the shift to the right. The $[A^-]$ increases by 0.543 mol/L as a result of the shift.				
	$[\ ]i$	0.840 − 0.543		0	0 + 0.543
Step 3.	$\Delta[\ ]$	$-y$		$+y$	$+y$
Step 4.	$[\ ]\rightleftharpoons$	$0.297 - y$		y	$0.543 + y$

Step 5. $K_a = 1.76 \times 10^{-4} = \dfrac{(y)(0.543 + y)}{0.297 - y}$

Let's approximate 0.543 >> y and 0.297 >> y:

$$1.76 \times 10^{-4} \simeq \frac{(y)(0.543)}{0.297}$$

$y \simeq 9.63 \times 10^{-5}$ (Both approximations are O.K.)

$$y = [H^+] = 9.63 \times 10^{-5}$$

$$pH = 4.02$$

Example 15.10 Calculate the change in pH after the addition of 0.220 mol HNO_3 to the buffer solution in Example 15.9.

Answer ΔpH = 0.47

Solution The strong acid shifts the equilibrium *left* by 0.220 mol/L, resulting in an increase of HA by 0.220 mol/L and a decrease of A^- by 0.220 mol/L.

Step 1.	equation	HA	$\rightleftharpoons$	H^+ +	A^-
Step 2.	[ ]i	0.297 + 0.220		0	0.543 − 0.220
Step 3.	Δ[ ]	$-y$		$+y$	$+y$
Step 4.	[ ]$\rightleftharpoons$	$0.517 - y$		y	$0.323 + y$

Step 5. $K_a = 1.76 \times 10^{-4} = \dfrac{y(0.323 + y)}{0.517 - y}$

With the approximation:

$$[H^+] = y = 2.81 \times 10^{-4}$$

$$pH = 3.55$$

From Example 15.9, pH = 4.02, therefore:

$$\Delta pH = 4.02 - 3.55 = 0.47$$

The pH change for the addition of 0.220 mol HNO_3 in 1.00 liter water instead of 1.00 liter of a buffer solution is

$$[H^+] = 0.220$$

$$pH = 0.66$$

$$\Delta pH = 7.00 - 0.66 = 6.34$$

A ΔpH of 0.47 compared to 6.34 shows that a buffer really resists a pH change!

If y is negligible compared to the concentrations of the weak acid and conjugate base at equilibrium (the 5% rule), then the pH of the solution at equilibrium can be calculated by using the Henderson-Hasselbalch equation:

$$pH = pK_a + \log \frac{[\text{base}]}{[\text{acid}]}$$

Using this equation to calculate the pH of the solution in Example 15.9,

$$pK_a\ (HCHO_2) = -\log (1.76 \times 10^{-4})$$

$$pK_a = 3.755$$

$$pH = 3.755 + \log \frac{0.543}{0.297}$$

$$= 3.755 + 0.262$$

$$= 4.017 \simeq 4.02$$

For Example 15.10, we have

$$pH = 3.75 + \log \frac{0.323}{0.517}$$

$$= 3.75 + (-0.20)$$

$$= 3.55$$

These values compare favorably with those values calculated in Examples 15.9 and 15.10.

Example 15.11 To prepare a solution with a pH of 10.00, what must be the ratio of $[HCO_3^-]$ to $[CO_3^{2-}]$?

Answer 1.78 to 1

Solution From Table 15.3, we select the K_2 of carbonic acid for the HCO_3^-/CO_3^{2-} system.

$$K_2 = 5.62 \times 10^{-11}$$
$$pK_2 = -\log(5.62 \times 10^{-11})$$
$$= 10.25$$

Knowing that the desired pH of the solution is 10.00 and that $pK_2 = 10.25$ we solve this problem using the Henderson-Hasselbalch equation,

$$\text{pH} = pK_2 + \log \frac{[CO_3^{2-}]}{[HCO_3^-]}$$

$$10.0 = 10.25 + \log \frac{[CO_3^{2-}]}{[HCO_3^-]}$$

$$\frac{[CO_3^{2-}]}{[HCO_3^{2-}]} = \text{antilog}(-0.25) = \frac{5.62 \times 10^{-1}}{1}$$

The ratio $[CO_3^{2-}]$ to $[HCO_3^-]$ must be 0.562 to 1 or $[HCO_3^-]/[CO_3^{2-}]$ is

$$\frac{1}{0.562} = \frac{1.78}{1} \quad \text{or} \quad 1.78 \text{ to } 1$$

Note: This means that if the $[HCO_3^-]$ is 0.420 mol/L, for a laboratory solution, then the $[CO_3^{2-}]$ must be

$$\left(\frac{1 \text{ mol/L } CO_3^{2-}}{1.78 \text{ mol/L } HCO_3^{2-}} \times 0.420 \text{ mol/L } HCO_3^-\right), \text{ or } 0.236 \text{ mol/L.}$$

Example 15.11A If 250 mL of 0.420 M HCO_3^- is prepared already, how many grams Na_2CO_3 should be added to have a buffer with a pH = 10.00? This is a very practical problem in preparing solutions for the laboratory.

Answer 6.25 g Na_2CO_3

Solution The number of moles of CO_3^{2-} added for the solution to have a pH = 10.00 is

$$\text{mol } CO_3^{2-} = 0.250 \text{ L} \times \frac{0.420 \text{ mol } HCO_3^-}{\text{L}} \times \frac{1 \text{ mol } CO_3^{2-}}{1.78 \text{ mol } HCO_3^-}$$

$$= 0.0590 \text{ mol } CO_3^{2-}$$

$$\text{mass of } Na_2CO_3 = 0.0590 \text{ mol } CO_3^{2-} \times \frac{1 \text{ mol } Na_2CO_3}{1 \text{ mol } CO_3^{2-}}$$

$$\times \frac{106 \text{ g } Na_2CO_3}{\text{mol } Na_2CO_3} = 6.25\text{g } Na_2CO_3$$

When you want a buffer with a certain pH range, choose one where the pH $= pK \pm 1$.

Example 15.12 A certain strain of bacteria can only grow in a buffer solution of pH 6.32. From the acids in Table 15.3, select the components for this buffer system.

Answer There are six possible systems.

Solution We select a weak acid/conjugate base pair in which the pK_a = pH ± 1 or one with a pK_a range of 5.32 to 7.32. From Table 15.3, we see that no monoprotic weak acid/conjugate base pairs are suitable. The following diprotic systems are:

pK_1 (H_2CO_3) = 6.37;	H_2CO_3/HCO_3^-
pK_2 ($H_2C_3H_2O_4$) = 5.69;	$HC_3H_2O_4^-/C_3H_2O_4^{2-}$
pK_2 (H_3PO_4) = 7.21;	$H_2PO_4^-/HPO_4^{2-}$
pK_2 ($H_2C_8H_4O_4$) = 5.4;	$HC_8H_4O_4^-/C_8H_4O_4^{2-}$
pK_2 ($H_2C_4H_4O_4$) = 5.61;	$HC_4H_4O_4^-/C_4H_4O_4^{2-}$
pK_2 (H_2SO_3) = 7.20;	HSO_3^-/SO_3^{2-}

Example 15.13 If the $HC_8H_4O_4^-/C_8H_4O_4^{2-}$ system were selected to prepare a buffer with a pH = 6.32 in Example 15.12. What is the $[HC_8H_4O_4^-]/[C_8H_4O_4^{2-}]$ ratio?

Answer $[HC_8H_4O_4^-]/[C_8H_4O_4^{2-}] = 0.12$

Solution The ionic equilibrium in solution is

$$HC_8H_4O_4^- \rightleftharpoons H^+ + C_8H_4O_4^{2-}$$

or more simply

$$HA^- \rightleftharpoons H^+ + A^{2-}$$

$$K_2 = \frac{[H^+][A^{2-}]}{[HA^-]} = 3.9 \times 10^{-6}$$

For a pH = 6.32, the desired

$$[H^+] = 10^{-6.32} = 4.79 \times 10^{-7}$$

$$\frac{[HA^-]}{[A^{2-}]} = \frac{[H^+]}{K_2} = \frac{4.79 \times 10^{-7}}{3.9 \times 10^{-6}} = 0.12$$

of from the Henderson-Hasselbalch equation

$$pH = pK_2 + \log \frac{[base]}{[acid]}$$

$$pK_2 = -\log (3.9 \times 10^{-6}) = 5.4$$

$$6.32 = 5.4 + \log \frac{[base]}{[acid]}$$

$$\frac{[base]}{[acid]} = \text{antilog}\,(6.32 - 5.4) = 8.31$$

$$\frac{[acid]}{[base]} = \frac{1}{8.31} = 0.12$$

SECTION 15.10 SLIGHTLY SOLUBLE SALTS

Objectives

- To describe the conditions necessary for calculating the solubility of "slightly soluble" salts.
- To calculate the solubility of a salt, given the K.
- To predict whether a salt precipitates, given the initial concentrations of the ions.

Focus

Like all other equilibrium constants, K, the constant for a slightly soluble salt depends on temperature. The equilibrium for the salt A_2B

$$A_2B(c) \rightleftharpoons 2A^+(aq) + B^{2-}(aq)$$

represents a solution, saturated with A^+ and B^- ions that are in equilibrium with solid A_2B. For this heterogeneous equilibrium,

$$K = [A^+]^2[B^{2-}]$$

The $[A_2B]$ does not appear in the equilibrium condition because the concentration of a solid is always constant (see text, Chapter 13.9).

Example 15.14 The molar solubility of A_2B is 1.82×10^{-6} mol/L. What is the equilibrium constant, K, for A_2B?

Answer 2.41×10^{-17}

Solution Since $[A_2B]$ is not in the equilibrium condition for K, we must look at the ionic concentrations of A^+ and B^{2-} in the solution. From the equation, 2 mol A^+ and 1 mol B^{2-} are in solution for each mole of A_2B that dissolves.

$$A_2B \rightleftharpoons 2A^+ + B^{2-}$$

If 1.82×10^{-6} mol A_2B dissolves per liter, then $2(1.82 \times 10^{-6})$ mol A^+ and 1.82×10^{-6} mol B^{2-} are present per liter of solution at equilibrium.

$$\begin{aligned} K &= [A^+]^2\,[B^{2-}] \\ &= [2(1.82 \times 10^{-6})]^2[1.82 \times 10^{-6}] \\ &= 2.41 \times 10^{-17} \end{aligned}$$

Example 15.15 Calculate the molar solubility of Ag_2CrO_4 in the presence of 0.0311 M Na_2CrO_4.

Answer 3.0×10^{-6} mol/L

Solution From Table 15.4, K of $Ag_2CrO_4 = 1.1 \times 10^{-12}$. Let's solve the problem as we did the weak acid equilibria.

Step 1.	equation	$Ag_2CrO_4 \rightleftharpoons$	$2Ag^+$ +	CrO_4^{2-}
Step 2.	$[\ ]i$		0	0.0311
Step 3.	$\Delta[\ ]$		$+2y$	$+y$

Step 4. $[\ \ ]\rightleftharpoons$ $\quad 2y \quad 0.0311 + y$

Step 5. $K = [Ag^+]^2 [CrO_4^{2-}] = 1.1 = 10^{-12}$

$K = (2y)^2 (0.0311 + y)$

Approximate $0.0311 >> y$.

$$1.1 \times 10^{-12} \simeq (2y)^2 (0.0311) = 0.124y^2$$

$$y \simeq 3.0 \times 10^{-6}$$

(The approximation is within the experimental error of 5%.)

$$2y = [Ag^+] = 6.0 \times 10^{-6} \text{ mol/L}$$

However, 2 mol/L Ag^+ are produced for every 1 mol/L of Ag_2CrO_4 that dissolves; therefore, the molar solubility of Ag_2CrO_4 is ½(6.0×10^{-6} mol/L) or 3.0×10^{-6} mol/L.

(*Note*: In calculations with slightly soluble salts, the concentration of an ion that originates from the salt is generally negligible compared to the concentration of the ion added from a second soluble salt. Therefore, $y \simeq 0$ when compared to the concentration of the second salt.)

We often use solubility product constants (also represented as K_{sp}) to determine whether a precipitate forms when the concentrations of two ions are given. The value for the reaction quotient, Q, is compared to the K. If $Q > K$, a precipitate forms until the product of the molar concentrations of the ions equals K. If $Q < K$, no precipitate forms because the solution is not yet saturated.

Example 15.16 A total of 0.00200 mol KI is added to 1 liter of a 1.00×10^{-3} M $Pb(NO_3)_2$ solution. Does a precipitate of PbI_2 form?

Answer No

Solution From Table 15.4, K for $PbI_2 = 1.39 \times 10^{-8}$.

$$PbI_2(c) \rightleftharpoons Pb^{2+} + 2I^-$$

In solution,

$$[Pb^{2+}] = 1.00 \times 10^{-3}$$

$$[I^-] = 2.00 \times 10^{-3}$$

$$Q = [Pb^{2+}][I^-]^2 = (1.00 \times 10^{-3})(2.00 \times 10^{-3})^2$$

$$Q = 4.00 \times 10^{-9} < K$$

Therefore, the solution is *not* saturated, and a precipitate does *not* form.

We also use solubility product constants to calculate the ionic concentration necessary for the precipitation of another ion.

Example 15.17 (a) What $[I^-]$ is required to form a saturated PbI_2 solution in Example 15.16?

(b) How many grams of KI are required?

(c) What happens in the solution if this amount of KI is exceeded?

Answer (a) 3.73×10^{-3} mol/L
(b) 0.619 g KI/L
(c) PbI_2 precipitates

Solution (a) $PbI_2(c) \rightleftharpoons Pb^{2+} + 2I^-$.

$$K = [Pb^{2+}][I^-]^2$$

At equilibrium, the $[Pb^{2+}] = 1.0 \times 10^{-3}$ and

$$[I^-] = \sqrt{\frac{K}{[Pb^{2+}]}} = \sqrt{\frac{1.39 \times 10^{-8}}{1.00 \times 10^{-3}}}$$

$$= 3.73 \times 10^{-3} \text{ mol/L for a saturated } PbI_2 \text{ solution}$$

(b) $$\text{grams KI/L} = 3.72 \times 10^{-3} \frac{\cancel{\text{mol I}^-}}{\text{L}} \times \frac{1 \cancel{\text{mol KI}}}{1 \cancel{\text{mol I}^-}} \times \frac{166 \text{ g KI}}{\cancel{\text{mol KI}}} = 0.619 \text{ g KI/L}.$$

(c) If more than 0.619 g KI/L is added, PbI_2 precipitates until there is equilibrium between the $PbI_2(c)$ and Pb^{2+} and I^- in solution.

Solubility product constants are used in the selective precipitation of two ions from a solution.

Example 15.18 A solution is 3.02×10^{-4} M $Pb(NO_3)$ and 1.18×10^{-4} M $Ba(NO_3)_2$.

(a) What concentration of SO_4^{2-} must be present in order for a precipitate to form?
(b) What is the precipitate?
(c) What is the molar concentration of Ba^{2+} when $PbSO_4$ begins to precipitate?
(d) What percentage of the Ba^{2+} precipitates from solution when Pb^{2+} begins to precipitate?

Answer (a) 9.15×10^{-17} mol/L
(b) $BaSO_4$
(c) 1.9×10^{-6} mol/L
(d) 98.3% of Ba^{2+} precipitated

Solution This could be a tough one, but let's go slowly. This is similar to the text, Example 33.

(a) First, we know from Table 15.4 that

$$BaSO_4 \; (K = 1.08 \times 10^{-10})$$

is less soluble than

$$PbSO_4 \; (K = 1.7 \times 10^{-8})$$

To determine which precipitates first, we calculate the $[SO_4^{2-}]$ needed for the precipitation of Ba^{2+} and Pb^{2+}.

$$BaSO_4(c) \rightleftharpoons Ba^{2+} + SO_4^{2-} \qquad PbSO_4(c) \rightleftharpoons Pb^{2+} + SO_4^{2-}$$

$$K = [Ba^{2+}][SO_4^{2-}] \qquad K = [Pb^{2+}][SO_4^{2-}]$$

$$[SO_4^{2-}] = \frac{1.08 \times 10^{-10}}{1.18 \times 10^{-4}} = 9.15 \times 10^{-7} \text{ mol } SO_4^{2-}\text{/L for } BaSO_4 \text{ precipitation.}$$

$$[SO_4^{2-}] = \frac{1.7 \times 10^{-8}}{3.02 \times 10^{-4}} = 5.6 \times 10^{-5} \text{ mol } SO_4^{2-}\text{/L for } PbSO_4 \text{ precipitation.}$$

Less SO_4^{2-} is required for Ba^{2+} precipitation; therefore,

(b) $BaSO_4$ precipitates first.

(c) When $PbSO_4$ begins to precipitate, the $[SO_4^{2-}] = 5.6 \times 10^{-5}$. The $[Ba^{2+}]$ at this point is less than its original 1.18×10^{-4} because the excess SO_4^{2-} pushes the $BaSO_4$ equilibrium farther to the left.

$$BaSO_4 \rightleftharpoons Ba^{2+} + SO_4^{2-}$$

At $[SO_4^{2-}] = 5.6 \times 10^{-5}$,

$$[Ba^{2+}] = \frac{K}{[SO_4^{2-}]} = \frac{1.08 \times 10^{-10}}{5.6 \times 10^{-5}} = 1.9 \times 10^{-6} \text{ mol/L}$$

(d) The $[Ba^{2+}]$ remaining when $[Pb^{2+}]$ begins to precipitate is (from part (c)) 1.9×10^{-6} mol/L. The $[Ba^{2+}]$ that precipitates equals the original $[Ba^{2+}]$ minus the remaining $[Ba^{2+}]$ or

$$1.18 \times 10^{-4} \text{ mol/L} - 1.9 \times 10^{-6} \text{ mol/L} = 1.16 \times 10^{-4} \text{ mol } Ba^{2+}\text{/L}$$

$$\%Ba^{2+} \text{ precipitated} = \frac{1.16 \times 10^{-4}}{1.18 \times 10^{-4}} \times 100 = 98.3\%$$

SECTION 15.11 COMPLEX IONS

Objectives

- To characterize a complex ion.
- To calculate equilibrium concentrations or constants from experimental data.

Focus

A **complex ion** consists of a transition metal ion bonded to a number of molecules or anions. The molecules or ions can dissociate from the metal ion to establish an equilibrium. For CdI_4^{2-},

$$CdI_4^{2-} \rightleftharpoons Cd^{2+} + 4I^-$$

$$K = \frac{[Cd^{2+}][I^-]^4}{[CdI_4^{2-}]}$$

K is called a dissociation constant.

Example 15.19 The deep blue $Cu(NH_3)_4^{2+}$ complex ion identifies the presence of Cu^{2+}. A solution is 2.31×10^{-2} M Cu^{2+}. What must be the $[NH_3]$ at equilibrium in order to have $[Cu^{2+}] = [Cu(NH_3)_4^{2+}]$?

Answer 8.1×10^{-4} mol/L

Solution $Cu(NH_3)_4^{2+} \rightleftharpoons Cu^{2+} + 4NH_3$

$$K = \frac{[Cu^{2+}][NH_3]^4}{[Cu(NH_3)_4^{2+}]} = 4.4 \times 10^{-13} \quad \text{(Table 15.5)}$$

For $[Cu^{2+}] = [Cu(NH_3)_4^{2+}]$ at equilibrium, then

$$4.4 \times 10^{-13} = [NH_3]^4$$

$$[NH_3] = 8.1 \times 10^{-4} \text{ mol/L}$$

QUESTIONS

Section 15.2

15.1 What is the $[H^+]$ for pure water at 25°C?

15.2 Calculate the $[H^+]$ of 3.20×10^{-4} M $Ba(OH)_2$, a strong base. Note that 1 mol $Ba(OH)_2$ produces 2 mol OH^-.

Section 15.3

15.3 The K_a for butyric acid, $HC_4H_7O_2$, is 1.5×10^{-5}. Calculate the pK_a.

15.4 The pH of a propionic acid solution is 4.20. Determine the $[H^+]$ and $[OH^-]$.

15.5 The K_w for water at 35°C is 2.06×10^{-14}. Calculate (a) the pK_w, (b) the $[H^+]$ and $[OH^-]$ in pure water at 35°C, (c) the pH of pure water at 35°C.

15.6 Calculate the pCl of a solution that contains 6.02×10^{-4} mol Cl^-/L solution.

Section 15.4

15.7 Calculate the pH at 25°C of (a) 0.00640 M HNO_3; (b) 0.00640 M KOH; (c) 0.00640 M $Ba(OH)_2$; (d) 4.18×10^{-3} M HI.

15.8 Calculate $[H^+]$ and $[OH^-]$ of a solution at 25°C having a (a) pH = 3.0; (b) pOH = 4.9; (c) pH = 6.1.

Section 15.5

15.9 Calculate the pH of the following solutions.

(a) 0.0177 M chloroacetic acid, $K_a = 1.40 \times 10^{-3}$. (Be careful on the approximation. Remember if the approximation of y is not less than 5% of the initial concentration of the weak acid or weak base, then you must solve by using the quadratic equation.)

(b) 3.31×10^{-1} M $HC_2H_3O_2$, $K_a = 1.75 \times 10^{-5}$

(c) 2.20×10^{-2} M NaCN, $K_b = 2.03 \times 10^{-5}$

Section 15.6

15.10 Calculate the K_a and pK_a for $(CH_3)_3NH^+$ in water.

15.11 Calculate the K_b and pK_b for NO_2^- in water.

15.12 Calculate the pH of a 0.0240 M hydroxylamine hydrochloride (or hydroxylammonium chloride), $HONH_3^+Cl^-$. (What is the conjugate of the ion affecting the pH of the solution?)

Section 15.7

15.13 Calculate the $[H^+]$, the succinate ion concentration, and the undissociated succinic acid concentration in 0.0200 M succinic acid, $H_2C_4H_4O_4$.

15.14 Calculate the basic ionization constant for the tartrate ion, $C_4H_4O_6^{2-}$.

15.15 Calculate the acidic ionization constant for $H_2NNH_3^+$.

Section 15.8

15.16 Calculate the H^+ and the benzoate ion concentration of a 0.0300 M benzoic acid solution. Compare this answer with Example 15.8. What is ΔpH? What can we say about the relative amounts of H^+ contributed by benzoic acid and HNO_3?

15.17 Calculate the H^+ and the benzoate ion concentration of a solution that is 0.0300 M benzoic acid and 0.400 M sodium benzoate. Compare this answer with Question 15.16 and Example 15.8.

Section 15.9

15.18 A mass of 2.03 g NaOH is added to 500 mL of 0.106 M propionic acid, $HC_3H_5O_2$. $K_a = 1.34 \times 10^{-5}$. Calculate the pH of the solution. Does this constitute a buffer system? Explain.

15.19 A volume of 500 mL of a solution is prepared that is 0.30 M in HCN and 0.30 M in NaCN.

(a) Calculate the pH of the solution.

(b) Calculate the pH after the addition of 0.080 mol HCl.

(c) Calculate the pH after the addition of 200 mL 0.040 M NaOH apart from part (b). Be careful in this problem because we have a dilution of the buffer solution. See this Study Guide, Section 13.11, for working this problem in terms of moles.

15.20 (a) What must be the $[HNO_2]/[NO_2^-]$ ratio to prepare a buffer with a pH of 4.00? Use the Henderson-Hasselbalch equation.

(b) If a 0.633 M $NaNO_2$ solution is already available, how many moles of HNO_2 are dissolved per liter?

Section 15.10

15.21 The solubility of $NiCO_3$ is 3.74×10^{-4} mol/L. What is the solubility product constant?

15.22 A volume of 600 mL of 2.07×10^{-4} M $MgSO_4$ is mixed with 300 mL of a solution at a pH of 10.0. Does $Mg(OH)_2$ precipitate? K of $Mg(OH)_2 = 1.8 \times 10^{-11}$. Remember to account for dilution.

15.23 (a) Calculate the molar solubility of CaF_2. K of $CaF_2 = 3.95 \times 10^{-11}$.

(b) How many moles of CaF_2 dissolve per liter in a 0.660 M NaF solution?

(c) What is the molar solubility of CaF_2 when the equilibrium $[Ca^{2+}]$ is adjusted to 0.660 M?

15.24 A solution contains 0.318 g Al^{3+} in 500 mL solution. Calculate the maximum pH that exists without precipitation of $Al(OH)_3$.

Section 15.11

15.25 A solution is 0.0300 M in $AuCl_4^-$ and 1.00 M in Cl^-. Calculate the $[Au^{3+}]$ at equilibrium. K of $AuCl_4^- = 5.0 \times 10^{-22}$. (Treat this calculation as a weak acid equilibrium.)

ANSWERS

15.1 $[H^+] = 1.00 \times 10^{-7}$

15.2 $[H^+] = 1.56 \times 10^{-11}$

15.3 $pK_a = 4.82$

15.4 $[H^+] = 6.31 \times 10^{-5}$
$[OH^-] = 1.58 \times 10^{-10}$

15.5 (a) $pK_w = 13.69$
(b) $[H^+] = [OH^-] = 1.44 \times 10^{-7}$
(c) pH = 6.84

15.6 pCl = 3.22

15.7 (a) 2.19
(b) 11.81
(c) 12.11
(d) 2.38

15.8 (a) $[H^+] = 1.0 \times 10^{-3}$; $[OH^-] = 1.0 \times 10^{-11}$
(b) $[H^+] = 7.94 \times 10^{-10}$; $[OH^-] = 1.26 \times 10^{-5}$
(c) $[H^+] = 7.94 \times 10^{-7}$; $[OH^-] = 1.26 \times 10^{-8}$

15.9 (a) pH = 2.36
(b) pH = 2.62
(c) pH = 10.83

15.10 $K_a = 1.55 \times 10^{-10}$
$pK_a = 9.81$

15.11 $K_b = 2.22 \times 10^{-10}$
$pK_b = 10.65$

15.12 pH = 3.83; $HONH_2$ (hydroxylamine)

15.13 $[H^+] = 1.14 \times 10^{-3}$ mol/L
$[X^{2-}] = [C_4H_4O_4^{2-}] = 2.47 \times 10^{-6}$ mol/L
$[H_2X] = [H_2C_4H_4O_4] = 1.89 \times 10^{-2}$ mol/L

15.14 $K_b = 2.20 \times 10^{-10}$

15.15 $K_a = 1.1 \times 10^{-8}$

15.16 $[H^+] = [X^-] \simeq 1.39 \times 10^{-3}$
$\Delta pH = 2.46$
The strong acid, HNO_3, contributes to most of the H^+, benzoic acid contributes very little to H^+ in solution.

15.17 $[H^+] = 4.85 \times 10^{-6}$ mol/L
$[C_7H_5O_2^-] \simeq 0.40$ mol/L
The presence of the benzoate ion reduces the amount of H^+ in solution.

15.18 pH = 6.24. Yes, the addition of OH^- shifts the reaction to the right, forming the conjugate base.

15.19 (a) pH = 9.31

(b) pH = 8.79

(c) pH = 9.35

15.20 (a) $\frac{[HNO_2]}{[NO_2^-]} = 0.224$

(b) 0.142 mol HNO_2

15.21 $K = 1.40 \times 10^{-7}$

15.22 $Q = 1.53 \times 10^{-13} < K = 1.8 \times 10^{-11}$
$Mg(OH)_2$ does not precipitate

15.23 (a) 2.14×10^{-4}

(b) 9.07×10^{-11}

(c) 3.87×10^{-6}

15.24 pH = 3.88

15.25 $[Au^{3+}] = 1.5 \times 10^{-23}$

Chapter 16

ANALYSIS OF IONS IN SOLUTIONS

SECTION 16.1 GRAVIMETRIC ANALYSIS

Objective

- To calculate the composition of a sample by analyzing its precipitate.

Focus

In order to analyze the composition of a substance gravimetrically, we must produce a precipitate of known composition. The precipitate is filtered, washed, dried, and then weighed on an analytical balance, one capable of weighing to ±0.0001 gram.

Example 16.1 A 1.4920 g sample of an impure hydrate of Ag_2TeO_4 is dissolved in water. Excess Cl^- is added from a NaCl solution, forming a white precipitate of AgCl. The washed and dried AgCl precipitate weighs 0.6296 g.

$$Ag_2TeO_4 + 2NaCl \rightarrow 2AgCl + Na_2TeO_4$$

(a) Calculate the mass of Ag in the sample.
(b) Determine the percent Ag_2TeO_4.

Answer (a) 0.4739 g Ag; (b) 59.97% Ag_2TeO_4

Solution (a) One mol Ag is present in each mol AgCl precipitate. Since we have not seen any calculations using dimensional analysis and a balanced equation for quite some time, we will go through this problem slowly (refer to Chapter 4).

(a) Given quantity: 0.6296 g AgCl
Desired quantity: g Ag
Conversion factor:

1 mol AgCl yields 1 mol Ag

$$1\,\cancel{\text{mol AgCl}} \times \frac{143.321\text{ g AgCl}}{1\,\cancel{\text{mol AgCl}}} \text{ yields } 1\,\cancel{\text{mol Ag}} \times \frac{107.868\text{ g Ag}}{1\,\cancel{\text{mol Ag}}}$$

The conversion factor to change g AgCl to g Ag is

$$\frac{107.868\text{ g Ag}}{143.321\text{ g AgCl}}$$

$$\text{desired} = \text{given} \times \text{conversion factor:}$$

$$\text{mass Ag} = 0.6296\ \cancel{\text{g AgCl}} \times \frac{107.868\ \text{g Ag}}{143.321\ \cancel{\text{g AgCl}}} = 0.4739\ \text{g Ag}$$

(b) The balanced equation shows

$$Ag_2TeO_4 \rightarrow 2AgCl$$

Given quantity: 0.6296 g AgCl

Desired quantity: g Ag_2TeO_4. From this, we calculate the percentage of Ag_2TeO_4 in the sample's total weight.

Conversion factor: From the balanced equation,

$$1\ \text{mol}\ Ag_2TeO_4\ \text{yields}\ 2\ \text{mol AgCl}$$

$$1\ \cancel{\text{mol}\ Ag_2TeO_4} \times \frac{407.34\ \text{g}\ Ag_2TeO_4}{1\ \cancel{\text{mol}\ Ag_2TeO_4}}\ \text{yields}\ 2\ \cancel{\text{mol AgCl}} \times \frac{143.321\ \text{g AgCl}}{1\ \cancel{\text{mol AgCl}}}$$

The conversion factor to convert from g AgCl to Ag_2TeO_4 is

$$\frac{407.34\ \text{g}\ Ag_2TeO_4}{2 \times 143.321\ \text{g AgCl}}$$

$$\text{desired} = \text{given} \times \text{conversion factor}$$

$$\text{mass}\ Ag_2TeO_4 = 0.6296\ \cancel{\text{g AgCl}} \times \frac{407.34\ \text{g}\ Ag_2TeO_4}{2 \times 143.321\ \cancel{\text{g AgCl}}}$$

$$= 0.8947\ \text{g}\ Ag_2TeO_4$$

Calculate the percentage Ag_2TeO_4 in the sample. The percentage of Ag_2TeO_4 in the sample is

$$\frac{0.8947\ \text{g}\ Ag_2TeO_4}{1.4920\ \text{g sample}} \times 100\% = 59.97\%\ Ag_2TeO_4$$

SECTION 16.2 VOLUMETRIC ANALYSIS AND TITRATIONS

Objectives

- To explain the technique of titration.
- To calculate unknowns, such as molarity, mass, and percent purity of a substance, by means of a titration.

Focus

Titration is an analytical technique whereby a measured volume of a standard solution (one of known concentration) is added to a solution of unknown concentration. A **buret** is used for the addition of a chemically equivalent amount (determined by the molar ratio in the balanced equation) of the standard solution to the unknown.

Example 16.2 A volume of 27.21 mL of 0.2241 M HNO_3 reacts with 50.00 mL $Ba(OH)_2$.

(a) What is the molarity of the $Ba(OH)_2$ solution?

(b) How many milligrams of $Ba(OH)_2$ are present in the solution?

Answer (a) 0.06098 M $Ba(OH)_2$; (b) 522.3 mg $Ba(OH)_2$

Solution (a) To know how the $Ba(OH)_2$ reacts with HNO_3, we need a balanced equation (always):

$$Ba(OH)_2 + 2HNO_3 \rightarrow Ba(NO_3)_2 + 2H_2O$$

To calculate the molarity of $Ba(OH)_2$, we need to find the number of mmoles of $Ba(OH)_2$ in the 50.00 mL because molarity is defined as mmol $Ba(OH)_2$/mL soln. Remember that mol/L or mmol/mL expresses molarity.

$$\text{Given quantity:}\quad 27.21\,\cancel{\text{mL}} \times \frac{0.2241 \text{ mmol } HNO_3}{\cancel{\text{mL}}} = 6.098 \text{ mmol } HNO_3$$

$$\text{Desired quantity:}\quad \text{M } Ba(OH)_2 \text{ or } \frac{\text{mmol } Ba(OH)_2}{\text{mL soln}}$$

Conversion factor: From the balanced equation,

1 mmol $Ba(OH)_2$ reacts with 2 mmol HNO_3 or

$$\frac{1 \text{ mmol } Ba(OH)_2}{2 \text{ mmol } HNO_3}$$

$$\text{mmol } Ba(OH)_2 = 6.098\,\cancel{\text{mmol } HNO_3} \times \frac{1 \text{ mmol } Ba(OH)_2}{2\,\cancel{\text{mmol } HNO_3}}$$

$$= 3.049 \text{ mmol } Ba(OH)_2$$

Finally,

$$M = \frac{3.049 \text{ mmol } Ba(OH)_2}{50.00 \text{ mL soln}} = 0.06098\ \frac{\text{mmol } Ba(OH)_2}{\text{mL soln}}$$

(b) Given quantity: 3.049 mmol $Ba(OH)_2$
Desired quantity: mg $Ba(OH)_2$
Conversion factor:

1 mmol $Ba(OH)_2$ = 171.3 mg $Ba(OH)_2$

$$\frac{171.3 \text{ mg } Ba(OH)_2}{1 \text{ mmol } Ba(OH)_2}$$

$$\text{mg } Ba(OH)_2 = 3.049\,\cancel{\text{mmol } Ba(OH)_2} \times \frac{171.3 \text{ mg } Ba(OH)_2}{1\,\cancel{\text{mmol } Ba(OH)_2}}$$

$$= 522.3 \text{ mg } Ba(OH)_2$$

Example 16.3 A 75.0 mg sample of impure $Ca(OH)_2$ dissolves in 50.00 mL of 0.0331 M HCl. 3.95 mL of 0.0430 M NaOH neutralizes the excess HCl. Calculate the percent $Ca(OH)_2$ in the sample.

$$Ca(OH)_2 + 2\,HCl \rightarrow CaCl_2 + 2H_2O$$

Answer 73.6% $Ca(OH)_2$

Solution We are looking for mass $Ca(OH)_2$ in the original sample. We find this by calculating the amount of H^+ (from HCl) it reacts with. The mmol H^+ reacting with the $Ca(OH)_2$ is determined by subtracting the excess mmol H^+ from the total mmol H^+ in which the $Ca(OH)_2$ is dissolved.

$$\text{Total mmol H}^+ \text{ added:} \quad 50.00 \cancel{\text{mL}} \times 0.0331 \frac{\text{mmol H}^+}{\cancel{\text{mL}}} = 1.66 \text{ mmol H}^+$$

Because there is a 1 to 1 mole ratio of OH^- to H^+ in the balanced equation, we can find excess mmol H^+ by calculating mmol OH^- in solution (from the neutralization by NaOH):

$$H^+ + OH^- \text{ (from NaOH)} \rightarrow H_2O$$

$$3.95 \cancel{\text{mL OH}^-} \times 0.0430 \frac{\text{mmol OH}^-}{\cancel{\text{mL OH}^-}} = 0.170 \text{ mmol OH}^-$$

Thus there is also 0.170 mmol H^+ excess. The mmol H^+ that reacts with $Ca(OH)_2$ in the sample is

$$\text{total mmol H}^+ - \text{excess mmol H}^+ = \text{mmol H}^+ \text{ that reacts with Ca(OH)}_2$$

$$1.66 \text{ mmol H}^+ - 0.170 \text{ mmol H}^+ = 1.49 \text{ mmol H}^+$$

We can now find the mass of $Ca(OH)_2$ in the sample, given 1.49 mmol H^+.

$$Ca(OH)_2 + 2H^+ \rightarrow Ca^{2+} + 2H_2O$$

Conversion factor:

$$1 \cancel{\text{mmol Ca(OH)}_2} \times \frac{74.1 \text{ mg Ca(OH)}_2}{\cancel{\text{mmol Ca(OH)}_2}} \quad \text{reacts with 2 mmol H}^+$$

or

$$\frac{74.1 \text{ mg Ca(OH)}_2}{2 \text{ mmol H}^+}$$

$$\text{mass Ca(OH)}_2 = 1.49 \cancel{\text{mmol H}^+} \times \frac{74.1 \text{ mg Ca(OH)}_2}{2 \cancel{\text{mmol H}^+}}$$

$$= 55.2 \text{ mg Ca(OH)}_2$$

$$\% \text{ purity} = \frac{55.2 \text{ mg Ca(OH)}_2}{75.0 \text{ mg sample}} \times 100\% = 73.6\%$$

SECTION 16.3 ACID-BASE INDICATORS

Objective

- To explain the chemical properties of an acid-base indicator in a titration.

Focus

An **indicator** is a weak acid (HIn) which has a different color from its conjugate base (In^-). The color of a solution containing the indicator depends on the $[HIn]/[In^-]$ ratio. If $[HIn]>>[In^-]$, the solution has the color of the acid, and if $[HIn]<<[In^-]$, the solution has the color of the base.

SECTION 16.4 TITRATION CURVES

Objectives

- To define and distinguish between the equivalence point and end point in a titration.
- To select an appropriate indicator for an acid-base titration.

Focus

The **equivalence point** in an acid-base titration is the point at which equivalent amounts (according to the balanced equation) of acid and base are present in the reacting mixture. It is sometimes called the *stoichiometric point* in a titration.

The **end point** in an acid-base titration is that point at which the indicator changes color. Ideally, we select an indicator that has an end point which coincides with the equivalence point in the reaction. To choose an indicator, first calculate the pH of the solution at the equivalence point and then select an indicator that has a color change range near the calculated pH.

At the equivalence point of a strong acid-strong base titration, the pH = 7.0 because $H^+ + OH^- \rightarrow H_2O$.

At the equivalence point of a weak acid-strong base titration, the pH > 7 because the conjugate base of the weak acid reacts with water:

$$B + H_2O \rightleftharpoons HB^+ + OH^-$$

The equilibrium condition is

$$K_b = \frac{[HB^+][OH^-]}{[B]} = \frac{K_w}{K_a(HB^+)}$$

At the equivalence point of a weak base-strong acid titration, the pH < 7 because the conjugate acid of the weak base reacts with water:

$$HB^+ + H_2O \rightleftharpoons B + H_3O^+$$

The equilibrium condition is

$$K_a = \frac{[B][H_3O^+]}{[HB^+]} = \frac{K_w}{K_b(B)}$$

Example 16.4 Calculate the pH at the equivalence point in the titration of 10.0 mmol pyridine, C_6H_5N, with 0.100 M HCl. The final volume of the solution is 95.7 mL. From text Table 15.3, $K_b = 1.78 \times 10^{-9}$. From text Table 16.1, select an appropriate indicator.

Answer pH = 3.12; methyl orange

Solution At the equivalence point, the conjugate *acid* of C_6H_5N, $C_6H_5NH^+$, is present. See Study Guide, Section 13.11, for a step-by-step problem-solving procedure.

Step 1.	equation	$C_6H_5NH^+$	+	H_2O	$\rightleftharpoons$	C_6H_5N	+	H_3O^+
Step 2.	[ ]*i*	$\frac{10.0\ \text{mmol}}{95.7\ \text{mL}}$				0		0
Step 3.	Δ[ ]	$-y$				$+y$		$+y$
Step 4.	[ ]$\rightleftharpoons$	$0.104 - y$				y		y

Step 5. $K_a = \frac{K_w}{K_b} = \frac{1.00 \times 10^{-14}}{1.78 \times 10^{-9}} = 5.62 \times 10^{-6} = \frac{(y)(y)}{0.104 - y}$

Assume $0.104 >> y$:

$$5.62 \times 10^{-6} = \frac{y^2}{0.104}$$

$$y = [H_3O^+] = 7.65 \times 10^{-4}$$

$$pH = 3.12$$

According to Table 16.1 an appropriate indicator is methyl orange, which has a pH range of 3 to 4.4.

SECTION 16.5 TITRATION EXAMPLES

Objective

- To use titration procedures to calculate unknown quantities of a substance.

Focus

Practicing on different types of titrations is the best way to solve for an unknown in a titration problem. The text illustrates six examples. Let's look at three more.

Example 16.5 A mass of 0.4936 g NaCl is dissolved and diluted to 100 mL. A volume of 36.2 mL of this solution precipitates the Hg_2^{2+} contained in 50.0 mL.

$$Hg_2^{2+} + 2Cl^- \rightarrow Hg_2Cl_2(c)$$

(a) Determine the molarity of the NaCl solution.
(b) Calculate the mass of Hg_2^{2+} in the 50.0 mL.

Answer (a) 8.45×10^{-2} M NaCl; (b) 614 mg Hg_2^{2+}

Solution

(a) Given quantity: $\frac{0.4935 \text{ g NaCl}}{0.100 \text{ L soln}}$

Desired quantity: $\frac{\text{mol NaCl}}{\text{L soln}}$

Conversion factor:

$$1 \text{ mol NaCl} = 58.44 \text{ g NaCl}$$

$$\text{M NaCl} = \frac{0.4935 \text{ g NaCl}}{0.100 \text{ L soln}} \times \frac{1 \text{ mol NaCl}}{58.44 \text{ g NaCl}}$$

$$= 8.45 \times 10^{-2} \frac{\text{mol NaCl}}{\text{L soln}}$$

(b) Given quantity: $36.2 \text{ mL soln} \times \frac{8.45 \times 10^{-2} \text{ mmol}}{\text{mL soln}} = 3.06 \text{ mmol NaCl}^-$

Desired quantity: mg Hg_2^{2+}

Conversion factor: From the balanced equation,

$$1 \text{ mmol } Hg_2^{2+} \text{ reacts with } 2 \text{ mmol } Cl^-$$

$$1\ \cancel{\text{mmol } Hg_2^{2+}} \times \frac{401.2 \text{ mg } Hg_2^{2+}}{1\ \cancel{\text{mmol } Hg_2^{2+}}} \text{ reacts with } 2 \text{ mmol } Cl^-$$

The conversion factor is

$$\frac{401.2 \text{ mg } Hg_2^{2+}}{2 \text{ mmol } Cl^-}$$

$$\text{mass } Hg_2^{2+} = 3.06\ \cancel{\text{mmol } Cl^-} \times \frac{401.2 \text{ mg } Hg_2^{2+}}{2\ \cancel{\text{mmol } Cl^-}}$$

$$= 614 \text{ mg } Hg_2^{2+}$$

Example 16.6 A volume of 16.7 mL of 0.231 M NaOH neutralizes the acid generated from the reaction of H_2SO_3 with a 25.0 mL solution of $KMnO_4$. The solution is boiled, decomposing the excess H_2SO_3 into H_2O and SO_2. Calculate the molarity of the $KMnO_4$ solution.

$$2MnO_4^- + 5H_2SO_3 \rightarrow 2Mn^{2+} + 4H^+ + 5SO_4^{2-} + 3H_2O$$

Answer 0.0772 M $KMnO_4$

Solution Boiling removes any acidity in the solution caused by the excess weak acid, H_2SO_3. If we determine the mol H^+ generated, we can calculate the mol MnO_4^- present from the balanced equation.

Given quantity: $16.7\ \cancel{\text{mL NaOH}} \times \frac{0.231 \text{ mmol NaOH}}{\cancel{\text{mL NaOH}}} = 3.86 \text{ mmol } OH^-$

Desired quantity: $\frac{\text{mmol } KMnO_4}{\text{mL soln}}$

Conversion factors: From the balanced equation

$$2 \text{ mmol } MnO_4^- \text{ yields } 4 \text{ mmol } H^+$$

and because $OH^- + H^+ \rightarrow H_2O$

$$1 \text{ mmol } H^+ \text{ reacts with } 1 \text{ mmol } OH^-$$

$$\text{mmol } MnO_4^- = 3.86\ \cancel{\text{mmol } OH^-} \times \frac{1\ \cancel{\text{mmol } H^+}}{1\ \cancel{\text{mmol } OH^-}} \times \frac{2 \text{ mmol } MnO_4^-}{4\ \cancel{\text{mmol } H^+}}$$

$$= 1.93 \text{ mmol } MnO_4^-$$

Because 1.93 mmol MnO_4^- is equivalent to 1.93 mmol $KMnO_4$, the molarity is

$$\text{M } KMnO_4 = \frac{1.93 \text{ mmol } KMnO_4}{25.0 \text{ mL soln}} = 0.0772 \frac{\text{mmol } KMnO_4}{\text{mL soln}}$$

Example 16.7 Calculate the volume of 0.239 M $H_2C_2O_4$ needed to reach the equivalence point in the titration of 50.0 mL 0.0772 M $KMnO_4$.

$$2MnO_4^- + 5H_2C_2O_4 + 6\,H^+ \rightarrow 2Mn^{2+} + 10CO_2 + 8H_2O$$

Answer 40.4 mL of 0.239 M $H_2C_2O_4$

Solution Use the balanced equation after converting to mol $KMnO_4$. Convert M to mmol/mL or mol/L to solve the problem.

Given quantity: $50.0\ \text{mL KMnO}_4 \times \dfrac{0.0772\ \text{mmol KMnO}_4}{\text{mL KMnO}_4} = 3.86\ \text{mmol KMnO}_4$

Desired quantity: mL $H_2C_2O_4$

Conversion factors:

2 mmol $KMnO_4$ reacts with 5 mmol $H_2C_2O_4$ or

$$\frac{5\ \text{mmol H}_2\text{C}_2\text{O}_4}{2\ \text{mmol KMnO}_4}$$

0.239 mmol $H_2C_2O_4$ are in 1 mL $H_2C_2O_4$ or

$$\frac{1\ \text{mL H}_2\text{C}_2\text{O}_4}{0.239\ \text{mmol H}_2\text{C}_2\text{O}_4}$$

$$\text{mL H}_2\text{C}_2\text{O}_4 = 3.86\ \text{mmol KMnO}_4 \times \frac{5\ \text{mmol H}_2\text{C}_2\text{O}_4}{2\ \text{mmol KMnO}_4} \times \frac{1\ \text{mL H}_2\text{C}_2\text{O}_4}{0.239\ \text{mmol H}_2\text{C}_2\text{O}_4} = 40.4\ \text{mL H}_2\text{C}_2\text{O}_4$$

SECTION 16.6 BALANCING OXIDATION-REDUCTION EQUATIONS

Objective

- To balance oxidation-reduction equations by the ion-electron method.

Focus

Balancing oxidation-reduction equations by inspection can be very difficult. It is made easier if the balancing is done in steps.

Example 16.8 Balance the equation

$$P_4 + NO_3^- \rightarrow H_3PO_4 + NO$$

Answer $8H_2O + 3P_4 + 20H^+ + 20NO_3^- \rightarrow 12H_3PO_4 + 20NO$

Solution

Step 1. Write two unbalanced partial equations. Look for the substances that change oxidation state. P changes from 0 in P_4 to +5 in H_3PO_4, and N changes from +5 in NO_3^- to +2 in NO.

$$\overset{0}{P_4} \rightarrow \overset{+5}{H_3PO_4} \qquad\qquad \overset{+5}{NO_3^-} \rightarrow \overset{+2}{NO}$$

Step 2. Balance each partial equation with respect to mass.

(a) First, balance the atoms that change oxidation state:

$P_4 \rightarrow 4H_3PO_4$ | $NO_3^- \rightarrow NO$

4 P atoms on each side | 1 N atom on each side

(b) Next, balance the oxygen atoms with an equal number of H_2O molecules (remember, this is an aqueous solution and H_2O is available). We do not balance the oxygen atoms with O_2 because it is not a reactant or a product. Do not change subscripts:

$16H_2O + P_4 \rightarrow 4H_3PO_4$ | $NO_3^- \rightarrow NO + 2H_2O$

16 O atoms on each side | 3 O atoms on each side

(c) Balance the hydrogen atoms with H^+ ions since H_2 is not a reactant or a product:

$16H_2O + P_4 \rightarrow 4H_3PO_4 + 20H^+$ | $4H^+ + NO_3^- \rightarrow NO + 2H_2O$

32 H atoms on each side | 4 H atoms on each side

Step 3. Balance the charge of each partial equation using only electrons—not any form of positive charge. Electrons, not protons, are lost and gained in an oxidation-reduction reaction.

There are 20 plus charges on the right hand side (from $20H^+$), so add $20e$ to the right side | There is a net charge of plus 3 on the left, so add $3e$ to the left side

$16\,H_2O + P_4 \rightarrow 4H_3PO_4 + 20H^+ + 20e^-$ | $3e^- + 4H^+ + NO_3^- \rightarrow NO + 2H_2O$

zero charge on each side | zero charge on each side

Step 4. A complete equation does not generate excess electrons (to avoid electrocution!); therefore, the total number of electrons lost by the oxidation of P_4 must equal those gained in the reduction of NO_3^-. Multiply the left equation by 3 and the right equation by 20:

$3(16H_2O + P_4 \rightarrow 4H_3PO_4 + 20H^+ + 20e^-)$ | $20(3e^- + 4H^+ + NO_3^- \rightarrow NO + 2H_2O)$

60 electrons lost | 60 electrons gained

Step 5. Add the two balanced partial equations:

$$48H_2O + 3P_4 + 60e^- + 80H^+ + 20NO_3^- \rightarrow 12H_3PO_1 + 60H^+ + 60e + 20NO + 40H_2O$$

Step 6. Cancel the substances common to each side of the equation: $60e^-$, $60H^+$, $40H_2O$.

$$8H_2O + 3P_4 + 20H^+ + 20NO_3^- \rightarrow 12H_3PO_4 + 20NO$$

It would be very difficult to balance this equation by inspection.

In an example where the reaction occurs in a basic solution, we balance the equation the same as we do for acidic solutions. However, we adjust the *final equation* for basicity by adding the same number of OH^- ions as there are H^+ in the balanced equation.

Example 16.9 Balance the equation

$$Cr^{3+} + O_2^{2-} \rightarrow CrO_4^{2-}$$

in a basic solution.

Answer $4OH^- + 2Cr^{3+} + 3O_2^{2-} \rightarrow 2CrO_4^{2-} + 2H_2O$

Solution

Step 1. Write two partial equations

$Cr^{3+} \rightarrow CrO_4^{2-}$

(Cr changes oxidation state from +3 to +6)

$O_2^{2-} \rightarrow H_2O$ (remember H_2O is always available!)

(O changes oxidation state from −1 to −2)

Step 2. (a) Balance the atoms changing oxidation state

$Cr^{3+} \rightarrow CrO_4^{2-}$ $\qquad$ $O_2^{2-} \rightarrow 2H_2O$

(b) Balance O atoms with H_2O:

$4H_2O + Cr^{3+} \rightarrow CrO_4^{2-}$ $\qquad$ $O_2^{2-} \rightarrow 2H_2O$

(c) Balance H atoms with H^+:

$4H_2O + Cr^{3+} \rightarrow CrO_4^{2-} + 8H^+$ $\qquad$ $4H^+ + O_2^{2-} \rightarrow 2H_2O$

Step 3. Balance charge with electrons:

$4H_2O + Cr^{3+} \rightarrow CrO_4^{2-} + 8H^+ + 3e^-$ $\qquad$ $2e^- + 4H^+ + O_2^{2-} \rightarrow 2H_2O$

(As a check on balancing electrons, remember that one partial equation *must* show oxidation, the other reduction.)

Step 4. Electrons lost = electrons gained:

$2(4H_2O + Cr^{3+} \rightarrow CrO_4^{2-} + 8H^+ + 3e^-)$ $\qquad$ $3(2e^- + 4H^+ + O_2^{2-} \rightarrow 2H_2O)$

Step 5. Add the balanced partial equations:

$$8H_2O + 2Cr^{3+} + 6e^- + 12H^+ + 3O_2^{2-} \rightarrow 2CrO_4^{2-} + 16H^+ + 6e^- + 6H_2O$$

Step 6. Cancel common substances appearing on both sides of the equation: $6H_2O$, $6e^-$, $12H^+$.

$$2H_2O + 2Cr^{3+} + 3O_2^{2-} \rightarrow 2CrO_4^{2-} + 4H^+$$

Step 7. Make the solution basic by adding $4OH^-$ to each side; in effect this neutralizes the $4H^+$.

$$4OH^- + 2H_2O + 2Cr^{3+} + 3O_2^{2-} \rightarrow 2CrO_4^{2-} + \underbrace{4H^+ + 4OH^-}_{\rightarrow 4H_2O}$$

and simplify by canceling same quantities ($2H_2O$'s):

$$4OH^- + 2Cr^{3+} + 3\,O_2^{2-} \rightarrow 2CrO_4^{2-} + 2H_2O$$

SECTION 16.7 EQUIVALENT WEIGHT AND NORMALITY

Objectives

- To define an equivalent of a substance.
- To determine the equivalent weight of acids, bases, and oxidizing and reducing agents.
- To calculate the normality of a solution.
- To determine the quantity of an unknown in a solution by volumetric analysis, using the principles of equivalent weight and normality.

Focus

One **equivalent** of a substance releases or accepts 1 mole of unit charge; one equivalent of an acid releases 1 mol H^+ and a base accepts 1 mol H^+; one equivalent of an oxidizing agent accepts 1 mol e^- and a reducing agent releases 1 mol e^-.

Two equivalents of an acid release 2 mol H^+; if their source is from 1 mole acid, there are 2 equivalents (eq) acid per mole acid or

$$2 \text{ eq acid/mol acid}$$

The **equivalent weight** is the *mass* that releases or accepts 1 mole of unit charge (H^+, OH^-, or e^-) and therefore has units of g/eq.

If an acid has a formula weight of 98.0 g/mol and there are 2 eq acid/mol acid, then the acid's equivalent weight is

$$\frac{98.0 \text{ g acid}}{\cancel{\text{mol acid}}} \times \frac{\cancel{\text{mol acid}}}{2 \text{ eq acid}} = \frac{49.0 \text{ g acid}}{\text{eq acid}}$$

The conditions of a chemical reaction determine the moles of unit charge a substance releases or accepts. Therefore, you need to know the reaction in which the substance participates before you can determine its equivalent weight.

Example 16.10 Determine the equivalent weight of $Ba(OH)_2$ in a reaction in which both OH^- ions are neutralized.

Answer 85.5 g/eq

Solution One mol $Ba(OH)_2$, or 171 g $Ba(OH)_2$, accepts 2 mol H^+ for neutralization; thus there are 2 eq $Ba(OH)_2$ in 1 mol $Ba(OH)_2$. Since the formula weight of $Ba(OH)_2$ is 171 g/mol, then the equivalent weight is

$$\text{eq wt } Ba(OH)_2 = \frac{171 \text{ g } Ba(OH)_2}{\cancel{\text{mol } Ba(OH)_2}} \times \frac{1 \cancel{\text{ mol } Ba(OH)_2}}{2 \text{ eq } Ba(OH)_2}$$

$$= \frac{85.5 \text{ g } Ba(OH)_2}{\text{eq } Ba(OH)_2}$$

This means that 85.5 g $Ba(OH)_2$ neutralizes 1 mol H^+ or 1 eq H^+.

Example 16.11 In Example 16.8, P_4 oxidizes to H_3PO_4 according to the partial oxidation equation:

$$16H_2O + P_4 \rightarrow 4H_3PO_4 + 20H^+ + 20e^-$$

Calculate the equivalent weight of P_4.

Answer 6.1948 g/eq

Solution One mol P_4, or 123.89 g, donates 20 mol e^-; thus there are 20 eq P_4 per mol P_4.

$$\text{eq wt } P_4 = \frac{123.89 \text{ g } P_4}{\cancel{\text{mol } P_4}} \times \frac{\cancel{\text{mol } P_4}}{20 \text{ eq } P_4} = \frac{6.1948 \text{ g } P_4}{\text{eq } P_4}$$

The **normality**, N, of the solution is the number of equivalents (milliequivalents) of acid, base, or oxidizing or reducing agent dissolved per liter (milliliter) of solution. The units of N are

$$N = \text{eq/L} \quad \text{or} \quad \text{meq/mL}$$

In any chemical reaction, 1 equivalent of a substance *always* reacts with 1 equivalent of another to produce 1 equivalent of each product.

Example 16.12 A volume of 29.3 mL of 0.611 N NaOH neutralizes 1.680 grams of a dry unknown acid. Calculate the equivalent weight of the acid.

Answer 93.8 g/eq

Solution First, we need to know what we are looking for; an acid's equivalent weight has units of g/eq or mg/meq. Since we know the number of grams of acid, we solve for equivalents of acid. We know that

$$1 \text{ eq acid} = 1 \text{ eq base} \quad \text{or}$$

$$1 \text{ meq acid} = 1 \text{ meq base}$$

Given the volume (mL) and normality (meq/mL) of the NaOH solution, we can solve for meq base:

$$\text{meq base} = 29.3 \cancel{\text{mL}} \times \frac{0.611 \text{ meq}}{\cancel{\text{mL}}} = 17.9 \text{ meq base}$$

$$17.9 \text{ meq base} = 17.9 \text{ meq acid} = 0.0179 \text{ eq acid}$$

$$\text{eq wt of acid} \left(\frac{\text{g}}{\text{eq}}\right) = \frac{1.680 \text{ g acid}}{0.0179 \text{ eq acid}} = \frac{93.8 \text{ g acid}}{\text{eq acid}}$$

Knowing the number of equivalents of H^+, OH^-, or e^- per mole of acid, base, or oxidizing or reducing agent converts molarity to normality.

$$N\left(\frac{\text{eq}}{\text{L}}\right) = \frac{\cancel{\text{mol}}}{\text{L}} \times \frac{\text{eq}}{\cancel{\text{mol}}} = \frac{\text{eq}}{\text{L}} \quad \text{or}$$

$$N = M \times \frac{\text{eq}}{\text{mol}}$$

Example 16.13 (a) What is the normality of the unknown acid in Example 16.12 when diluted to produce 200 mL solution?

(b) If 1 mole of the acid releases 2 mol H^+, what is its molarity?

Answer (a) 0.0896 eq acid/L soln; (b) 0.0448 mol acid/L soln

Solution (a) Use the definition of normality: eq/L or meq/mL.

$$N\left(\frac{eq}{L}\right) = \frac{1.680\ \text{g acid}}{0.200\ \text{L soln}} \times \frac{1\ \text{eq acid}}{93.8\ \text{g acid}} = 0.0896\ \frac{\text{eq acid}}{\text{L soln}}$$

(b) There are 2 eq acid per mol acid because 1 mol acid produces 2 mol H^+

$$M\ \text{acid} = \frac{0.0896\ \text{eq}}{\text{L soln}} \times \frac{1\ \text{mol acid}}{2\ \text{eq}} = 0.0448\ \frac{\text{mol acid}}{\text{L soln}}$$

SECTION 16.8 QUALITATIVE INORGANIC ANALYSIS

Objective

- To explain how pH, complex ion formation, buffer solutions, and amphoterism are used to selectively precipitate various "insoluble" salts.

Focus

Most cations are separated into five groups (text, Figure 16.3). Understand the principles of the separation, but do not memorize the table. See if you can answer Question 16.23.

Example 16.14 A solution is prepared in which $[Cu^{2+}] = [Ni^{2+}] = 1.0 \times 10^{-3}$ M. What is the range of $[S^{2-}]$ that selectively separates the two ions as the insoluble sulfide salts? $K_{CuS} = 1.3 \times 10^{-36}$ and $K_{NiS} = 1.1 \times 10^{-21}$ (text, Table 15.3).

Answer $1.3 \times 10^{-33} < [S^{2-}] < 1.1 \times 10^{-18}$

Solution $CuS(c) \rightleftharpoons Cu^{2+} + S^{2-}$ $\qquad$ $NiS(c) \rightleftharpoons Ni^{2+} + S^{2-}$

$$K = [Cu^{2+}][S^{2-}] = 1.3 \times 10^{-36} \qquad K = [Ni^{2+}][S^{2-}] = 1.1 \times 10^{-21}$$

$$[S^{2-}] = \frac{K}{[Cu^{2+}]} = \frac{1.3 \times 10^{-36}}{1.0 \times 10^{-3}} = 1.3 \times 10^{-33} \qquad [S^{2-}] = \frac{K^2}{[Ni^{2+}]} = \frac{1.1 \times 10^{-21}}{1.0 \times 10^{-3}} = 1.1 \times 10^{-18}$$

These are the $[S^{2-}]$ needed to precipitate each salt. NiS precipitates when the $[S^{2-}] = 1.1 \times 10^{-18}$. So, as long as $1.3 \times 10^{-33} < [S^{2-}] < 1.1 \times 10^{-18}$, CuS precipitates and NiS does not.

Example 16.15 If the source of S^{2-}, Example 16.14, is from the dissociation of H_2S, calculate the pH at which NiS begins to precipitate in a 0.10 M H_2S solution

$$H_2S \rightleftharpoons 2H^+ + S^{2-} \qquad K = 6.3 \times 10^{-22} \qquad \text{(text, Table 15.3)}$$

Answer pH = 2.12

Solution NiS precipitates when $[S^{2-}] = 1.1 \times 10^{-18}$. The H_2S equilibrium in which H^+ appears is

$$H_2S \rightleftharpoons 2H^+ + S^{2-}$$

$$K = 6.3 \times 10^{-22} = \frac{[H^+]^2[S^{2-}]}{H_2S} = \frac{[H^+]^2(1.1 \times 10^{-18})}{0.10}$$

$$[H^+]^2 = 5.73 \times 10^{-5}$$

$$[H^+] = 7.57 \times 10^{-3}$$

$$pH = 2.12$$

If the pH is 2.12 or greater (according to LeChatelier's principle, less H^+ indicates more S^{2-}), NiS precipitates; if the pH is less than 2.12, NiS does not precipitate but CuS does.

Example 16.16 How are the following cation pairs separated in a solution? See text, Figure 16.3.

(a) Ag^+ and Mn^{2+}
(b) Sn^{4+} and Co^{2+}
(c) Ba^{2+} and Mg^{2+}

Answer (a) Cl^-; (b) S^{2-}/acid solution; (c) CO_3^{2-}

Solution (a) Add Cl^-; AgCl is insoluble and $MnCl_2$ is not.
(b) Add S^{2-} in acidic solution; SnS_2 precipitates and CoS does not.
(c) Add $(NH_4)_2CO_3$ and NH_3; $BaCO_3$ precipitates and $MgCO_3$ is soluble.

QUESTIONS

Section 16.1

16.1 A 0.8774 g sample of an iron ore dissolves in acid. The iron precipitates as the carbonate yielding 0.8184 g $FeCO_3$. Calculate the percentage of Fe.

16.2 A 16.30 g sample of high S coal, heated in excess air, oxidizes the S to SO_2. The SO_2 that is passed through a CaO solution produces insoluble $CaSO_3$:

$$CaO + SO_2 \rightarrow CaSO_3$$

If 2.432 g $CaSO_3$ precipitates, what percent of S is in the coal?

Section 16.2

16.3 A 25.00 mL solution of $HC_2H_3O_2$ is neutralized by 20.92 mL of 0.1443 M NaOH. Calculate the molarity of the $HC_2H_3O_2$ solution.

16.4 Calculate the volume of 0.662 M HCl needed for the neutralization of 60.0 mL of 0.291 M KOH.

16.5 A 23.66 mL volume of 0.0224 M H_2SO_4 neutralizes 50.0 mL of a NaOH solution. How many milligrams of NaOH are in solution?

16.6 The Cl^- in a 2.3324 g sample of $FeCl_2$ precipitates with 0.4111 M $AgNO_3$. K_2CrO_4

is the indicator in the titration. When the red color of Ag_2CrO_4 appears, the buret reading is 44.77 mL. At the start of the titration, the buret reading was 1.00 mL. What is the percentage of Cl^- in the sample?

16.7 Calculate the volume of 0.331 M Na_2SO_4 required to precipitate (a) the Pb from 20.0 mL of 0.322 M $Pb(NO_3)_2$:

$$Pb^{2+} + SO_4^{2-} \rightarrow PbSO_4(c)$$

(b) 0.3844 g Ba^{2+}:

$$Ba^{2+} + SO_4^{2-} \rightarrow BaSO_4(c)$$

Section 16.3

16.8 At a pH of 6, what color is a solution containing phenolphthalein? Refer to the text, Table 16.1.

16.9 What color is a solution containing the methyl orange indicator at a pH of 6?

Section 16.4

16.10 A 25.00 mL solution of KOH requires 16.79 mL of 0.1033 M HNO_3 to reach the equivalence point in the titration. Calculate the pH at this point. From Table 16.1, select an appropriate indicator.

16.11 A 25.00 mL solution of HNO_2 is titrated to the equivalence point with 16.79 mL of 0.0919 M KOH. Calculate the pH at the equivalence point. K_a for HNO_2 is 4.5×10^{-4}. Select an appropriate indicator.

Section 16.5

16.12 A limestone rock weighing 0.8440 g requires a volume of 48.31 mL of 0.1044 M HCl to reach the equivalence point. Calculate the percentage of $CaCO_3$ in the rock:

$$CaCO_3(c) + 2HCl \rightarrow CaCl_2 + H_2O + CO_2$$

16.13 (a) A 1.288 g sample of dried Na_2CO_3 is dissolved and diluted with water to 500 mL. What is the molarity of the solution? (b) 35.33 mL of this solution precipitates the chromium from 400 mL of a $Cr(NO_3)_2$ solution. How many milligrams of Cr^{2+} are present?

$$Cr^{2+} + CO_3^{2-} \rightarrow CrCO_3(c)$$

(c) What is the molarity of the $Cr(NO_3)_2$ solution?

16.14 For the reaction in water,

$$4Zn + AsO_4^{3-} + 11H^+ \rightarrow 4Zn^{2+} + AsH_3 + 4H_2O$$

What volume of 0.3889 M of H^+ dissolves 210.0 mg Zn?

Section 16.6

16.15 Balance the following equations in acidic solutions:

(a) $PbO_2 + I^- \rightarrow Pb^{2+} + I_2$

(b) $NO_2 + UO_2^+ \rightarrow NO + UO_2^{2+}$

(c) $ClO_3^- + V^{2+} \rightarrow V^{3+} + ClO_2$

(d) $S + Ni \rightarrow Ni^{2+} + H_2S$

(e) $VO^{2+} + Np^{3+} \rightarrow V^{3+} + Np^{4+}$

16.16 Balance the following equations in basic solutions:

(a) $NO_3^- + Cr^{3+} \rightarrow CrO_4^{2-} + NO_2^-$

(b) $NiO_2 + Fe \rightarrow Fe^{2+} + Ni^{2+}$

(c) $Br_2 + NO \rightarrow NO_3^- + Br^-$

(d) $HO_2^- + S^{2-} \rightarrow S + OH^-$

(e) $PuO_2^+ + CuI \rightarrow Pu^{4+} + Cu^{2+} + I^-$

Section 16.7

16.17 A 200 mL solution of 0.00481 N NaOH neutralizes 83.9 mL of an acid. What is the normality of the acid?

16.18 How many equivalents are in 38.32 mL of 0.114 N HCl?

16.19 A volume of 42.66 mL of 0.1066 N KOH neutralizes 0.6921 g mandelic acid. What is the acid's equivalent weight?

16.20 What is the equivalent weight of K_2SO_3 in this reaction?

$$K_2SO_3 + HC_2H_3O_2 \rightarrow KC_2H_3O_2 + KHSO_3$$

16.21 Find the normality of a 200 mL solution containing 2.22 g K_2CO_3. The solution is used for the reaction

$$H_3PO_4 + K_2CO_3 \rightarrow K_2HPO_4 + H_2O + CO_2$$

(*Hint*: How many equivalents of H^+ does 1 mole of K_2CO_3 accept?)

16.22 What is the normality of 0.182 M $H_2C_2O_4$ in the following reactions?

(a) $NaOH + H_2C_2O_4 \rightarrow NaHC_2O_4 + H_2O$

(b) $H_2C_2O_4 \rightarrow 2CO_2 + 2H^+ + 2e^-$

Section 16.8

16.23 Indicate a solution that separates the following. Write the reaction that represents the separation.

(a) Ag^+ and Cu^{2+}

(b) Cu^{2+} and Ba^{2+}

(c) Cd^{2+} and Fe^{3+}

(d) Zn^{2+} and Mg^{2+}

(e) $Fe(OH)_3$ and $Al(OH)_3$

16.24 Calculate the minimum pH at which FeS precipitates from a 0.001 M Fe^{2+} solution saturated with H_2S.

$$[H_2S] = 0.10 \text{ for a saturated } H_2S \text{ solution.}$$

$$K = 1.6 \times 10^{-19} \text{ for FeS; } K = 6.3 \times 10^{-22} \text{ for } H_2S \rightleftharpoons 2H^+ + S^{2-}.$$

16.25 What is the total mass of precipitate formed by adding excess 1.00 M $(NH_4)_2CO_3$

to 20.00 mL of an ammonia solution containing 0.010 M Ba^{2+}, 0.010 M Ca^{2+}, and 0.010 M Mg^{2+}? (See text, Table 15.4, for solubility product constants.)

16.26 A solution is 0.010 M Hg^{2+} and 0.010 M Mn^{2+}.

(a) Calculate the $[S^{2-}]$ needed to precipitate each.

(b) What is the $[Hg^{2+}]$ when MnS begins to precipitate?

(c) Does either of the ions precipitate in a saturated H_2S solution ($[H_2S] = 0.10$) at a pH = 1.5?

$$K = 6.3 \times 10^{-22} \quad \text{for } H_2S \rightleftharpoons 2H^+ + S^{2-}$$

See text Table 15.4 for solubility product constants.

ANSWERS

16.1 44.96% Fe

16.2 3.98% S

16.3 0.1208 M $HC_2H_3O_2$

16.4 26.4 mL 0.662 M HCl

16.5 42.4 mg NaOH

16.6 27.35% Cl^-

16.7 (a) 19.5 mL 0.331 M Na_2SO_4

(b) 8.46 mL 0.331 M Na_2SO_4

16.8 colorless

16.9 yellow

16.10 pH = 7; bromthymol blue, neutral red

16.11 pH = 7.96; bromthymol blue, neutral red, or phenolphthalein

16.12 29.9% $CaCO_3$

16.13 (a) 0.02430 M Na_2CO_3

(b) 44.64 mg Cr^{2+}

(c) 0.002146 M $Cr(NO_3)_2$

16.14 22.71 mL 0.3889 M H^+

16.15 (a) $PbO_2 + 2I^- + 4H^+ \rightarrow Pb^{2+} + I_2 + 2H_2O$

(b) $NO_2 + 2UO_2^+ + 2H^+ \rightarrow NO + 2UO_2^{2+} + H_2O$

(c) $ClO_3^- + V^{2+} + 2H^+ \rightarrow V^{3+} + ClO_2 + H_2O$

(d) $S + Ni + 2H^+ \rightarrow Ni^{2+} + H_2S$

(e) $VO^{2+} + Np^{3+} + 2H^+ \rightarrow V^{3+} + Np^{4+} + H_2O$

16.16 (a) $3NO_3^- + 2Cr^{3+} + 10\,OH^- \rightarrow 2CrO_4^{2-} + 3NO_2^- + 5H_2O$

(b) $NiO_2 + Fe + 2H_2O \rightarrow Fe^{2+} + Ni^{2+} + 4OH^-$

(c) $3Br_2 + 2NO + 8\,OH^- \rightarrow 2NO_3^- + 6\,Br^- + 4H_2O$

(d) $HO_2^- + S^{2-} + H_2O \rightarrow S + 3\,OH^-$

(e) $PuO_2 + CuI + 2H_2O \rightarrow Pu^{4+} + Cu^{2+} + I^- + 4OH^-$

16.17 0.0115 N acid

16.18 0.00437 equivalent

16.19 152.2

16.20 158.26 g/eq

16.21 0.161 N K_2CO_3

16.22 (a) 0.182 N $H_2C_2O_4$

(b) 0.364 N $H_2C_2O_4$

16.23 (a) HCl; $Ag^+ + Cl^- \rightarrow AgCl(c)$

(b) H_2S/0.3 M H^+; $Cu^{2+} + S^{2-} \rightarrow CuS(c)$

(c) H_2S/0.3 M H^+; $Cd^{2+} + S^{2-} \rightarrow CdS(c)$

(d) H_2S/NH_4^+, NH_3; $Zn^{2+} + S^{2-} \rightarrow ZnS(c)$

(e) OH^-, excess; $Al(OH)_3(c) + OH^- \rightarrow Al(OH)_4^-$

16.24 pH = 3.20

16.25 59.5 mg ($MgCO_3$ is considered soluble.)

16.26 (a) $[S^{2-}] = 6 \times 10^{-51}$ for HgS

$[S^{2-}] = 4.6 \times 10^{-12}$ for MnS

(b) $[Hg^{2+}] = 1.3 \times 10^{-41}$ mol/L

(c) $[S^{2-}] = 6.3 \times 10^{-20}$ at pH = 1.5; only HgS precipitates.

Chapter 17

ELECTROCHEMISTRY

SECTION 17.1 INTRODUCTION

Objective

• To explain qualitatively how an oxidation-reduction reaction occurs in an aqueous solution.

Focus

Ions conduct current by migrating through solution, picking up electrons (reduction) at one electrode and disposing of them (oxidation) at another.

Electrons conduct current in a metal wire that externally connects the two electrodes.

The two categories of electrochemical reactions are:

1. In an **electrolytic cell**, an electric current from an external power source produces a chemical reaction. Examples are charging a car battery or a calculator.
2. In a **galvanic cell**, a chemical reaction produces electric current. Examples are the use of a flashlight and a calculator.

Electrolytic cells are studied in the next section.

SECTION 17.2 ELECTRODE REACTIONS

Objectives

• To define the anode and the cathode in terms of oxidation and reduction processes.

• To define a faraday and an ampere.

• To calculate one unknown for an electrolysis reaction, given all but one of the following: Faradays, current, mass, moles, volume for a gas, charge, amperes, time, and/or ionic charge.

Focus

Oxidation occurs at the **anode**, reduction occurs at the **cathode**.

One electron has a charge of 1.60×10^{-19} coulomb (C); the charge of 1 mole of electrons is

$$\frac{\text{C}}{\text{mol } e^-} = 1.60 \times 10^{-19} \frac{\text{C}}{\cancel{e^-}} \times \frac{6.02 \times 10^{23} \cancel{e^-}}{\text{mol } e^-}$$

$$= \frac{9.65 \times 10^4 \text{ C}}{\text{mol } e^-}$$

Therefore 1 mol e^- = 96,500 C

= 1 **faraday** (by definition), $\mathfrak{F}$

= $6.02 \times 10^{23}\ e^-$

Current is the movement of charge or electrons through a wire. An **ampere** (A), the measurement of current, is the passage of 1 coulomb of charge by a given point in 1 second.

$$1 \text{ A} = \frac{1 \text{ C}}{\text{sec}}$$

Know each of these units so that they may be used in problems involving electrolytic processes.

Example 17.1 How much time (minutes) is required to deposit 67.3 g Cr from a $Cr_2(SO_4)_3$ solution if a current of 600 A is used?

Answer 10.4 minutes

Solution As with nearly all problems, we must know the chemical reaction that is occurring. Chromium in the +3 oxidation state in $Cr_2(SO_4)_3$ is reduced to chromium metal.

$$Cr^{3+} + 3e^- \rightarrow Cr$$

From this equation we see that 3 $\mathfrak{F}$, or 3 mol e^-, are required to electroplate 1 mol Cr.

For 67.3 g Cr:

$$\mathfrak{F} = 67.3 \cancel{\text{g Cr}} \times \frac{\cancel{\text{mol Cr}}}{52.0 \cancel{\text{g Cr}}} \times \frac{3\ \mathfrak{F}}{\cancel{\text{mol Cr}}}$$

$$= 3.88\ \mathfrak{F}$$

With 3.88 $\mathfrak{F}$, the total charge, C, passing through the cell is

$$\text{C} = 3.88 \cancel{\mathfrak{F}} \times \frac{96{,}500 \text{ C}}{\cancel{\mathfrak{F}}}$$

$$= 3.74 \times 10^5 \text{ C}$$

Using a current of 600 A or 600 C/sec, then

$$\text{time (min)} = 3.74 \times 10^5 \cancel{\text{C}} \times \frac{1 \cancel{\text{sec}}}{600 \cancel{\text{C}}} \times \frac{1 \text{ min}}{60 \cancel{\text{sec}}}$$

$$= 10.4 \text{ min}$$

or:

Given quantity: 67.3 g Cr
Desired quantity: time (sec)
Conversion factors: 52.0 g Cr = 1 mol Cr
1 mol Cr = 3 $\mathcal{F}$
96,500 C = 1 $\mathcal{F}$
600 C = 1 sec

$$\text{time (min)} = 67.3\,\cancel{\text{g Cr}} \times \frac{1\,\cancel{\text{mol Cr}}}{52.0\,\cancel{\text{g Cr}}} \times \frac{3\,\cancel{\mathcal{F}}}{1\,\cancel{\text{mol Cr}}} \times \frac{96{,}500\,\cancel{\text{C}}}{\cancel{\mathcal{F}}} \times \frac{1\,\cancel{\text{sec}}}{600\,\cancel{\text{C}}} \times \frac{1\text{ min}}{60\,\cancel{\text{sec}}}$$

$$= 10.4\text{ min}$$

Example 17.2 A weight loss of 12.57 grams occurs over a 6.00 hour period at a titanium anode that uses a current of 4.70 amperes. Calculate the oxidation state of the titanium ion in the molten solution.

Answer Ti^{4+}

Solution The number of moles of electrons (or faradays) that pass through the cell, per mole of titanium ion formed, determines its oxidation state.

$$Ti \rightarrow Ti^{n+} + ne^-$$

Given quantity: $4.70\text{ A} = 4.70\,\frac{\text{C}}{\text{sec}}$

Desired quantity: $\mathcal{F}$

Conversion factors:

$$1\text{ hr} = 3600\text{ sec}$$

$$1\,\mathcal{F} = 96{,}500\text{ C}$$

$$\mathcal{F} = \frac{4.70\,\cancel{\text{C}}}{\cancel{\text{sec}}} \times \frac{3600\,\cancel{\text{sec}}}{\cancel{\text{hr}}} \times 6.00\,\cancel{\text{hr}} \times \frac{\mathcal{F}}{96{,}500\,\cancel{\text{C}}}$$

$$= 1.05\,\mathcal{F} = 1.05\text{ mol }e^-$$

$$\text{mol Ti} = 12.57\,\cancel{\text{g Ti}} \times \frac{\text{mol Ti}}{47.90\,\cancel{\text{g Ti}}} = 0.262\text{ mol Ti}$$

$$n = \frac{\text{mol }e^-}{\text{mol Ti}} = \frac{1.05\text{ mol }e^-}{0.262\text{ mol Ti}}$$

$$= 4.0\,\frac{\text{mol }e^-}{\text{mol Ti}}$$

Therefore, the Ti on the anode must be oxidized to Ti^{4+} in the melt.

SECTION 17.3 GALVANIC CELLS

Objectives

- To describe a galvanic cell in terms of ion and electron flow, and the substances involved in the oxidation-reduction reaction.
- To set up an operative galvanic cell.

Focus

A **galvanic cell** converts chemical energy into electrical energy by directing the transfer of electrons in an oxidation-reduction reaction through an external circuit.

The reaction that occurs in a galvanic cell is spontaneous; ΔG is negative. A barrier separates the substances that spontaneously oxidize and reduce in a galvanic cell. This barrier, which can be a salt bridge or a porous cup, allows ions to flow, but prevents the spontaneous mixing of the two substances and therefore forces the electrons to pass through an external circuit. These moving electrons can provide work.

Figure 17.1 shows the important features of a galvanic cell. At the metallic anode, oxidation occurs; electrons are removed from a substance in solution. This removal of electrons causes a buildup of positive charge in the solution, but it remains electrically neutral because of an *inflow of negative ions (anions)* into the anode chamber and/or an *outflow of positive ions (cations)* into the cathode chamber. Likewise reduction occurs at the cathode; electrons are accepted from the external circuit by a substance in solution. This decreases the positive charge in the cathode chamber. The same ion flow, however, maintains neutrality; for every electron entering the cathode chamber, one unit of negative charge in the form of an ion either leaves through the barrier or one unit of positive charge in the form of a cation enters it.

When two different substances are placed in solution on the two sides of the barrier in the galvanic cell, the one which has more negative ΔG for reduction is reduced at the cathode. This forces the reaction for the substance with a more positive ΔG for reduction to be reversed so that it is oxidized at the anode.

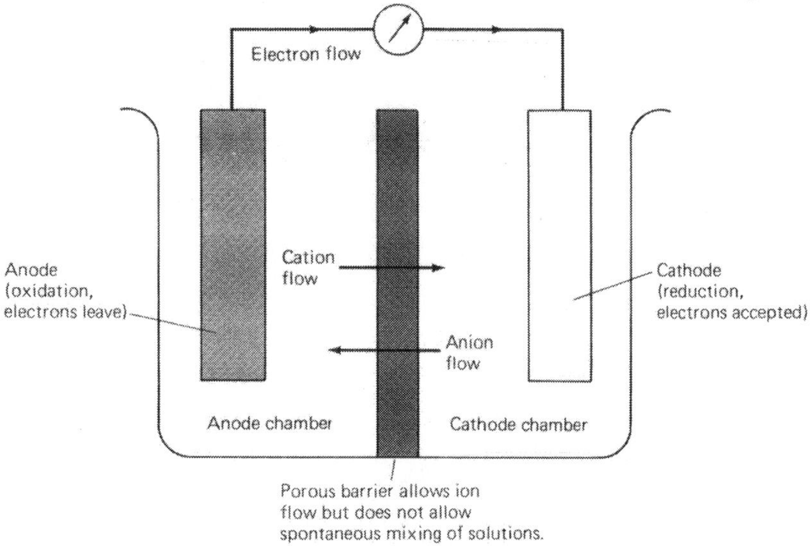

Figure 17.1 The movement of anions, cations, and electrons in a galvanic cell.

Example 17.3 Two solutions, a $NiCl_2$ solution with a Ni electrode and a $CdCl_2$ solution with a Cd electrode, are placed on opposite sides of the porous barrier (Figure 17.2). Of the two possible reduction reactions

$$Ni^{2+} + 2e^- \rightarrow Ni$$

$$Cd^{2+} + 2e^- \rightarrow Cd$$

the Ni^{2+} reduction is more favorable, a more negative ΔG. Designate the anode, cathode, the direction of Cl^-, Ni^{2+}, and Cd^{2+} flow, and the direction of electron flow in the galvanic cell.

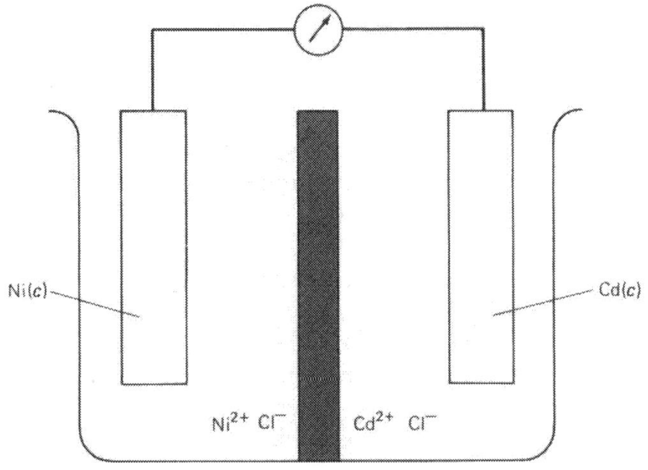

Figure 17.2 Set up of Cd | Cd^{2+} | Ni^{2+} | Ni galvanic cell.

Answer Since the Ni^{2+} reduction reaction has a more negative ΔG than does the Cd^{2+} reduction reaction, Ni^{2+} is reduced at the *cathode*. Cd metal therefore is oxidized at the *anode*. Cd^{2+} generated at the anode moves out of the anode chamber and/or Cl^- moves into the anode chamber to maintain electrical neutrality. Similarly, Ni^{2+} is removed (plated as Ni metal on the cathode) from the cathode chamber; this causes the exit of Cl^- from the cathode chamber and/or the inflow of Cd^{2+} from the anode chamber to maintain neutrality (see Figure 17.3).

SECTION 17.4 CONVENTIONAL NOTATION FOR CELLS

Objectives

- To write a simplified cell diagram for a galvanic cell.
- To write the partial and net equations for the reaction in a cell, given the cell diagram.

Focus

In the simplified cell notation for a galvanic cell, remember that | represents an interface between two phases or solutions and || represents a salt bridge.

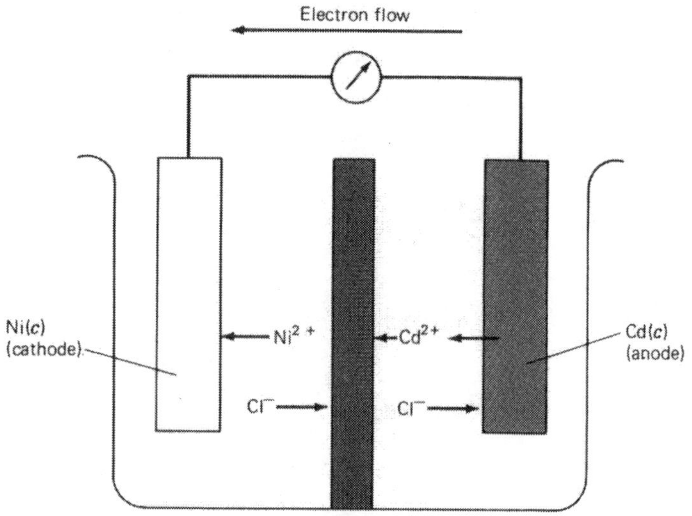

Figure 17.3 Movement of cation, anion, and electron flow in a galvanic cell.

For the reaction

$$A(c) + B(c) \rightarrow A^+(aq) + B^-(aq)$$

the cell representation is

A | A^+ (aq)	‖	B^- (aq) | B
Anode	barrier	Cathode
Oxidation		Reduction
Left		Right
$A \rightarrow A^+ + e^-$		$B + e^- \rightarrow B^-$

If there is no salt bridge, the cell is represented as

$$A \mid A^+(aq) \mid B^-(aq) \mid B.$$

Example 17.4 What reaction occurs in the cell?

$$Pb \mid Pb(NO_3)_2(aq) \mid AgNO_3(aq) \mid Ag$$

Answer $Pb(c) \rightarrow Pb^{2+}(aq) + 2e^-$; $Ag^+(aq) + e^- \rightarrow Ag(c)$

$$Pb(c) + 2Ag^+(aq) \rightarrow Pb^{2+}(aq) + 2Ag(c)$$

Solution Pb and Ag change oxidation states. In the cell description, the anode is on the left and the cathode on the right. Therefore, Pb is oxidized

$$Pb(c) \rightarrow Pb^{2+}(aq) + 2e^-$$

and Ag is reduced

$$Ag^+(aq) + e^- \rightarrow Ag(c)$$

The sum of these reactions is the cell reaction (after balancing for charge):

$$Pb(c) + 2Ag^+(aq) \rightarrow Pb^{2+}(aq) + 2Ag(c)$$

Example 17.5 Write the cell notation for the following reaction:

$$Zn(c) + ClO_4^-(aq) + 2H^+(aq) \rightarrow Zn^{2+}(aq) + ClO_3^-(aq) + H_2O$$

Answer $Zn \mid Zn^{2+}(aq) \parallel ClO_4^-(aq), ClO_3^-(aq), H^+(aq) \mid Pt$ or, without a salt bridge

$$Zn \mid Zn^{2+}(aq) \mid ClO_4^-(aq), ClO_3^-(aq)\ H^+(aq) \mid Pt$$

Solution On the left side of the cell, zinc is oxidized (zinc increases in oxidation state from 0 to 2+).

$$Zn(c) \rightarrow Zn^{2+}(aq) + 2e^-$$

For the right side, $ClO_4^-(aq)$ is reduced:

$$2e^- + 2H^+(aq) + ClO_4^-(aq) \rightarrow ClO_3^-(aq) + H_2O$$

A salt bridge or a porous cup is used to separate the two solutions. For the reduction side of the galvanic cell we use a nonreactive metallic electrode to conduct the electrons. Pt or C can serve this purpose.

SECTION 17.5 ELECTRICAL WORK

Objective

- To explain electrical work in terms of electrode potential and charge.

Focus

The quantity of electrical work that a galvanic cell produces is measured by

$$\text{Work} = \left(\begin{array}{l}\text{Potential difference be-}\\ \text{tween the reactions}\\ \text{occurring (reduction and}\\ \text{oxidation) at the}\\ \text{electrodes}\end{array}\right) \times \left(\begin{array}{l}\text{The quantity of}\\ \text{charge that passes}\\ \text{through the cell}\end{array}\right)$$

$$\text{Work (J)} = \quad \text{J/C} \quad \times \quad \text{C}$$

A **joule** is a unit of work and energy. A joule/coulomb, equal to a **volt**, is the electrical energy available from the cell per coulomb of electrical charge.

The **potential difference** is called the **electromotive force (emf)** of the cell. The *standard* emf, $\mathcal{E}^0$, is measured when the substances are in their standard states: at 25°C, 1 mol/L for soluble solutes, and 1 atm pressure for gases.

Example 17.6 The potential difference for the galvanic cell

$$Zn \mid Zn^{2+}(1.0\ M) \parallel Sn^{4+}(1.0\ M), Sn^{2+}\ (1.0\ M) \mid Pt$$

is 0.90 volt. Calculate the quantity of work the cell generates per mol Zn oxidized.

Answer 1.74×10^5 J

Solution The cell reaction is

Anode:	$Zn(c) \rightarrow Zn^{2+}\ (1.0\ M) + 2e^-$
Cathode:	$Sn^{4+}\ (1.0\ M) + 2e^- \rightarrow Sn^{2+}\ (1.0\ M)$
Cell reaction:	$Zn(c) + Sn^{4+}\ (1.0\ M) \rightarrow Zn^{2+}\ (1.0\ M) + Sn^{2+}\ (1.0\ M)$

Two $\mathcal{F}$ or 2 mol e^- pass through the cell. The quantity of available work then is

$$\text{work (J)} = 0.90\ \frac{\text{J}}{\cancel{\mathcal{C}}} \times \frac{96{,}500\ \cancel{\mathcal{C}}}{\cancel{\mathcal{F}}} \times 2\cancel{\mathcal{F}}$$

$$= 1.74 \times 10^5\ \text{J}$$

SECTION 17.6 MEASUREMENT OF ELECTROMOTIVE FORCE

Objectives

- To explain how a potentiometer is used in measuring the potential difference of a cell.
- To understand the significance of the emf measured by the potentiometer.

Focus

A **potentiometer** is an instrument that produces an emf *opposite* to the spontaneous flow of electrons from the anode to the cathode in a galvanic cell. The emf that stops this spontaneous current flow is the measured emf of the cell.

The emf measured by the potentiometer at zero current flow is the reversible emf of the cell and is used to calculate the maximum or ideal amount of electrical work a cell produces. The actual work, however, is less because of the cell's resistance to electron and ion flow.

SECTION 17.7 FREE ENERGY AND ELECTROMOTIVE FORCE

Objective

- To calculate the maximum amount of available work a galvanic cell generates.

Focus

The maximum amount of **electrical work** that a galvanic cell produces is

$$w_{max} = w_{rev} = -\Delta G \quad \text{or,}$$

$$\Delta G = -w_{max}$$

From Section 17.5, we can determine the quantity of work from a galvanic cell by knowing the potential difference, $\mathcal{E}$, and the quantity of charge passing through the cell, n. See this Study Guide, Example 17.6.

$$w_{max} = n\mathcal{F}\mathcal{E}$$

where n = number of faradays or moles of electrons
$\mathcal{F}$ = 96,500 C/$\mathcal{F}$
$\mathcal{E}$ = potential difference of the cell, J/C
Therefore, $\Delta G = -n\mathcal{F}\mathcal{E} = -w_{max} = -w_{rev}$

For standard conditions of the solutes and/or gases in the cell reaction

$$\Delta G^0 = -n\mathcal{F}\mathcal{E}^0$$

When using these equations, remember that w_{max} and ΔG depend upon the coefficients of the balanced equation for the cell reaction—the more moles that react, the greater are the values for ΔG and w_{max}. However, $\mathcal{E}$ is a potential difference between two electrodes with units of joules/coulomb: if more joules of energy are produced, then more coulombs of charge are passed through the cell. This maintains the joule/coulomb ratio constant ($\mathcal{E}$ is independent of the balanced equation).

Example 17.7 (a) Calculate ΔG^0 and w_{rev} for

$$O_2(g)(1\text{ atm}) + 4H^+(1.0\text{ M}) + 4\,Fe^{2+}(1.0\text{ M}) \rightarrow 4Fe^{3+}(1.0\text{ M}) + 2H_2O$$

for which $\mathcal{E}^0 = 0.460$ V.

(b) Determine $\mathcal{E}^0$ and ΔG^0 for the reaction

$$2O_2(g)(1\text{ atm}) + 8H^+(1.0\text{ M}) + 8Fe^{2+}(1.0\text{ M}) \rightarrow 8Fe^{3+}(1.0\text{ M}) + 4H_2O$$

Answer (a) $\Delta G^0 = -178$ kJ; $w_{rev} = +178$ kJ

(b) $\mathcal{E}^0 = 0.460$ V; $\Delta G^0 = -355$ kJ

Solution

(a) $\Delta G^0 = -n\mathcal{F}\mathcal{E}^0$

$n = 4\mathcal{F}$ or 4 mol e^- because of

$4Fe^{2+} \rightarrow 4Fe^{3+} + 4e^-$ and/or

$O_2 + 4H^+ + 4e^- \rightarrow 2H_2O$

$\mathcal{E}^0 = 0.460\text{ V} = 0.460\text{J/C}$

$$\Delta G^0 = -4\not{\mathcal{F}} \times \frac{96{,}500\not{C}}{\not{\mathcal{F}}} \times \frac{0.460\text{J}}{\not{C}}$$

$$= -1.78 \times 10^5\text{ J} = -178\text{ kJ}$$

$$w_{rev} = -\Delta G^0 = -(-178\text{ kJ}) = 178\text{ kJ}$$

(b) $\mathcal{E}^0 = 0.460\text{ V} = 0.460$ J/C. This value is independent of the coefficients of the balanced equation.

$n = 8\,\mathcal{F}$ or 8 mol e^- because of

$8Fe^{2+} \rightarrow 8Fe^{3+} + 8e^-$ and/or

$2O_2 + 8H^+ + 8e^- \rightarrow 4H_2O$

$$\Delta G^0 = -8\not{\mathcal{F}} \times \frac{96{,}500\not{C}}{\not{\mathcal{F}}} \times 0.460\ \frac{\text{J}}{\not{C}}$$

$$= -3.55 \times 10^5\text{ J} = -355\text{ kJ}$$

SECTION 17.8 THE SIGN OF AN ELECTRODE

Objective

- To identify the signs of the electrodes in galvanic and electrolytic cells.

Focus

Always remember that oxidation occurs at the anode and reduction at the cathode, regardless of the type of cell.

For a *galvanic cell* the anode is said to be negative because it is the source of electrons in the external circuit and the cathode is positive because it "pulls in" the electrons.

Galvanic cell:	anode (−)	oxidation
	cathode (+)	reduction

For an *electrolytic cell*, the anode is positive because electrons are pulled away from the anode by the external power source and the cathode is negative because electrons are pushed into it.

Electrolytic cell:	anode (+)	oxidation
	cathode (−)	reduction

SECTION 17.9 HALF-CELL POTENTIALS

Objective

- To predict $\mathcal{E}^0_{cell}$, given the $\mathcal{E}^0$ values for the reduction and oxidation half-cell reactions.

Focus

The text, Table 17.1, lists the *standard* reduction potentials, $\mathcal{E}^0$, for a number of half-cells. Notice that for each half-cell, the potential for reduction relative to the standard H_2 electrode is expressed in volts (joules/coulomb). The standard H_2 electrode is defined at 25°C as

$$2H^+(1.0\ M) + 2e^- \rightleftharpoons H_2(g)(1\ atm) \qquad \mathcal{E}^0 = 0.00\ V$$

The half-cell reactions with positive potentials are more likely to be reduced than H^+ ($2H^+ + 2e^- \rightarrow H_2$), but those with negative potentials are less likely to be reduced than H^+ *or* more likely to be oxidized relative to H_2. Become familiar with Table 17.1, but do not memorize it.

The standard cell potential, $\mathcal{E}^0_{cell}$, is determined by adding the two half-cell potentials. We change the sign of $\mathcal{E}^0$ for the anode (oxidation) reaction and add it to $\mathcal{E}^0$ for the cathode (reduction) reaction. To get $\mathcal{E}^0_{anode}$ (oxidation) we must change the sign of the value listed in Table 17.1 because the table gives standard *reduction* ($\mathcal{E}^0_{cathode}$) potentials. When a reaction is reversed, the sign of $\mathcal{E}^0$ must also be reversed.

$$\mathcal{E}^0_{cell} = \mathcal{E}^0_{cathode} + (-\mathcal{E}^0_{anode})$$

If the cell reaction is written in the direction of spontaneous electron flow, or if the simplified cell notation is written properly with the anode on the left and the cathode on the right, then $\mathcal{E}^0_{cell}$ is always a positive potential and $\Delta G^0 = -n\mathcal{F}\mathcal{E}^0$ is negative. The reaction is thermodynamically spontaneous as written.

Example 17.8 (a) Determine $\mathcal{E}^0_{cell}$ at 25°C for the cell:

$$\text{Pt} \mid V^{3+}(1.0\ M), V^{2+}(1.0\ M) \parallel Au^{3+}(1.0\ M) \mid \text{Au}$$

(b) Write the cell reaction.
(c) Calculate the maximum amount of electrical work that the cell produces.
(d) What is the standard free energy change for the cell reaction, ΔG^0?

Answer (a) $\mathcal{E}^0 = 1.68$ V
(b) $Au^{3+} + 3V^{2+} \rightarrow 3V^{3+} + Au$
(c) $w_{max} = 486$ kJ
(d) $\Delta G^0 = -486$ kJ

Solution (a) To determine the $\mathcal{E}^0_{cell}$ let us first rewrite each half-cell reaction with the corresponding reduction potential (from Table 17.1). This is a check to make sure that the simplified cell notation is correct as written—with the anode on the left and the cathode on the right.

$$Au^{3+} + 3e^- \rightleftharpoons Au \qquad \mathcal{E}^0 = +1.42\ V$$
$$V^{3+} + e^- \rightleftharpoons V^{2+} \qquad \mathcal{E}^0 = -0.255\ V$$

Since the $Au^{3+} \mid Au$ half-cell has a higher reduction potential, reduction of Au^{3+} occurs at the cathode (Au electrode). As a result, oxidation *must* occur at the Pt anode where V^{2+} oxidizes to V^{3+}. Therefore, the V^{3+}, V^{2+} half-cell appears on the left and the $Au^{3+} \mid Au$ on the right. The given cell notation is correct.

Write down each half-cell equation and the corresponding reduction potential as it actually occurs in the cell. For the cathode reaction,

$$Au^{3+} + 3e^- \rightarrow Au \qquad \mathcal{E}^0_{cathode} = +1.42\ V$$

For the anode reaction, the half-cell equation is reversed and its potential is the negative of what appears in Table 17.1:

$$V^{2+} \rightarrow V^{3+} + e^- \qquad -\mathcal{E}^0_{anode} = -(-0.255\ V) = +0.255\ V$$

Therefore, $\mathcal{E}^0_{cell}$ is the sum of the half-cell potentials for each half-cell reaction as each occurs in the cell.

$$\mathcal{E}^0_{cell} = \mathcal{E}^0_{cathode} + (-\mathcal{E}^0_{anode})$$
$$= 1.42\ V + (+0.255\ V) = 1.68\ V$$

Watch the signs!
(b) The cell reaction is the sum of the two half-cell reactions:

Cathode	$Au^{3+} + 3e^- \rightarrow Au$
Anode	$3(V^{2+} \rightarrow V^{3+} + e^-)$
	$Au^{3+} + 3V^{2+} \rightarrow 3V^{3+} + Au$

Remember, the charge must be balanced; $3e^-$ are gained in the reduction of Au^{3+} and therefore $3e^-$ must be lost in the oxidation of V^{2+}. To balance electrons, we must multiply the oxidation reaction by 3.

(c) The w_{max} of the cell, with $3\mathcal{F}$ or 3 mol e^- passing through the cell, is

$$\begin{aligned} w_{max} &= n\mathcal{F}\mathcal{E} \\ &= 3\mathcal{F} \times \frac{96{,}500\ C}{\mathcal{F}} \times \frac{1.68\ J}{C} \\ &= 4.86 \times 10^5\ J = 486\ kJ \end{aligned}$$

(d) $$\begin{aligned} \Delta G &= -w_{max} \\ &= -486\ kJ \end{aligned}$$

Example 17.9 Given the half-cell reactions and potentials:

$$Sn^{2+} + 2e^- \rightleftharpoons Sn \qquad \mathcal{E}^0 = -0.1346\ V$$
$$VO^{2+} + 2H^+ + e^- \rightleftharpoons V^{3+} + H_2O \quad \mathcal{E}^0 = +0.36\ V$$

(a) Write the simplified cell diagram for a galvanic cell using these two half-cell reactions.

(b) Write the half-cell reactions occurring in the cell with the corresponding potentials.

(c) Calculate $\mathcal{E}^0_{cell}$.

Answer (a) $Sn \mid Sn^{2+}(aq) \parallel VO^{2+}(aq), H^+(aq), V^{3+}(aq) \mid$ Pt or C

(b) $VO^{2+} + 2H^+ + e^- \rightleftharpoons V^{3+} + H_2O \qquad \mathcal{E}^0_{cathode} = +0.36\ V$

$Sn \rightleftharpoons Sn^{2+} + 2e^- \qquad -\mathcal{E}^0_{anode} = -(-0.1364\ V) = +0.1364\ V$

(c) $\mathcal{E}^0_{cell} = 0.50\ V$

Solution (a) Since the VO^{2+}, V^{3+} half-cell reaction has a higher $\mathcal{E}^0$, this reaction occurs at the cathode and is written on the right. Since the $Sn^{2+} \mid Sn$ half-cell reaction occurs at the anode, it is written on the left.

$Sn \mid Sn^{2+} \parallel VO^{2+}, H^+, V^{3+} \mid$ nonreactive electrode such as C or Pt

(b) For the cathode reaction, write the VO^{2+}, V^{3+} reaction as shown in Table 17.1:

$$VO^{2+} + 2H^+ + e^- \rightleftharpoons V^{3+} + H_2O \qquad \mathcal{E}^0_{cathode} = +0.36\ V$$

For the anode reaction, reverse the $Sn^{2+} \mid Sn$ half-cell reaction and change the sign of $\mathcal{E}^0$:

$$Sn \rightleftharpoons Sn^{2+} + 2e^- \qquad -\mathcal{E}^0_{anode} = -(-0.1364\ V) = +0.1364\ V$$

(c) $$\begin{aligned} \mathcal{E}^0_{cell} &= \mathcal{E}^0_{cathode} + (-\mathcal{E}^0_{anode}) \\ &= 0.36\ V + (+0.1364\ V) \\ &= 0.50\ V \end{aligned}$$

SECTION 17.10 THE NERNST EQUATION

Objective

- To determine the emf of a galvanic cell when the reactants and/or products are *not* in their standard states.

Focus

The **Nernst equation**, derived from thermodynamics, calculates the emf of a galvanic cell when the reactants and/or products are not in their standard states. In the Nernst equation,

$$\mathcal{E}_{cell} = \mathcal{E}^0_{cell} - \frac{RT}{n\mathrm{F}} \ln Q$$

$$= \mathcal{E}^0_{cell} - \frac{2.303RT}{n\mathrm{F}} \log Q$$

$\mathcal{E}_{cell}$ is the experimental cell potential, $\mathcal{E}^0_{cell}$ is the standard cell potential, $\mathcal{F}$ is Faraday's constant, n is the number of moles of electrons, and Q is the reaction quotient. At 25°C (298.15 K),

$$\frac{2.303RT}{\mathcal{F}} = 2.303 \frac{\left(8.314 \frac{\mathrm{J}}{(\mathrm{mol\,K})}\right)(298.15\,\mathrm{K})}{96{,}500\ \mathrm{C}/\mathcal{F}}$$

$$= \frac{0.0592\ \mathrm{J}\ \mathcal{F}}{(\mathrm{mol\ C})} = \frac{0.0592\ (\mathrm{volt}\ \mathcal{F})}{\mathrm{mol}}$$

Learn the Nernst equation (at 25°C)

$$\mathcal{E}_{cell} = \mathcal{E}^0_{cell} - \frac{0.0592}{n} \log Q$$

The equations for w_{max} and ΔG (Section 17.7) are also applicable for electrical work and the free energy change of the cell reaction even when the substances are not in their standard states. Merely substitute $\mathcal{E}_{cell}$ for $\mathcal{E}^0_{cell}$.

$$w_{max} = n\mathcal{F}\mathcal{E}_{cell}$$

$$\Delta G = -n\mathcal{F}\mathcal{E}_{cell}$$

Example 17.10 (a) Calculate the standard cell potential at 25°C for the cell:

Hg | $HgCl_4^{2-}$ (0.010 M), Cl^- (10^{-7} M) | CrO_4^{2-} (3.0×10^{-5} M) | Ag_2CrO_4(c) | Ag.

(b) Write the standard cell reaction.
(c) Calculate the value of the reaction quotient, Q.
(d) Calculate the cell potential at 25°C. Interpret the answer.
(e) Determine the maximum electrical work the cell produces when 1 mol Hg is consumed or produced.

Answer (a) $\mathcal{E}^0_{cell} = 0.07$ V
(b) $Ag_2CrO_4 + Hg + 4Cl^- \rightarrow 2Ag + CrO_4^{2-} + HgCl_4^{2-}$
(c) 3.0×10^{21}
(d) $\mathcal{E}_{cell} = -0.57$ V
Because the ionic concentrations are significantly different from standard (1 M) concentrations in solution, the negative cell potential means that the reaction is not spontaneous as written. (See Section 17.13.)
(e) $w_{max} = 110$ kJ for the *reverse* cell reaction.

Solution (a) Write the two cell reactions and their corresponding standard potentials. Remember to reverse the direction of the anode reaction *and* reverse the sign of $\mathcal{E}^0$

for the standard anode potential (see Table 17.1).

$$Hg + 4Cl^- \rightleftharpoons HgCl_4^{2-} + 2e^- \qquad -\mathcal{E}^0_{anode} = -(+0.38\ V) = -0.38\ V$$

$$Ag_2CrO_4 + 2\,e^- \rightarrow 2Ag + CrO_4^{2-} \qquad \mathcal{E}^0_{cathode} = +0.4463\ V$$

$\mathcal{E}^0_{cell}$ is the sum of the two standard potentials or

$$\begin{aligned}\mathcal{E}^0_{cell} &= \mathcal{E}^0_{cathode} + (-\mathcal{E}^0_{anode})\\ &= +0.4463\ V + (-0.38\ V) = 0.07\ V\end{aligned}$$

(b) Before using the Nernst equation, write the cell reaction to avoid error in setting up the reaction quotient. The cell reaction is the sum of the two half-cell reactions:

$$Ag_2CrO_4(c) + Hg(c) + 4Cl^-(aq) \rightarrow 2Ag(c) + CrO_4^{2-}(aq) + HgCl_4^{2-}(aq)$$

(c) The reaction quotient is the product of the molar concentrations of the products divided by that of the reactants, with each concentration raised to the power of the coefficient in the balanced equation, or

$$Q = \frac{[CrO_4^{2-}][HgCl_4^{2-}]}{[Cl^-]^4}$$

The concentrations of Ag_2CrO_4, Hg, and Ag are pure substances (not a part of the solution) and therefore do not appear as part of the reaction quotient

$$Q = \frac{(3.0 \times 10^{-5})(0.010)}{(10^{-7})^4} = 3.0 \times 10^{21}$$

(d) $\mathcal{E}_{cell} = \mathcal{E}^0_{cell} - \dfrac{0.0592}{n} \log Q$

For this reaction, 2 mol e^- are exchanged; $n = 2$.

$$\begin{aligned}\mathcal{E}_{cell} &= 0.07\ V - \frac{0.0592}{n} \log 3.0 \times 10^{21}\\ &= 0.07\ V - \frac{0.0592}{2}(21.48)\\ &= 0.07\ V - 0.64\ V\\ &= -0.57\ V\end{aligned}$$

This negative $\mathcal{E}$ value means that the cell reaction does *not* occur spontaneously in the direction from left to right as written. Rather the spontaneous reaction is from right to left (see Section 17.13). This can be explained if we apply Le Chatelier's principle. The concentrations of both $HgCl_4^{2-}$ and CrO_4^{2-} are less than 1 M (less than standard). This decrease in product concentration forces the reaction to the right. However, the concentration of Cl^- is only 10^{-7} M. This is *more* significantly different from standard concentrations than are the concentrations of $HgCl_4^{2-}$ and CrO_4^{2-}; the reaction is therefore driven to the left.

(e) For the *reverse* cell reaction, with the given concentrations of ions,

$$2Ag + CrO_4^{2-} + HgCl_4^{2-} \rightarrow Ag_2CrO_4 + Hg + 4Cl^-$$

$$\mathcal{E}_{cell} = +0.57\ V$$

and

$$
\begin{aligned}
w_{max} &= n\mathcal{F}\mathcal{E} \\
&= 2\mathcal{F} \times \frac{96{,}500\ C}{\mathcal{F}} \times \frac{0.57\ J}{C} \\
&= 1.1 \times 10^5\ J \\
&= 110\ kJ
\end{aligned}
$$

SECTION 17.11 MEASUREMENT OF CONCENTRATIONS

Objectives

- To describe a glass electrode.
- To calculate the pH of an unknown concentration from a measured $\mathcal{E}_{cell}$.

Focus

A **glass electrode** is one in which a potential difference is established between the H^+ inside the electrode and the H^+ in solution. The potential, $\mathcal{E}'$, is measured relative to a reference electrode, most commonly a saturated calomel electrode,

$$Hg_2Cl_2(c) + 2e^- \rightarrow 2Hg + 2Cl^-$$

where the Cl^- is from a saturated KCl solution. The emf of this cell (at 25°C) is

$$\mathcal{E}_{cell} = \mathcal{E}' - \frac{2.303RT}{n} \log[H^+]$$

or, more simply,

$$\mathcal{E}_{cell} = \mathcal{E}' + 0.0592\ pH$$

where $\mathcal{E}'$ for the instrument is standardized with a buffer solution of known H^+ concentration. Thus, we can use the Nernst equation to calculate the pH of a solution.

Actually, we can measure the molar concentration of *any* ion by determining the emf of a cell containing the ion and by applying the Nernst equation.

Example 17.11 $\mathcal{E}_{cell}$ = 0.80 volt for the cell

$$Au \mid Au(CN)_2^- (0.010\ M),\ CN^- (1.0\ M) \mid Br^- (?\ M) \mid AgBr(c) \mid Ag$$

Determine $[Br^-]$ and pBr at 25°C.

Answer $[Br^-]$ = 0.63 mol/L

pBr = 0.20

Solution From Table 17.1 and the given cell notation, we have two half-cell reactions with the corresponding $\mathcal{E}^0$ potentials:

$$Au + 2CN^- \rightarrow Au(CN)_2^- + e^- \qquad -\mathcal{E}^0_{anode} = -(-0.60\ V) = +0.60\ V$$

$$AgBr + e^- \rightarrow Ag + Br^- \qquad \mathcal{E}^0_{cathode} = +0.0713\text{ V}$$

$$\mathcal{E}^0_{cell} = \mathcal{E}_{cathode} + (-\mathcal{E}_{anode})$$
$$\mathcal{E}^0_{cell} = 0.0713\text{ V} + (+0.60\text{ V}) = +0.67\text{ V}$$

Now, write the cell reaction by adding the two half-cell reactions:

$$AgBr + Au + 2CN^- \rightarrow Ag + Br^- + Au(CN)_2^-$$

Substituting the molar concentrations into the Nernst equation,

$$\mathcal{E}_{cell} = \mathcal{E}^0_{cell} \frac{0.0592}{n} \log Q,$$

where $\mathcal{E}_{cell} = 0.80$ V, $\mathcal{E}^0_{cell} = 0.67$ V, and $n = 1$.

$$0.80\text{ V} = 0.67\text{ V} - \frac{0.0592}{1} \log \frac{[Br^-][Au(CN)_2^-]}{[CN^-]^2}$$

$$0.80\text{ V} = 0.67\text{ V} - 0.0592 \log \frac{[Br^-][0.010]}{[1.0]^2}$$

$$\frac{0.80\text{ V} - 0.67\text{ V}}{-0.0592\text{ V}} = \log[Br^-] + \log \frac{[0.010]}{[1.0]^2}$$

$$-2.20 = \log [Br^-] + (-2.0)$$

$$-2.20 + 2.0 = \log [Br^-]$$

$$-0.20 = \log [Br^-]$$

$$pBr = 0.20$$

$$[Br^-] = \text{antilog}(-0.20) = 0.63\text{ mol/L}$$

Example 17.12 Assuming no change in $[CN^-]$ or $[Au(CN)_2^-]$, what is the $[Br^-]$ when $\mathcal{E}_{cell} = 0$ in Example 17.10?

Answer $[Br^-] = 2.0 \times 10^{13}$ mol/L

Solution Use the Nernst equation, where $\mathcal{E}_{cell} = 0$.

$$0 = 0.67 - \frac{0.0592}{1} \log \frac{[Br^-][0.010]}{[1.0]^2}$$

$$\frac{-0.67}{-0.0592} = \log [Br^-] + \log 0.010$$

$$= \log [Br^-] + (-2.0)$$

$$+11.3 + 2.0 = 13.3 = \log [Br^-]$$

$$[Br^-] = \text{antilog}(13.3) = 2.0 \times 10^{13}\text{ mol/L}$$

Examples 17.10 and 17.11 indicate that an unknown concentration of any ion, including the H^+ (text, Example 17.14), can be determined by measuring the potential of a suitable cell and by knowing all other ionic concentrations in solution.

SECTION 17.12 ELECTROMOTIVE FORCE AND EQUILIBRIUM

Objective

• To calculate an equilibrium constant from given standard reduction potentials.

Focus

At equilibrium $Q = K$, therefore the equilibrium constant for a cell reaction can be determined from the equation

$$\log K = \frac{n\mathcal{F}\mathcal{E}^0}{2.303RT}$$

$$\text{at } 25°\text{C}, \log K = \frac{n\mathcal{E}^0}{0.0592}$$

Example 17.13 For the cell reaction $Ag_2CrO_4 + Hg + 4Cl^- \rightarrow 2Ag + CrO_4^{2-} + HgCl_4^{2-}$ $\mathcal{E}^0_{cell} = 0.070$ V. See Study Guide Example 17.10. Calculate K (25°C).

Answer $K = 2.3 \times 10^2$

Solution $n = 2$ for the reaction. Substitute the known values into the equation

$$\log K = \frac{n\mathcal{E}^0}{0.0592}$$

$$\log K = \frac{(2)(0.070 \text{ J/C})}{0.0592 \text{ J/C}} = 2.36$$

$$K = \text{antilog } (0.36 + 2) = 2.3 \times 10^2$$

We can determine the K for *any* reaction if the reaction can be resolved into two half-cell reactions. The $\mathcal{E}^0$ potential for each half-cell reaction is then used to calculate $\mathcal{E}^0_{cell}$ for the theoretical galvanic cell. The equation

$$\log K = \frac{n\mathcal{E}^0_{cell}}{0.0592}$$

can then be used to calculate the equilibrium constant.

Example 17.14 From the following data, calculate the K_a for HF; the B/L (Brönsted-Lowry) equation is

$$HF + H_2O \rightleftharpoons H_3O^+ + F^-$$

The half-cell reaction and potential for F_2 is

$$F_2 + 2H^+ + 2e^- \rightleftharpoons 2HF \qquad \mathcal{E}^0 = 3.03 \text{ V}$$

Answer $K_a = 3.9 \times 10^{-6}$

Solution First, notice that the cell reaction shows HF as a reactant, but the given half-cell reaction shows HF as a product. We must reverse the partial reaction and also reverse the sign of $\mathcal{E}^0$:

$$2HF \rightleftharpoons F_2 + 2H^+ + 2e^- \qquad -\mathcal{E}^0_{anode} = -3.03\ V$$

Second, choose another half-cell reaction that will give the desired reactant and products shown in the cell reaction when the two half-cell reactions are added together. A suitable partial reaction (text Table 17.1) is

$$F_2 + 2e^- \rightleftharpoons 2F^- \qquad \mathcal{E}^0_{cathode} = +2.87V$$

The cell reaction, equal to the weak acid dissociation of HF in water, is written:

$$F_2 + 2e^- \rightleftharpoons 2F^- \qquad \mathcal{E}^0_{cathode} = +2.87V$$

$$2\ HF \rightleftharpoons F_2 + 2H^+ + 2e^- \qquad -\mathcal{E}_{anode} = -3.03\ V$$

$$2\ HF \rightleftharpoons 2H^+ + 2F^-$$

$$\mathcal{E}^0_{cell} = 2.87\ V + (-3.03\ V) = -0.16\ V$$

$$\log K = \frac{n\mathcal{E}^0}{0.0592} = \frac{2(-0.16)}{0.0592} = -5.41$$

$$K = \text{antilog}\,(-5.41) = 3.9 \times 10^{-6}$$

SECTION 17.13 PREDICTING THE DIRECTION OF A REACTION

Objective

- To predict the spontaneous direction for an oxidation-reduction reaction, given the $\mathcal{E}^0_{cell}$.

Focus

For solutions in which the ionic concentration range is between 0.01 M and 10 M, the log term in the Nernst equation does not greatly change $\mathcal{E}^0_{cell}$. In this concentration range, the $\mathcal{E}^0_{cell}$ can tell us whether or not the cell reaction is spontaneous.

If $\mathcal{E}^0_{cell} > 0$, the reaction progresses from left to right. ΔG^0 is negative ($\Delta G^0 = -n\mathcal{F}\mathcal{E}^0$).

If $\mathcal{E}^0_{cell} < 0$, the reaction progresses from right to left. ΔG^0 is positive.

In Example 17.13 (Study Guide), $\mathcal{E}^0_{cell} = -0.16$ V; the dissociation of HF in an aqueous solution is not favored. On the other hand, the cell reaction in Example 17.1 is spontaneous because $\mathcal{E}^0_{cell}$ is positive. The changes in ionic concentrations from the standard 1 mol/L had little affect on $\mathcal{E}^0_{cell}$ because they were within the 0.01 mol/L to 10 mol/L range. However, in Example 17.10 the ionic concentrations fall outside this range. In this case $\mathcal{E}^0_{cell}$ was *not* a good indicator for predicting the direction of the cell reaction.

SECTION 17.14 STRENGTHS OF OXIDIZING AND REDUCING AGENTS

Objective

- To predict the relative strengths of oxidizing and reducing agents using Table 17.1.

Focus

In Table 17.1, F_2 is the strongest oxidizing agent because F_2 has the greatest potential to gain electrons and be reduced to F^-. Li metal is the strongest reducing agent because it has the least tendency to exist in the reduced form; it prefers to donate an electron to form Li^+. Interpret Table 17.1 as follows.

$Li^+ + e^- \rightleftharpoons Li$

Increasing strength of oxidizing agent (↓) | Increasing strength of reducing agent (↑)

$F_2 + 2e^- \rightleftharpoons 2F^-$

In an oxidation-reduction reaction, the stronger oxidizing and reducing agents produce weaker oxidizing and reducing agents as products.

This is analogous to the Brönsted-Lowry theory of acids and bases (Chapter 14). A strong acid produces a weak conjugate base, and a strong base produces a weak conjugate acid.

HCl	+	H_2O	→	H_3O^+	+	Cl^-
$acid_1$		$base_2$		$acid_2$		$base_1$
(strong)		(strong)		(weak)		(weak)

Because HCl is a stronger acid than H_3O^+ and H_2O is a stronger base than Cl^-, the reaction proceeds to the right. Similarly, for oxidizing and reducing agents, a strong oxidizing agent produces a weak reducing agent, and a strong reducing agent produces a weak oxidizing agent. Recall that an oxidizing agent causes other substances to lose electrons, thereby gaining electrons. Strong oxidizing agents have a strong affinity for electrons. Strong reducing agents, on the other hand, want to get rid of their electrons (be oxidized).

F_2	+	2 Na	→	2 Na^+	+	2 F^-
oxidizing $agent_1$		reducing $agent_2$		oxidizing $agent_2$		reducing $agent_1$
(strong)		(strong)		(weak)		(weak)

F_2 is a stronger oxidizing agent than Na^+, and Na is a stronger reducing agent than F^-; therefore, the reaction proceeds to the right.

You can read Table 17.1 and write the balanced cell equation using the **counterclockwise rule** which states: When reduction potentials for half-reactions are listed in order of increasing values, the half-reaction with the larger $\mathcal{E}^0$ (lower in the table) goes to the right, the half-reaction with the smaller $\mathcal{E}^0$ (higher in the table) goes to the left.

Example 17.15 Does the reaction

$$Hg^{2+} + 2Fe^{2+} \rightarrow Hg + 2Fe^{3+}$$

occur spontaneously?

Answer Yes

Solution In Table 17.1, the two half-cell reactions appear in the following order:

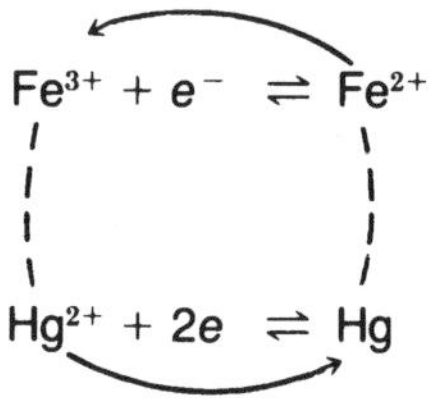

Spontaneous counterclockwise rule

The counterclockwise rule says that the $Hg^{2+} \rightarrow Hg$ and $Fe^{2+} \rightarrow Fe^{3+}$ reactions are spontaneous.

SECTION 17.15 DECOMPOSITION POTENTIAL

Objective

- To calculate the applied potential required for the onset of electrolysis in an electrolytic cell (called the decomposition potential).

Focus

The **decomposition potential** for electrolysis is determined by calculating the $\mathcal{E}_{cell}$ as if it were a galvanic cell. If $\mathcal{E}_{cell}$ is negative, the reaction is *not* spontaneous. The calculated emf (J/C) indicates the degree of nonspontaneity.

In order for the cell reaction to go as written with a negative $\mathcal{E}_{cell}$, an emf equivalent to the negative $\mathcal{E}_{cell}$ must be applied. This is the decomposition potential, that is, the emf necessary for the reaction to go as written. For example, if $\mathcal{E}_{cell} = -0.66$ volt for the cell reaction, then *at least* 0.66 volt must be supplied for electrolysis to occur (for the cell reaction to go as written).

Example 17.16 Calculate the decomposition potential for the cell reaction

$$Pb^{2+}(1.0\ M) + Hg \rightarrow Hg^{2+}(1.0\ M) + Pb$$

Answer 0.977 V

Solution Set up the two half-cell reactions with the corresponding standard potentials:

$$Pb^{2+}(aq) + 2e^- \rightleftharpoons Pb(c) \qquad \mathcal{E}^0_{cathode} = -0.1263\ V$$

$$Hg(c) \rightleftharpoons Hg^{2+}(aq) + 2e^- \qquad -\mathcal{E}^0_{anode} = -(+0.851\ V) = -0.851\ V$$

$$\mathcal{E}^0_{cell} = \mathcal{E}^0_{cathode} + (-\mathcal{E}^0_{anode}) = -0.1263\ V + (-0.851\ V) = -0.977\ V$$

Therefore a minimum emf of 0.977 volt must be supplied to the electrolytic cell for the cell reaction to go as written.

SECTION 17.16 PREDICTING ELECTRODE REACTIONS

Objective

- To predict the most likely product(s) at the anode and cathode in an electrolysis reaction.

Focus

When electrical energy is supplied to an electrolytic cell, the substance most easily oxidized (most negative reduction potential) oxidizes at the anode and the substance most easily reduced (most positive reduction potential) reduces at the cathode.

In the electrolysis of an aqueous solution, generally three possible reactions can occur at the anode:

1. Oxidation of anion(s): $X^{n-} \rightarrow X + ne^-$.
2. Oxidation of H_2O: $2H_2O \rightarrow O_2 + 4H^+ + 4e^-$.
3. Oxidation of a metallic electrode: $M \rightarrow M^{n+} + ne^-$.

and generally two can occur at the cathode:

1. Reduction of cation(s): $A^{n+} + ne^- \rightarrow A$.
2. Reduction of H_2O: $2H_2O + 2e^- \rightarrow H_2 + 2OH^-$.

In some systems, other reactions may occur at either electrode, but these are the most common. The actual half-cell reactions and the corresponding standard potentials are shown in Table 17.1.

Example 17.17 For the electrolysis of an aqueous $CdBr_2$ solution in HBr using nonreactive electrodes,

(a) Write the equations for the possible reactions occurring at each electrode.

(b) Calculate the decomposition potential required for the onset of electrolysis. Assume $\mathcal{E} \simeq \mathcal{E}^0$.

Answer (a) Anode: $2Br^- \rightarrow Br_2 + 2e^-$; Cathode: $2H^+ + 2e^- \rightarrow H_2$

(b) 1.08 volts

Solution

Possible anode reactions

(nonreactive electrode)

Anion:	$2Br^- \rightarrow Br_2 + 2e^-$	$\mathcal{E}^0 = +1.08$ V*
H_2O:	$2H_2O \rightarrow O_2 + 4H^+ + 4e^-$	$\mathcal{E}^0 = +1.229$ V

* The anode reaction is the one with the more negative reduction potential, the reaction least likely to occur as a reduction (or the one most likely to occur as an oxidation). The oxidation of Br^- occurs at the anode.

Possible cathcde reactions

Cation:	$Cd^{2+} + 2e^- \rightarrow Cd$	$\mathcal{E}^0 = -0.4026$
Cation:	$2H^+ + 2e^- \rightarrow H_2$	$\mathcal{E}^0 = 0.00$ V*
H_2O:	$2H_2O + 2e^- \rightarrow H_2 + 2OH^-$	$\mathcal{E}^0 = -0.8277$ V

* The cathode reaction is the one with the more positive reduction potential. The reduction of H^+ occurs at the cathode. The cell reaction is

$$2Br^- + 2H^+ \rightarrow H_2 + Br_2$$

$$\mathcal{E}^0_{\text{cathode}} = 0.00 \text{ V}$$

$$-\mathcal{E}^0_{\text{anode}} = -(+1.08 \text{ V}) = -1.08 \text{ V}$$

$$\mathcal{E}^0_{\text{cell}} = \mathcal{E}^0_{\text{cathode}} + (-\mathcal{E}^0_{\text{anode}})$$

$$\mathcal{E}^0_{\text{cell}} = 0.00 \text{ V} + (-1.08 \text{ V}) = -1.08 \text{ V}$$

The decomposition potential is 1.08 volts.

SECTION 17.17 SOME PRACTICAL CELLS

Objectives

- To categorize the various types of practical galvanic cells and to explain how they work.
- To consider some advantages and disadvantages of H_2/O_2 fuel cells and a H_2 energy base.

Focus

Be familiar with the various types of practical galvanic cells.

- **Single-use** or **primary cells**: dry cell or common flashlight battery, alkaline dry cell, and the Hg cell.
- **Storage cells**: lead-acid cell and the alkaline "Nicad" cell.
- **Fuel cells**: H_2/O_2.

QUESTIONS

Section 17.2

17.1 How many faradays are needed to electroplate 38.7 g Cu from a $Cu(NO_3)_2$ solution?

17.2 A current of 0.600 A passes through a molten K_2RuCl_6 for 45 minutes. Calculate the mass of ruthenium (atomic weight = 101) that is electroplated.

17.3 Calculate the current in amperes required to produce 15.0 L Cl_2 at STP in 30 minutes from the electrolysis of molten NaCl.

Section 17.3

17.4 Using a $CuSO_4$ solution, a strip of Cu metal, a $Cr_2(SO_4)_3$ solution, and a strip of Cr metal, construct an operative galvanic cell. ΔG^0 for the reduction of Cu^{2+} is more negative than for Cr^{3+}. Label the anode, cathode, direction of electron flow, and the direction of Cu^{2+}, Cr^{3+}, and SO_4^{2-} flow.

17.5 Write the anode and cathode reactions and the cell reaction that occur in each cell:

(a) $Ni(c) \mid NiCl_2(aq) \parallel Hg_2Cl_2(c) \mid Hg(\ell)$

(b) $Fe(c) \mid Fe(NO_3)_2(aq) \parallel Cu(NO_3)_2(aq) \mid Cu(c)$

17.6 Write the cell notation for the following reaction:

$$3Sn(c) + 2HCrO_4^-(aq) + 14H^+(aq) \rightarrow 3Sn^{2+}(aq) + 2Cr^{3+}(aq) + 8H_2O$$

Section 17.8

17.7 In a galvanic cell oxidation occurs at the (anode, cathode) which is the (positive, negative) electrode.

Section 17.9

17.8 Determine $\mathcal{E}^0_{cell}$ for the galvanic cell

$$Ni|Ni^{2+}\ (1.0\ M) \parallel Pb^{2+}\ (1.0\ M) \mid Pb$$

17.9 In the galvanic cell $Sn \mid Sn^{2+}$ (1.0 M) $\parallel Cu^{2+}$ (1.0 M) $\mid$ Cu, the (Sn, Cu) electrode increases in mass for the cell as written.

17.10 Calculate ΔG^0 for the galvanic cell

$$Cd \mid Cd^{2+}\ (1.0\ M) \parallel HgCl_4^{2-}\ (1.0\ M),\ Cl^-\ (1.0\ M) \mid Hg$$

17.11 $\mathcal{E}^0_{cell}$ for the galvanic cell

$$M \mid M^{3+}\ (1.0\ M) \parallel Hg^{2+}\ (1.0\ M) \mid Hg$$

is 0.110 V. Calculate $\mathcal{E}^0$ for the $M^{3+} \mid M$ half-cell reaction.

Section 17.10

17.12 Refer to Example 17.8. Which of the following changes increase the cell potential?

(a) increasing the Sn^{2+} concentration

(b) increasing the V^{3+} concentration

(c) increasing the pH in the cathode chamber

(d) increasing the VO^{2+} concentration

(e) increasing the size of the Sn anode

17.13 Given the electrode potentials

$$Cr^{3+} + e^- \rightleftharpoons Cr^{2+} \qquad \mathcal{E}^0 = -0.41 \text{ V}$$

$$ClO_3^- + 3H^+ + 2e^- \rightleftharpoons HClO_2 + H_2O \qquad \mathcal{E}^0 = 1.21 \text{ V}$$

and the galvanic cell

$$Pt \mid Cr^{3+}\ (10^{-2}\ M),\ Cr^{2+}\ (10^{-6}\ M) \parallel ClO_3^-\ (10^{-3}\ M),\ H^+\ (pH = 4),\ HClO_2\ (1.0\ M) \mid Pt$$

(a) Calculate $\mathcal{E}^0_{cell}$.

(b) Write the standard cell reaction.

(c) Calculate $\mathcal{E}_{cell}$.

(d) What is ΔG of the galvanic cell? Is the cell reaction spontaneous as written?

17.14 Determine $\mathcal{E}_{cell}$ for the galvanic cell

$$Ag \mid Ag^+\ (2.00 \times 10^{-4}\ M) \parallel Fe^{3+}\ (2.00\ M),\ Fe^{2+}\ (1.00 \times 10^{-6}\ M) \mid Pt$$

Section 17.11

17.15 For the galvanic cell

$$Mn \mid Mn^{2+}\ (xM) \parallel Cr^{3+}\ (1.0 \times 10^{-7}\ M) \mid Cr$$

the $\mathcal{E}_{cell}$ is 0.40 V. What is the Mn^{2+} concentration?

$$Mn^{2+} + 2e^- \rightleftharpoons Mn \qquad \mathcal{E}^0 = -1.18 \text{ V}$$

$$Cr^{3+} + 3e^- \rightleftharpoons Cr \qquad \mathcal{E}^0 = -0.74 \text{ V}$$

17.16 A pH meter is equipped with a glass electrode and an internal standard reference calomel electrode with a potential, $\mathcal{E}'$, of 0.283 V. This reference calomel electrode has a half-cell reaction

$$2e^- + Hg_2Cl_2(c) \rightarrow 2Hg(\ell) + 2Cl^-(1.0\ M)$$

If the measured potential for a solution with an unknown pH is 0.598 V, what is the pH of the solution?

Section 17.12

17.17 Calculate K for the cell reactions in Questions 17.8 and 17.13.

17.18 Using Tables 17.1 and 17.2, calculate K of $PbSO_4$.

$$PbSO_4 \rightleftharpoons Pb^{2+} + SO_4^{2-}$$

17.19 What are the values of $\mathcal{E}_{cell}$, ΔG, and w_{max} when a galvanic cell reaction has reached equilibrium?

Section 17.13

17.20 Which of the following reactions are likely to occur spontaneously? Calculate $\mathcal{E}^0_{cell}$ for each.

(a) $Cd + 2H^+ \rightarrow Cd^{2+} + H_2$

(b) $HgBr_4^{2-} + 4I^- \rightleftharpoons HgI_4^- + 4Br^-$

(c) $3Pt + 2MnO_4^- + 4H_2O \rightleftharpoons 3Pt^{2+} + 2MnO_2 + 8OH^-$

(d) $5Pt + 2MnO_4^- + 16H^+ \rightleftharpoons 5Pt^{2+} + 2Mn^{2+} + 8H_2O$

Section 17.14

17.21 Which is the stronger oxidizing agent?

(a) Pb^{2+} or Ni^{2+}

(b) HgI_4^- or $HgBr_4^-$

(c) O_2 (acid solution) or O_2 (basic solution)

(d) MnO_4^- (acid solution) or MnO_4^- (basic solution)

17.22 Which is the stronger reducing agent?

(a) Ni or Sn

(b) Ag or Au

(c) Zn or Na

17.23 (a) Does Zn displace H^+ from an aqueous solution?

(b) Will Pb displace Ag^+ from an aqueous solution?

(c) Will Cl_2 oxidize Pt?

Section 17.15

17.24 Calculate the decomposition potential for the reactions

(a) $Pt + I_2 \rightarrow Pt^{2+} + 2I^-$

(b) $2Na^+ + 2Cl^- \rightarrow 2Na + Cl_2$

Section 17.16

17.25 When an emf is applied to a solution containing the following ions at approximate equal concentrations, which ion electroplates first?

(a) Cu^{2+} or Fe^{2+}

(b) Zn^{2+} or Pb^{2+}

(c) Mg^{2+} or Hg^{2+}

17.26 An aqueous solution of ZnI_2 is electrolyzed. Assume $\mathcal{E} \simeq \mathcal{E}^0$. (a) What is the predominant reaction at each electrode using nonreactive electrodes? What is the decomposition potential for the reaction?

(b) What is the predominant reaction at each electrode using Zn electrodes? What is the decomposition potential for the reaction?

Section 17.17

17.27 Complete the following table.

	Dry Cell	Alkaline Dry Cell	Mercury Cell
Anode reaction			
Cathode reaction			
Electrolyte			
Emf			

17.28 Complete the following table.

	Lead-Acid Battery	Nicad Battery
Electrolyte		
Discharge anode reaction		
Discharge cathode reaction		
Discharge cell reaction		
Discharge emf		
Charge anode reaction		
Charge cathode reaction		
Charge cell reaction		

17.29 How does the density of the electrolyte in the lead-acid battery determine whether or not a cell is charged or discharged?

17.30 The anode reaction for the CH_4 fuel cell is

$$CH_4 + 2H_2O \rightarrow CO_2 + 8H^+ + 8e^-$$

Assuming O_2 to be the "oxidizer," write the cell reaction for the fuel cell.

ANSWERS

17.1 1.22$\mathfrak{F}$

17.2 0.424 g Ru

17.3 71.8 amperes

17.4 See Figure 17.4.

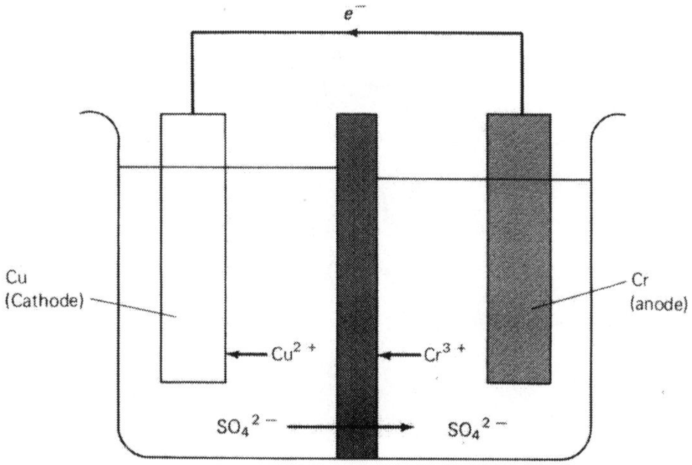

Figure 17.4 Answer to question 17.4.

17.5 (a) $Ni(c) + Hg_2Cl_2(c) \rightarrow Ni^{2+}(aq) + Hg(\ell) + 2Cl^-(aq)$
(b) $2Fe(c) + 3Cu^{2+}(aq) \rightarrow 2\,Fe^{3+}(aq) + 3Cu(c)$

17.6 $Sn(c) \mid Sn^{2+}(aq) \parallel HCrO_4^-(aq), Cr^{3+}(aq), H^+(aq) \mid Pt$

17.7 anode, negative

17.8 $\mathcal{E}^0_{cell} = 0.12$ V

17.9 Cu

17.10 $\Delta G^0 = -151$ kJ

17.11 $\mathcal{E}^0\,(M^{3+} \mid M) = 0.741$ V

17.12 (d) only

17.13 (a) 1.62 V
(b) $ClO_3^- + 3H^+ + 2Cr^{2+} \rightarrow HClO_2 + H_2O + 2Cr^{3+}$
(c) 0.94 V
(d) -180 kJ; yes

17.14 0.56 V

17.15 $[Mn^{2+}] = 4.83 \times 10^{-4}$ mol/L

17.16 pH = 5.32

17.17 1.13×10^4; 5.37×10^{54}

17.18 1.70×10^{-8}

17.19 all equal zero

17.20 (a) yes, $\mathcal{E}^0 = +0.4026$ V
(b) yes, $\mathcal{E}^0 = +0.25$ V
(c) no, $\mathcal{E}^0 = -0.6$ V
(d) yes, $\mathcal{E}^0 = +0.3$ V

17.21 (a) Pb^{2+}
(b) $HgBr_4^{2-}$
(c) O_2 (acidic solution)
(d) MnO_4^- (acidic solution)

17.22 (a) Ni; (b) Ag; (c) Na

17.23 (a) yes; (b) yes; (c) yes

17.24 (a) +0.7 V; (b) +4.071 V

17.25 (a) Cu^{2+}; (b) Pb^{2+}; (c) Hg^{2+}

17.26 (a) Anode: $I_2 \rightarrow 2I^- + 2e^-$ $\quad -\mathcal{E}^0 = -0.535$ V
Cathode: $Zn^{2+} + 2e^- \rightarrow Zn$ $\quad \mathcal{E}^0 = -0.7678$ V
$\mathcal{E}^0 = -1.298$ V

Decomposition potential = 1.298 V

(b) Anode: $Zn \rightarrow Zn^{2+} + 2e^-$ $-\mathcal{E}^0 = -(-0.7628\ V)$

Cathode: $Zn^{2+} + 2e^- \rightarrow Zn$ $\mathcal{E}^0 = -0.7628\ V$

$\mathcal{E}^0 = 0.00\ V$

Decomposition potential = 0.00 V

17.27

	Dry Cell	Alkaline Dry Cell	Mercury Cell
Anode reaction	$Zn \rightarrow Zn^{2+} + 2e^-$	$Zn \rightarrow Zn^{2+} + 2e^-$	$Zn + 2OH^- \rightarrow ZnO + 2e^-$
Cathode reaction	$NH_4^+ + MnO_2 + e^- \rightarrow Mn(O)OH + NH_3$	$NH_4^+ + MnO_2 + e^- \rightarrow Mn(O)OH + NH_3$	$HgO + H_2O + 2e^- \rightarrow Hg + 2OH^-$
Electrolyte	H_2O, NH_4Cl	KOH	KOH
Emf	1.48 V	1.48 V	1.3 V

17.28

	Lead-Acid Battery	Nicad Battery
Electrolyte	H_2SO_4	KOH
Discharge anode reaction	$Pb + HSO_4^- \rightarrow PbSO_4 + H^+ + 2e^-$	$Cd + 2OH^- \rightarrow CdO + H_2O + 2e^-$
Discharge cathode reaction	$PbO_2 + 3H^+ + HSO_4^- + 2e^- \rightarrow PbSO_4 + 2H_2O$	$Ni_2O_3 + 3H_2O + 2e^- \rightarrow 2Ni(OH)_2 + 2OH^-$
Discharge cell reaction	$Pb + PbO_2 + 2H^+ + 2HSO_4^- \rightarrow 2PbSO_4 + 2H_2O$	$Cd + Ni_2O_3 + 2H_2O \rightarrow CdO + 2Ni(OH)_2$
Discharge emf	2 V	1.2 V
Charge anode reaction	$PbSO_4 + H^+ + 2e^- \rightarrow Pb + HSO_4^-$	$CdO + 2H_2O + 2e^- \rightarrow Cd + 2OH^-$
Charge cathode reaction	$PbSO_4 + 2H_2O \rightarrow PbO_2 + 3H^+ + HSO_4^- + 2e^-$	$2Ni(OH)_2 + 2OH^- \rightarrow Ni_2O_3 + 3H_2O + 2e^-$
Charge cell reaction	$2PbSO_4 + 2H_2O \rightarrow Pb + PbO_2 + 2H^+ + 2HSO_4^-$	$CdO + 2Ni(OH)_2 \rightarrow Cd + Ni_2O_3 + 2H_2O$

17.29 The solution in a charged cell has a higher density from the H_2SO_4 than that in the discharged cell which has more H_2O.

17.30 $CH_4 + 2O_2 \rightarrow CO_2 + 2H_2O$

Chapter 18

CHEMICAL KINETICS

SECTION 18.1 DETERMINATION OF CONCENTRATION CHANGES WITH TIME

Objective

- To explain the concept of reaction rate.

Focus

We determine the **rate of a chemical reaction** by observing the change in the concentration of a reactant or product as a function of time. This *change* may be a color change, a pressure change (if one of the substances is a gas), a temperature change, or a pH change.

For the hydrolysis of butyl chloride,

$$C_4H_9Cl(\ell) + H_2O(\ell) \rightarrow C_4H_9OH(\ell) + HCl(\ell)$$

we can determine the reaction rate by measuring the change in pH (a function of the production of HCl) with time.

Example 18.1 Explain how we determine the rate of growth of a child as a function of time. What is the general shape of the curve?

Answer $\dfrac{\text{centimeters of growth}}{\text{month}}$

Solution A baby's length is always measured at birth. Measure the height at the end of each year. Record and plot the data. The curve should have the shape shown in Figure 18.1.

The rate of a reaction generally decreases with time. For example, when a matchhead burns, it ignites rapidly. The burning rate slows down, however, because the concentration of the chemicals on the matchhead decreases.

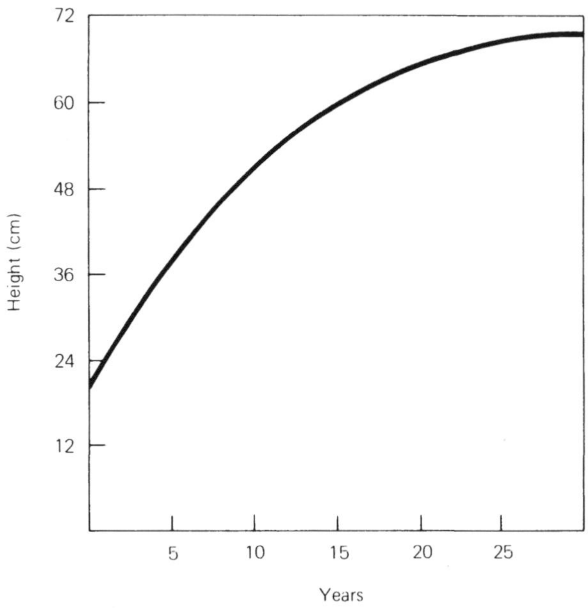

Figure 18.1 Plot of child's growth against time.

SECTION 18.2 CONDITIONS AFFECTING REACTION RATES

Objectives

- To list various factors that affect the rate of a chemical reaction.
- To determine the rate equation for a chemical reaction.
- To determine the order of a reaction.

Focus

Five factors affect the rate of a chemical reaction.

1. **The nature of the reactants**. Substances have different chemical properties and therefore react at different rates. For example, Fe reacts more rapidly with O_2 than does Au.
2. **The concentration of reactants**. The effect of concentration on reaction rate is expressed quantitatively by a *rate equation*. The rate equation for the reaction $A + B \rightarrow AB$, is

$$rate = k[A]^x[B]^y$$

   x and y indicate the *order* of the reaction with respect to [A] and [B] and can only be determined from experimental data. If doubling the molar concentration of A doubles the reaction rate, then the reaction is first order with respect to A or $x = 1$. If [A] is doubled and the reaction rate increases by a factor of 4, then the reaction is second order with respect to A and $x = 2$. The order of a reaction is often a whole number, but zero, fractional, and even negative orders are known.

Example 18.2 The reaction,

$$R_2 + 2T \rightarrow 2RT$$

given the following set of data at 50°C:

Expt. No.	$[R_2]$	$[T]$	Rate $= \frac{\text{change in } [R_2]}{\text{change in time}} \left(\frac{\text{mol } R_2}{\text{L sec}}\right)$
1	0.010	0.030	2.0×10^{-2}
2	0.020	0.030	16.0×10^{-2}
3	0.020	0.060	32.0×10^{-2}

Determine the rate equation for the reaction. Find the order of the reaction.

Answer rate $= k[R_2]^3[T]$; fourth order

Solution The general form of this reaction's rate equation is

$$\text{rate} = k[R_2]^x[T]^y$$

Expt. No.	$[R_2]$	$[T]$	Rate $= \frac{\text{change in } [R_2]}{\text{change in time}} \left(\frac{\text{mol } R_2}{\text{L sec}}\right)$
1	0.010 ⤵ ×2	0.030	2.0×10^{-2} ⤵ ×8
2	0.020	0.030 ⤵ ×2	16.0×10^{-2} ⤵ ×2
3	0.020	0.060	32.0×10^{-2}

x determination: To find the value of x, the order of the reaction with respect to R_2, compare experiments 1 and 2 in which the concentration of T remains constant and therefore does not affect reaction rate. The rate equation has the shortened form:

$$\text{rate} = k[R_2]^x \times \text{constant}$$

The rate increases by a factor of 8 when $[R_2]$ increases by a factor of 2 or

$$(\times 8) = k\ [\times 2]^x \times \text{constant}$$

Mathematically how is a factor of 2 made equal to a factor of 8? To maintain the equality, x must equal 3. Therefore, the reaction is third order with respect to R_2.

y determination: Compare experiments 2 and 3 where $[R_2]$ is constant. The rate equation is

$$\text{rate} = k \times \text{constant} \times [T]^y$$

From the data, the rate increases by a factor of 2 when the concentration of T increases by a factor of 2, or

$$(\times 2) = k \times \text{constant} \times [\times 2]^y$$

y can only equal 1 to maintain the equality; the reaction is first order with respect to T.

Therefore, the rate equation is rate $= k[R_2]^3[T]^1$, which is a fourth-order reaction.

Recognize that the absolute concentrations of the reactants and the absolute reaction rates are unimportant in deriving the rate equation. We consider only the *change* in concentrations and how this change affects the *change* in the rate.

We calculate the rate constant, k, by substituting the molar concentrations of the reactants and the reaction rate for a given experiment into the rate equation. Rate constants are constant only as long as the temperature does not change.

Example 18.3 Calculate the rate constant, k, for the reaction in Example 18.2.

Answer $k = 6.7 \times 10^5\ L^3/(mol^3\ sec)$

Solution Substituting the data from experiment 1 into the rate equation,

$$\text{rate} = k[R_2]^3[T]$$

$$2.0 \times 10^{-2}\ \frac{\text{mol}}{\text{L sec}} = k[0.010]^3[0.030]$$

$$k = 6.7 \times 10^5\ \frac{\text{L}^3}{(\text{mol}^3\ \text{sec})}$$

Substituting data from experiments 2 and 3 into the rate equation gives the same value for k.

The balanced equation for the reaction can tell us the *relative rates* at which each substance reacts or is produced.

Example 18.4 For the reaction

$$X_2 + 3Z \rightarrow X_2Z_3$$

suppose that X_2 reacts at the rate of 3.0×10^{-6} mol/(L sec). What is the rate at which Z is reacting?

Answer 9.0×10^{-6} mol Z/(L sec)

Solution According to the balanced equation, for each mol X_2 reacting, three mol Z also reacts. Therefore, the reaction rate with respect to Z

$$= 3.0 \times 10^{-6}\ \frac{\cancel{\text{mol X}_2}}{\text{L sec}} \times \frac{3\ \text{mol Z}}{1\ \cancel{\text{mol X}_2}}$$

$$= 9.0 \times 10^{-6}\ \text{mol Z/(L sec)}$$

Now let us return to the other three factors that affect reaction rate.

3. **Particle size in heterogeneous reactions.** For heterogeneous reactions occurring at the interface between two phases, the larger the surface area at the interface, the faster is the reaction. For example, HCl(*aq*) reacts very slowly with marble, a natural form of $CaCO_3$, but rapidly with marble dust.
4. **Temperature.** A rise in temperature invariably increases reaction rate; generally, for every 10°C temperature rise, the rate doubles or triples.
5. **Catalyst.** We will discuss this in section 18.7.

SECTION 18.3 HALF-LIFE FOR FIRST-ORDER REACTIONS

Objectives

- To explain the relationship between the half-life and the kinetic stability of first-order reactions.
- To calculate the time needed for a reactant to decrease to a certain concentration.

Focus

The rate of a *first-order* reaction is frequently expressed in terms of its **half-life**, the time required for one-half of a given quantity of reactant to be consumed. The longer the half-life, the greater is the kinetic stability of the reactant. Remember that the kinetic stability of a reaction refers to the rate at which the reaction occurs.

The relationship between the initial quantity or concentration of a reactant, q_0, and its quantity or concentration, q_t, after time t has elapsed for a first-order reaction is

$$\log \frac{q_0}{q_t} = \frac{kt}{2.303}$$

Memorize this equation and know how to work the equation to solve for q_o, q_t, and t. q_0 and q_t have no required units because the units cancel in the "log" term. Therefore, q_0 and q_t may have units of moles, mol/L, grams, volume at STP, number of atoms or molecules, and so on.

Example 18.5 For the reaction $3A + B_2 \rightarrow A_3B_2$, the experimental rate equation is rate $= k[B_2]$, where $k = 2.66 \times 10^{-3}\ \text{sec}^{-1}$. If 17.0 g B_2 are placed in the reaction vessel, how long will it take to reduce the amount of B_2 to 5.00 grams?

Answer 4.61×10^2 sec

Solution Because the reaction is first order, (with respect to $[B_2]$), we can use the equation

$$\log \frac{q_0}{q_t} = \frac{kt}{2.303}$$

In the problem we have given,

q_0 = 17.0 g, the amount of B_2 initially

q_t = 5.00 g, the amount of B_2 after time, t, the time that we solve for

$k = 2.66 \times 10^{-3}\ \text{sec}^{-1}$

$$\log \frac{17.0}{5.00} = 0.532 = \frac{2.66 \times 10^{-3}\ \text{sec}^{-1} \times t}{2.303}$$

$$t = \frac{(2.303)(0.532)}{2.66 \times 10^{-3}\ \text{sec}^{-1}} = 4.61 \times 10^2\ \text{sec}$$

The half-life of a reactant is the time required for one-half of it to react. At $t_{1/2}$, q_t equals one-half of q_0 in the equation. Thus the log term becomes

$$\log \frac{1}{1/2} \quad \text{or} \quad \log \frac{2}{1} = 0.301$$

$$\text{and } 0.301 = \frac{kt_{1/2}}{2.303}$$

The shorter the half-life, the more kinetically *un*stable is the reactant and the faster the reaction.

$$t_{1/2} = \frac{(2.303)(0.301)}{k} = \frac{0.693}{k}$$

For calculating the $t_{1/2}$ (half-life) of a reaction from its rate constant, memorize this equation.

Example 18.6 Calculate the half-life of B_2 in Example 18.5.

Answer 260 sec

Solution We use the equation

$$t_{1/2} = \frac{0.693}{k}$$

$$k = 2.66 \times 10^{-3}\ \text{sec}^{-1}$$

$$t_{1/2} = \frac{0.693}{2.66 \times 10^{-3}\ \text{sec}^{-1}} = 260\ \text{sec}$$

Therefore, given any amount of B_2, in 260 sec only one-half of the original amount remains.

Example 18.7 Calculate the mass of B_2 in Example 18.5 left after 1000 seconds have elapsed in the reaction.

Answer 1.19 g

Solution The equations to use are

$$\log \frac{q_0}{q_t} = \frac{kt}{2.303} \quad \text{or} \quad q_t = q_0 \times 0.5^{t/t_{1/2}}$$

and solve for q_t. All other terms are known.

$$\frac{q_0}{q_t} = \frac{17.0\ \text{g}}{q_t} = \text{antilog}\ \frac{kt}{2.303} = \text{antilog}\left(\frac{2.66 \times 10^{-3}\ \text{sec}^{-1} \times 1000\ \text{sec}}{2.303}\right)$$

$$= \text{antilog}\ 1.155 = 14.29$$

$$q_t = \frac{17.0\ \text{g}}{14.29} = 1.19\ \text{g}$$

Or since $t_{1/2} = 260$ sec,

$$q_t = 17.0\ \text{g} \times 0.5^{1000\ \text{sec}/260\ \text{sec}}$$

$$q_t = 1.19\ \text{g}$$

SECTION 18.4 COLLISION THEORY OF REACTION RATES

Objective

- To identify the three requirements for an effective collision between reactant molecules to form product molecules.

Focus

The three requirements are:

1. Molecules must collide with one another; therefore the frequency of collisions is important. The greater the frequency of collisions, the faster is the reaction rate.
2. Molecules must collide with enough kinetic energy so that when kinetic energy converts to internal energy, bonds in the reactant molecules are weakened. The energy required in a collision for bond rupture to occur is called the **energy of activation**, E_a.
3. Reactant molecules in the collision must be oriented so that there is a favorable exchange of atoms to produce the new molecules. This orientation factor for the colliding molecules is called the **entropy of activation**, $\Delta S\ddagger$. The greater the $\Delta S\ddagger$, the larger is the number of preferred orientations for a reaction and the faster is the reaction. A large $\Delta S\ddagger$ means that the required arrangement of the colliding molecules is nonspecific for the formation of products; any arrangement is desirable for the rapid formation of products. A small or even negative $\Delta S\ddagger$ indicates that the arrangement of the colliding molecules is very specific—if the molecules do not collide with the right orientation, products will not form.

Example 18.8 The E_a and $\Delta S\ddagger$ for two sets of reactions are determined at 25°C. Which reaction in each set proceeds at a faster rate? Assume all reactions to be kinetically *un*stable.

	E_a (kJ/mol)	$\Delta S\ddagger$ (J/mol K)
Set A		
1. $2A + B_2 \rightarrow A_2B_2$	66.7	+21.0
2. $A + 2C \rightarrow AC_2$	14.2	+21.0
Set B		
1. $B_2 + 2C \rightarrow 2BC$	44.7	+14.0
2. $B_2 + D \rightarrow BD + B$	44.7	+38.9

Answer Set A, reaction 2; Set B, reaction 2

Solution Set A: Reaction 2 has a lower energy of activation; that is, less energy is required to break the bonds in the reactant molecules.

Set B: Reaction 2 has a higher entropy of activation; that is, the atoms in the transition state need not be as specifically oriented as in reaction 1. Less specificity in the orientation means the reaction is faster.

SECTION 18.5 TRANSITION STATE THEORY OF REACTION RATES

Objective

- To identify the energy factors required for reactants to proceed through the transition state to form products.

Focus

Understanding the nature of transition states and its energy factors helps chemists understand chemical reaction rates.

In order for reactant molecules to reach the **transition state**, the bonds breaking between atoms in the reactant molecules must be of the approximate strength as are the bonds forming in the product molecules. The enthalpy needed for reactant molecules to reach the transition state, $\Delta H\ddagger$ (**enthalpy of activation**), is

$$\Delta H\ddagger = H_{ts} - H_r$$

where H_{ts} is the enthalpy of the transition state molecules and H_r is the enthalpy of the reactant molecules. For most reactions,

$$\Delta H\ddagger \simeq E_a$$

When the transition state proceeds to the formation of products, energy is released and the final stable state of enthalpy for the product molecules is H_p. The energy absorbed or released in the overall reaction is given by the equation

$$\Delta H = H_p - H_r$$

Generally, the higher the H_{ts}, the slower is the reaction; the reactants are kinetically more stable. If the enthalpy of the products, H_p, is less than that of the reactants, H_r, the reaction is exothermic, regardless of the magnitude of H_{ts}. If H_p is greater than H_r, the reaction is endothermic. This is summarized in Figure 18.2.

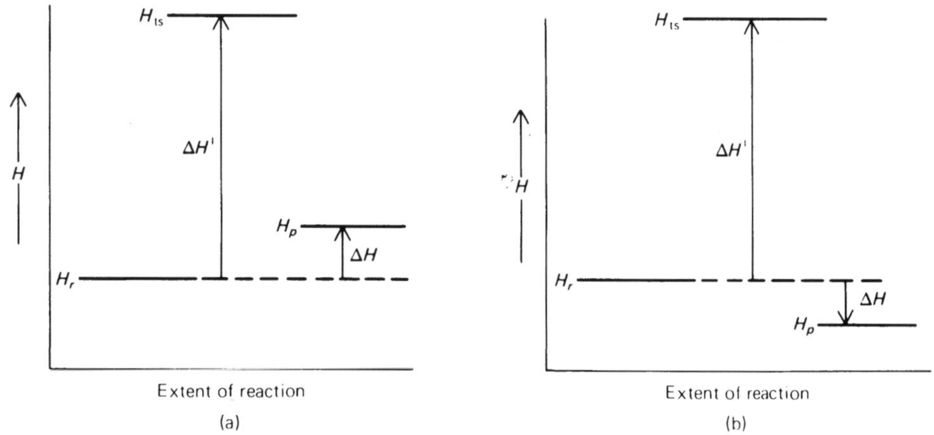

Figure 18.2 Enthalpy diagram for an (a) endothermic reaction; (b) exothermic reaction.

Example 18.9 From Figure 18.3, determine each of the following.

(a) H_r = ________ kJ
(b) H_p = ________ kJ
(c) H_{ts} = ________ kJ
(d) $\Delta H\ddagger$ = ________ kJ
(e) ΔH = ________ kJ
(f) The enthalpy of activation for the reverse reaction = ________ kJ
(g) Is the reaction exothermic or endothermic?

Answer (a) 24.0 kJ
(b) 32.0 kJ
(c) 48.0 kJ
(d) (48.0 − 24.0) kJ = 24.0 kJ
(e) (32.0 − 24.0) kJ = 8.0 kJ
(f) (48.0 − 32.0) kJ = 16.0 kJ
(g) endothermic

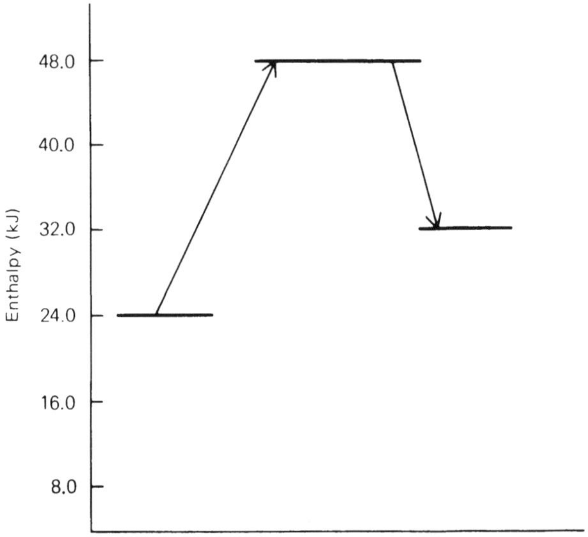

Figure 18.3 Enthalpy diagram, Example 18.9.

Example 18.10 Using the data in Example 18.8, calculate the kinetic stability, $\Delta G\ddagger$, for the four reactions at 25°C. Assume $\Delta E \simeq \Delta H\ddagger$.

Answer Set A.1: $\Delta G\ddagger$ = +60.4 kJ/mol
Set A.2: $\Delta G\ddagger$ = +7.9 kJ/mol
Set B.1: $\Delta G\ddagger$ = +40.5 kJ/mol
Set B.2: $\Delta G\ddagger$ = +33.1 kJ/mol

Solution

A.1: $\Delta G\ddagger = \Delta H\ddagger - T\Delta S\ddagger$

$$= 66.7\ \frac{\text{kJ}}{\text{mol}} - 298\ \cancel{\text{K}} \left(\frac{0.021\ \text{kJ}}{\text{mol}\ \cancel{\text{K}}}\right)$$

$$= +60.4\ \text{kJ/mol}$$

$$\text{A.2:} \quad \Delta G\ddagger = \frac{14.2 \text{ kJ}}{\text{mol}} - 298 \text{ K} \left(\frac{0.021 \text{ kJ}}{\text{mol K}}\right)$$
$$= +7.9 \text{ kJ/mol}$$

$$\text{B.1:} \quad \Delta G\ddagger = 44.7 \frac{\text{kJ}}{\text{mol}} - 298 \text{ K} \left(\frac{0.014 \text{ kJ}}{\text{mol K}}\right)$$
$$= +40.5 \text{ kJ/mol}$$

$$\text{B.2:} \quad \Delta G\ddagger = 44.7 \frac{\text{kJ}}{\text{mol}} - 298 \text{ K} \left(\frac{0.0389 \text{ kJ}}{\text{mol K}}\right)$$
$$= +33.1 \text{ kJ/mol}$$

Reactions A.2 and B.2 are kinetically less stable because $\Delta G\ddagger$ for each reaction in their respective set is more negative.

SECTION 18.6 MECHANISM OF A REACTION

Objectives

- To recognize a correct possible mechanism for a reaction, given the experimental rate equation.
- To predict the rate equation from a suggested mechanism.

Focus

The **mechanism** of a chemical reaction is the proposed step-by-step sequence by which reactants form products. Frequently, chemical reactions occur in several consecutive steps, each proceeding at its own rate. The rate of the overall reaction, however, proceeds no faster than its slowest step. After all, how fast can a team of horses go? No faster than the slowest horse!

Chemists propose a mechanism on the basis of all available kinetic data for a chemical reaction, particularly the data used in determining the rate equation.

> The proposed mechanism is based upon several after-the-fact observations dealing with the reaction. In this case, a chemist faces nearly the same dilemmas as a detective trying to determine the finer details of a murder. The detective knows the "before" and "after" but not the "how" or "who." Various leads are followed (more experiments), which may reveal additional facts (presence or absence of chemical intermediates in the reaction). After rechecking each step in the investigation, the detective proposes an analysis of the "how" and "who" consistent with the available facts. This is the *mechanism* of the murder. For a chemist, the proposed mechanism is the one most consistent with present experimental facts, but continued research may show that the mechanism is not entirely correct.

Each step in a mechanism has its own rate equation, but the one for the slowest step must be consistent with the experimentally determined rate equation.

Example 18.11 Write the rate equation for each of the following steps in the proposed mechanism for the reaction:

$$2A + B_2 \rightarrow 2AB$$

Step 1. $A + A \rightarrow A_2$

Step 2. $A_2 + B_2 \rightarrow A_2B + B$

Step 3. $A_2B + B \rightarrow 2AB$

Answer

Step 1: rate $= k_1[A]^2$
Step 2: rate $= k_2[A_2][B_2]$
Step 3: rate $= k_3[A_2B][B]$

Solution The rate equation for each step is proportional to the product of the molar concentration of each reactant. The exponent of the reactant in the rate equation for each step is the same as its coefficient in that equation. (Remember that this is not true for the overall reaction.) k_1, k_2, and k_3 designate the proportionality constant for each step.

Example 18.12 The experimental rate equation for the reaction $2A + B_2 \rightarrow 2AB$ in Example 18.11 is

$$\text{rate} = k[A]^2[B_2]$$

What step in the proposed mechanism, given in Example 18.11, is the rate determining step?

(*Note*: It is pure coincidence that the coefficients in the balanced equation are the same as the exponents of the reactants in the rate equation.)

Answer Step 2

Solution From the mechanism and the balanced equation, A_2, A_2B, and B are reaction intermediates and do not appear in the experimental rate equation.

1. Step 1 does *not* agree with the experimental rate equation, so step 1 is assumed to be a fast, reversible reaction.

$$A + A \rightleftharpoons A_2 \quad \text{(fast)}$$

The equilibrium condition is

$$K_1 = \frac{[A_2]}{[A]^2}$$

2. Step 2 also does not agree with the experimental rate equation. It shows an intermediate, $[A_2]$, not shown in the experimental rate equation. Notice, however, that by substituting the equilibrium conditions in step 1,

$$[A_2] = K_1[A]^2$$

into the rate equation for step 2, we have

$$\text{rate} = k_2[A_2][B_2] = k_2(K_1[A]^2)[B_2]$$

k_2K_1 equals the rate constant, k, in the experimental rate equation. The rate equation for step 2 agrees with the experimental rate equation. Therefore, step 2 is the slowest step in the mechanism.

SECTION 18.7 CATALYSIS

Objective

- To understand how catalysts work.

Focus

Homogeneous and heterogeneous (contact) **catalysts** increase reaction rate without undergoing any *resultant* change in chemical composition. A catalyst *may* change its chemical form in one or more of the several steps in a reaction mechanism; but at the end of the reaction, it has the same chemical composition as at the beginning.

A catalyst increases reaction rate by altering a reaction's mechanism in a way that lowers the activation energy, E_a or $\Delta H‡$. Remember that *catalysts are specific*; that is, a catalyst may lower the activation energy for one chemical reaction and have no effect on the rate of a very similar reaction. In addition, a change in a catalyst can result in a change in products because of the altered mechanism caused by the catalyst. For example, the products formed by the reaction between carbon monoxide and hydrogen depend on the catalysts as shown below:

$$CO + H_2 \xrightarrow{Ni} CH_4 + H_2O$$

$$CO + 2H_2 \xrightarrow{ZnO/Cr_2O_3} CH_3OH$$

SECTION 18.8 CHAIN REACTIONS

Objectives

- To examine the mechanism of chain reactions.
- To know the function of inhibitors or poisons.

Focus

Chain reactions occur more rapidly than other reactions because of the formation of a very reactive reaction intermediate called a **free radical**. A free radical, having a single, unpaired electron, reacts quickly to achieve stability. In a chain reaction, however, one free radical is generated for each one consumed. Because free radicals are so reactive, their lifetime is very short.

Recognize these three basic steps in the mechanism for a chain reaction.

1. **Chain-initiation:** the generation of a free radical.
2. **Chain-propagation:** the reaction of one free radical to form a product and another free radical.
3. **Chain-termination:** the reaction of two free radicals to produce a neutral molecule.

Many light-sensitive reactions occur because absorbed photons generate free radicals. Many secondary compounds produced in photochemical smog are due to the generation and subsequent reaction of free radicals. The photolytic activity of sunlight produces these reactive free radical intermediates.

Chemical inhibitors are subdivided into two types:

1. The catalytic poison inhibits the effectiveness of a catalyst by physically covering the reactive surface. For example, lead from leaded gasoline covers the platinum catalyst in an auto's catalytic convertor, thereby reducing the efficiency for the conversion of unburned hydrocarbons to carbon dioxide and water.
2. Other inhibitors react with a free radical, thereby preventing the generation of a second free radical. These compounds are sometimes called "free radical scavengers."

As the text points out, inhibitors are necessary in some reactions to reduce the probability of particularly bad side reactions from occurring.

QUESTIONS

Section 18.1

18.1 The rate of reaction for $A_2 + B_2 \rightarrow 2AB$ is observed by noting the loss in color intensity with time. The color intensity is proportional to the A_2 molar concentration. From the plot, determine the rate of A_2 disappearance at $t = 0$, at $t = 4$ sec, and at $t = 10$ sec (see Figure 18.4).

Remember the reaction rate can be calculated from the slope of the tangent drawn at time, t:

$$\frac{\text{change in } y}{\text{change in } x} \quad \text{or} \quad \frac{\text{change in } [A_2]}{\text{change in time}}$$

18.2 What is a convenient method for determining the rate of the reaction between NH_4^+ and NO_2^-?

$$NH_4^+(aq) + NO_2^-(aq) \rightarrow N_2(g) + 2H_2O$$

Express the rate expression mathematically.

Section 18.2

18.3 For the reaction, $A + 3B + 2C \rightarrow 2X + 4Y$, the following set of rate data is obtained in the laboratory:

[A]	[B]	[C]	rate $\left(\frac{\text{mol A}}{\text{L sec}}\right)$
1.0×10^{-2}	1.0×10^{-4}	5.0×10^{-3}	3.0×10^{-4}
1.0×10^{-2}	5.0×10^{-5}	5.0×10^{-3}	1.5×10^{-4}
1.0×10^{-2}	1.0×10^{-4}	1.0×10^{-2}	3.0×10^{-4}
1.0×10^{-1}	1.0×10^{-4}	5.0×10^{-3}	3.0×10^{-2}

(a) What is the order with respect to A?
What is the order with respect to B?
What is the order with respect to C?

(b) What is the rate equation for the reaction?

(c) What is the value of the rate constant?

(d) What is the rate of the reaction with the following concentrations?
[A] = 6.6×10^{-1}; [B] = 3.7×10^{-5}; [C] = 1.1×10^{-12}.

18.4 In the reaction,

$$2NO + Br_2 \rightarrow 2NOBr,$$

what is the rate of disappearance of Br_2 at the time when NOBr is forming at the rate of 3.79×10^{-3} mol/(L sec)?

18.5 List four factors studied in this section that increase the rate of a chemical reaction.

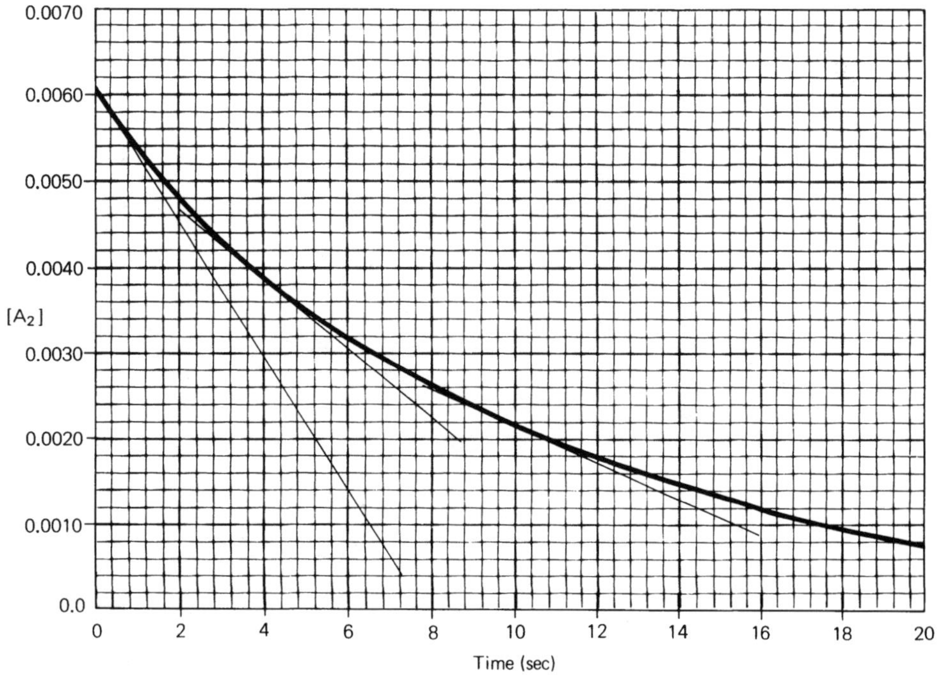

Figure 18.4 Plot of concentration of A_2 against time for the reaction $A_2 + B_2 \rightarrow 2AB$.

Section 18.3

18.6 For a series of first-order reactions, the rate constants are:
Reaction 1. $k = 0.67\ sec^{-1}$
Reaction 2. $k = 0.47\ sec^{-1}$
Reaction 3. $k = 0.92 \times 10^{-3}\ sec^{-1}$

(a) Which reaction has the shortest half-life?

(b) Which reaction is kinetically most stable?

18.7 The rate constant for the first-order decomposition of azomethane, $CH_3N = NCH_3$, is 40.8 min^{-1}.

(a) Calculate the amount of azomethane left from 30.0 mg after 3.70 secs.

(b) Calculate the time (in seconds) required for the 30.0 mg of azomethane to be reduced to 3.70 mg.

18.8 If the half-life for a radioactive isotope (first-order reaction) is 12.3 years, how much of a 100.0 g sample remains after one half-life? after two half-lives? after six half-lives?

Section 18.4

18.9 Gas molecules under normal conditions of temperature and pressure undergo about 10^{32} collisions per liter per second. A "rule of thumb" is that the rate of a chemical reaction doubles for every 10°C temperature rise. Of the three factors necessary for an effective collision, which one most likely accounts for this observation?

Section 18.5

18.10 Match the following with Figure 18.5.

(a) H_r (reactants) ________

(b) H_p (products) ________

(c) $\Delta H\ddagger$ (activation energy forward) ________

(d) H_{ts} (transition state) ________

(e) $\Delta H\ddagger$ (activation energy, backward) ________

(f) ΔH (heat of reaction) ________

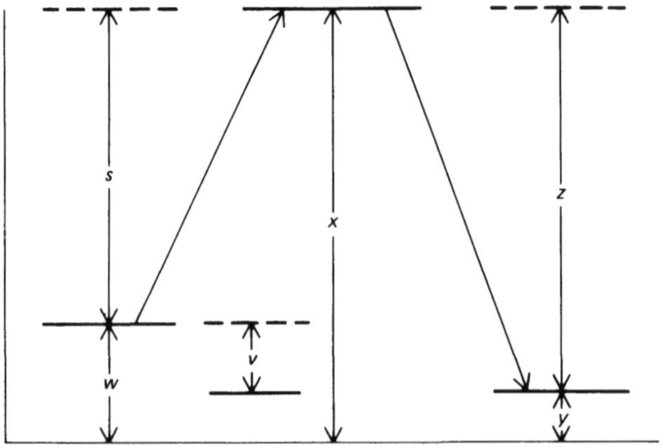

Figure 18.5 Enthalpy diagram, question 18.10.

18.11 The $\Delta H\ddagger$ of a forward reaction is 15.0 kJ and ΔH is −26.0 kJ. What is $\Delta H\ddagger$ for the reverse reaction?

Section 18.6

18.12 H_2O_2 decomposes in the presence of I^-:

$$2H_2O_2 \xrightarrow{I^-} 2H_2O + O_2$$

The mechanism is a two-step process:

$$H_2O_2 + I^- \rightarrow H_2O + IO^- \quad \text{(slow, reversible)}$$

$$IO^- + H_2O_2 \rightarrow H_2O + O_2 + I^- \quad \text{(fast)}$$

What is the rate law for the reaction?

Section 18.6

18.13 The experimental rate equation for the reaction

$$AB + CB_2 \rightarrow AB_2 + CB$$

is

$$\text{rate} = k[CB_2]^2$$

Which mechanism agrees with the rate equation?

(a) $AB + CB_2 \rightarrow AB_2 + CB$

(b) $2CB_2 \rightleftharpoons C_2B_4$ (fast, reversible)
$C_2B_4 + AB \rightarrow AB_2 + C_2B_3$ (slow)
$C_2B_3 + AB \rightarrow AB_2 + 2CB$ (fast)

(c) $2CB_2 \rightarrow CB_3 + CB$ (slow)
$CB_3 + AB \rightarrow CB_2 + AB_2$ (fast)

(d) $2CB_2 \rightleftharpoons 2CB + B_2$ (fast, reversible)
$2AB + B_2 \rightarrow 2AB_2$ (slow)

Determine the rate equation for each mechanism and compare each to the experimental rate equation.

Section 18.7

18.14 Explain why a finely divided solid catalyst is more effective in increasing reaction rate than the same mass of the catalyst in granular form.

18.15 How does a homogeneous catalyst differ in catalytic activity from a heterogeneous catalyst?

ANSWERS

18.1 $t = 0$ sec; $\dfrac{-0.00077 \text{ mol/L}}{\text{sec}}$

$t = 4$ sec; $\dfrac{-0.00028 \text{ mol/L}}{\text{sec}}$

$t = 10$ sec; $\dfrac{-0.00021 \text{ mol/L}}{\text{sec}}$

18.2 Observing the amount of N_2 gas evolved with time,

$$\frac{\text{change in } [N_2]}{\text{change in time}}$$

18.3 (a) 2; 1; 0

(b) rate $= k[A]^2[B]^1[C]^0 = k[A]^2[B]$

(c) $k = 3.0 \times 10^4$ L^2/(mol^2 sec)

(d) rate $= 4.8 \times 10^{-1}$ mol A/(L sec)

18.4 1.90×10^{-3} mol/(L sec)

18.5 nature of the reactants

concentration of the reactants

particle size in heterogeneous reactions

temperature of the reaction

presence of a catalyst

18.6 (a) Reaction 1, $t_{1/2} = 1.03$ sec

(b) Reaction 3, $t_{1/2} = 75.3$ sec

18.7 (a) 2.42 mg

(b) 3.08 sec

18.8 50.0 g; 25.0 g; 1.56 g

18.9 The increase in kinetic energy of the molecules causes a greater weakening of the bonds when the molecules collide.

18.10 (a) w; (b) y; (c) s; (d) x; (e) z; (f) v

18.11 +41 kJ

18.12 rate $= k[H_2O_2][I^-]$

18.13 (c)

(a) rate $= k[AB][CB_2]$

(b) rate $= k[CB_2]^2[AB]$

(c) rate $= k[CB_2]^2$

(d) rate $= k[CB_2]^2[AB]^2$

18.14 A finely divided catalyst has more surface area that is in contact with reactants. Therefore, more bonds are weakened per unit time and the reaction rate is greater.

18.15 A homogeneous catalyst enters into the chemical reaction and even changes chemical form but is regenerated in a later step. A heterogeneous catalyst provides a surface upon which reactants are absorbed and react; it does not change chemical form in the reaction.

Chapter 19

NUCLEAR CHEMISTRY

SECTION 19.1 RADIOACTIVITY

Objectives

- To identify the properties of α- and β- particles and γ-rays.
- To explain radioactive decay.
- To write balanced equations for nuclear reactions.

Focus

Thus far we have not considered the nucleus of the atom in any detail. We have focused our attention on what happens outside the nucleus; the chemistry of acids and bases, oxidation and reduction, kinetics, solubility, and so on, involves the behavior of valence electrons.

The nucleus acts as the atom's control center, much like the control tower at a very busy airport. The nucleus "controls" the energy states for electrons through its strong electrostatic attraction for the negative electrons. The strength of this attraction determines which electrons are lost, shared, or gained in chemical reactions.

In this section we look at the natural stability of some nuclei. For reasons still unknown, some nuclei spontaneously rearrange their protons and neutrons (collectively called *nucleons*) to form a more stable nuclear configuration. In the process the unstable nucleus emits a particle and/or energy. This is **radioactive decay**. Review the properties of α-particles, β-particles and γ-rays outlined in the text, Table 19.1. Remember that

- α-particles are high-energy He^{2+} ions.
- β-particles are high-energy electrons emitted from the nucleus.
- γ-rays are high-energy photons (light waves having a very short wavelength).

Radioactive nuclei spontaneously emit particles and/or energy without absorbing any energy. A radioactive decay process is represented by a nuclear equation in which the mass number sum (protons plus neutrons) and the atomic number sum of the nuclear reactants (the sum of the positive charge) equals the mass number sum and the atomic number sum of the nuclear products.

Example 19.1 What isotope forms as a result of the α-decay of ${}^{207}_{84}Po$?

Answer ${}^{203}_{82}Pb$

Solution From Table 19.1, an α-particle has a mass number of 4 and an atomic number of 2, ${}^{4}_{2}He$; therefore, the product nucleus has a mass number of 203 (207 − 4) and an atomic number of 82 (84 − 2). Pb is the element with an atomic number of 82. The product nucleus is ${}^{203}_{82}Pb$.

$${}^{207}_{84}Po \rightarrow {}^{203}_{82}Pb + {}^{4}_{2}He$$

Mass number balance: $207 = 203 + 4$
Atomic number balance: $84 = 82 + 2$

SECTION 19.2 THE INTERCONVERTIBILITY OF MATTER AND ENERGY: NUCLEAR ENERGY

Objectives

- To describe mass-energy interconversions in chemical and nuclear reactions.
- To calculate ΔE from a Δm in a reaction.
- To predict the penetrating effect of α-particles, β-particles, and γ-rays.

Focus

The energy absorbed or evolved in any reaction originates from a change in mass between reactants and products. When energy is absorbed, energy is converted into mass and the mass within the system increases; when energy is evolved, mass is converted into energy and the mass decreases.

In the conventional chemical reactions we have studied so far, this change in mass, Δm, is smaller than any highly sophisticated laboratory instrument can measure. In the text, Example 2, the combustion of 400 g gasoline causes a Δm of only -2.3×10^{-7} g, which is a *very* negligible quantity compared to the initial 400 g.

In nuclear reactions, however, Δm is significant, meaning that ΔE is also significant. For a **nuclear reaction**, small mass changes between reactants and products produce large amounts of energy. This is the principal source of nuclear energy. The equation for the mass-energy interconversion is

$$\Delta E = \Delta mc^2$$

Example 19.2 Calculate the quantity of energy released in the radioactive decay of 1 mol ${}^{210}_{84}Po$ by α decay.

$${}^{210}_{84}Po \rightarrow {}^{206}_{82}Pb + {}^{4}_{2}He$$

${}^{210}_{84}Po$	${}^{206}_{82}Pb$	${}^{4}_{2}He$
209.9829	205.9745	4.0026

molar masses, g/mol

Answer -5.2×10^{11} J/mol

Solution To solve for Δm, use as many significant figures as possible because many times the Δm does not appear until the 3rd or 4th decimal place. For this reaction

$$\Delta m = \Sigma m_{\text{products}} - \Sigma m_{\text{reactants}}$$
$$\Delta m = (205.9745 + 4.0026) - 209.9829$$
$$= -0.0058 \text{ g/mol}$$

To convert grams to joules, first convert grams to kilograms, and then multiply by c^2, the speed of light squared, $(3.0 \times 10^8 \text{ m/sec})^2$. Then convert to joules, 1 kg m²/sec² = 1 J.

$$\Delta E = \Delta mc^2$$
$$\Delta E = \left(-0.0058 \frac{\text{g}}{\text{mol}}\right)\left(\frac{\text{kg}}{10^3 \text{ g}}\right)\left(3.0 \times 10^8 \frac{\text{m}}{\text{sec}}\right)\left(\frac{\text{J}}{1 \frac{\text{kg m}^2}{\text{sec}^2}}\right)$$
$$= -5.2 \times 10^{11} \text{ J/mol}$$

Mass/energy interconversions have been observed experimentally with the simultaneous appearance of an electron, ${}_{-1}^{0}\beta$, and a positron (antielectron, ${}_{+1}^{0}\beta$). A γ-ray photon produces both the electron and the positron. When an electron and positron combine, they annihilate one another to form two γ-ray photons. This same annihilation occurs when any matter combines with its antimatter.

The penetrating power of radioactive emissions depends on its mass and energy. Alpha particles are relatively large and cannot penetrate more than a few sheets of paper; a sheet of aluminum metal can stop β-particles; and several feet of lead or concrete can stop γ-rays. Nuclear reactor walls have lead salts mixed with the reinforced concrete to absorb some of the γ-radiation.

SECTION 19.3 NUCLEAR STRUCTURE

Objective

- To compare the structure of the nucleus with the electronic structure of the atom.

Focus

According to Hund's rule and the Pauli exclusion principle (Chapter 7), the energy state of any electron in any atom is determined by a set of four quantum numbers, n, ℓ, m_ℓ, and m_s. Although these numbers come from the application of quantum mechanics, they fit very well with experiment.

When we treat the nucleus the same way, a set of four nuclear quantum numbers indicates the energy state of each nucleon. Hund's rule is, however, applied separately to protons and neutrons. For example, the ${}_{1}^{2}H$ nucleus has 1 proton and 1 neutron; the proton and the neutron are unpaired. Like an unpaired electron, an unpaired nucleon possesses a magnetic moment.

Nuclear stability is greater for nuclei with 2, 8, 20, 28, 50, 82, and 126 protons and/or neutrons. This is similar to the atomic stability for atoms with 2, 10, 18, 36, 54, and 86 electrons (the noble gases). When a nucleon deexcites from a higher to a lower energy level, a photon (a γ ray) is emitted. This resembles the electron's deexcitation resulting in the emission of a photon. The photon from a nuclear deexcitation, however, has a much higher energy than one from an electronic deexcitation.

Hence, we can speak of *shells* for protons and neutrons, as well as electrons, each shell having an energy with a maximum number of particles arranged according to Hund's rule and the Pauli exclusion principle.

Example 19.3 Which nuclei have a "magic number" of protons and/or neutrons? (a) $^{40}_{20}$Ca; (b) $^{120}_{50}$Sn; (c) $^{208}_{82}$Pb; (d) $^{50}_{23}$V

Answer (a) $^{40}_{20}$Ca; (b) $^{120}_{50}$Sn; (c) $^{208}_{82}$Pb

Solution

$^{40}_{20}$Ca has 20 protons and (40 − 20 =) 20 neutrons;

$^{40}_{20}$Ca has a magic number of protons *and* neutrons (extra stable).

$^{120}_{50}$Sn has 50 protons and (120 − 50 =) 70 neutrons;

$^{120}_{50}$Sn has a magic number of protons.

$^{208}_{82}$Pb has 82 protons and (208 − 82 =) 126 neutrons;

$^{208}_{82}$Pb has two magic numbers.

$^{50}_{23}$V has 23 protons and (50 − 23 =) 27 neutrons; there are no magic number of nucleons in $^{50}_{23}$V.

Remember that nuclear stability is greater for a magic number of protons and/or neutrons but not their sum. In $^{50}_{23}$V, the mass number 50 (the sum of the number of protons and neutrons) is not considered a stability factor.

SECTION 19.4 THE STABILITY OF NUCLEI

Objectives

- To calculate the total binding energy for a nucleus and the binding energy per nucleon.
- To explain the stability of a nucleus according to the π-meson theory.

Focus

The **nuclear binding energy** (B.E.) is the energy necessary to break a nucleus into its protons and neutrons. This is similar to the total energy required to break the bonds of a molecule into its atoms. However, the nuclear binding energy is several million times greater than a bond energy.

Example 19.4 Calculate the binding energy and the binding energy per nucleon of $^{56}_{26}$Fe (55.93493 g/mol) and $^{235}_{92}$U (235.04394 g/mol). The molar mass of $^{1}_{1}$H is 1.00783 g/mol and of $^{1}_{0}n$ is 1.00867 g/mol. In which nucleus is a nucleon more stable?

Answer

	B.E.	B.E./nucleon
$^{56}_{26}Fe$	-4.76×10^{13} J/mol	-1.41×10^{-12} J/nucleon
$^{235}_{92}U$	-1.72×10^{14} J/mol	-1.22×10^{-12} J/nucleon

In the ^{56}Fe nucleus each nucleon is more stable.

Solution Binding energy is calculated from ΔE = B.E. = Δmc^2. Δm is the difference between the atomic weight of the $^{56}_{26}Fe$ atom and the sum of the atomic weights of 26 protons and electrons (or 26 H atoms) and (56 − 26, or) 30 neutrons.

$$26\,{}^{1}_{1}\mathrm{H} \quad + \quad 30\,{}^{1}_{0}n \quad \rightarrow \quad {}^{56}_{26}\mathrm{Fe}$$
$$26(1.00783\ \mathrm{g/mol}) + 30(1.00867\ \mathrm{g/mol}) \rightarrow 55.93493\ \mathrm{g/mol}$$

$$\Delta m = 55.93493 - (26.20358 + 30.26010)$$
$$= -0.52875\ \mathrm{g/mol}$$
$$\mathrm{B.E.} = \Delta mc^2 = \frac{-0.52875\ \cancel{\mathrm{g}}}{\mathrm{mol}} \times \frac{\mathrm{kg}}{10^3\ \cancel{\mathrm{g}}} \times \left(\frac{3.0 \times 10^8\ \mathrm{m}}{\mathrm{sec}}\right)^2$$
$$= -4.76 \times 10^{13}\ \mathrm{J/mol}$$

Since each $^{56}_{26}Fe$ atom has 56 nucleons or 1 mol Fe has

$$6.02 \times 10^{23}\ \frac{\cancel{\mathrm{nuclei}}}{\mathrm{mol}} \times 56\ \frac{\mathrm{nucleons}}{\cancel{\mathrm{nucleus}}}\text{, the B.E./nucleon is}$$

$$\frac{-4.61 \times 10^{13}\ \mathrm{J}/\cancel{\mathrm{mol}}}{(6.02 \times 10^{23} \times 56)\ \dfrac{\mathrm{nucleons}}{\cancel{\mathrm{mol}}}} = -1.41 \times 10^{-12}\ \frac{\mathrm{J}}{\mathrm{nucleon}}$$

Likewise, for $^{235}_{92}U$, Δm is the difference between the atomic mass of the $^{235}_{92}U$ atom and the sum of the masses of 92 H atoms and 143 neutrons.

$$92\,{}^{1}_{1}\mathrm{H} \quad + \quad 143\,{}^{1}_{0}n \quad \rightarrow \quad {}^{235}_{92}\mathrm{U}$$
$$92(1.00783\ \mathrm{g/mol}) + 143(1.00867\ \mathrm{g/mol}) \rightarrow 235.04394\ \mathrm{g/mol}$$

$$\Delta m = 235.04394 - (92.72036 + 144.23981)$$
$$= -1.9162\ \mathrm{g/mol}$$
$$\mathrm{B.E.} = -1.9162\ \frac{\cancel{\mathrm{g}}}{\mathrm{mol}} \times \frac{\mathrm{kg}}{10^3\ \cancel{\mathrm{g}}} \times \left(3.00 \times 10^8\ \frac{\mathrm{m}}{\mathrm{sec}}\right)^2$$
$$= -1.72 \times 10^{14}\ \mathrm{J/mol}$$

Each $^{235}_{92}U$ atom has 235 nucleons or 1 mol ^{235}U

has $6.02 \times 10^{23}\ \dfrac{\cancel{\mathrm{nuclei}}}{\mathrm{mol}} \times 235\ \dfrac{\mathrm{nucleons}}{\cancel{\mathrm{nucleus}}}$. The B.E./nucleon is

$$\frac{-1.72 \times 10^{14}\ \mathrm{J}/\cancel{\mathrm{mol}}}{(6.02 \times 10^{23} \times 235)\ \dfrac{\mathrm{nucleons}}{\cancel{\mathrm{mol}}}} = -1.22 \times 10^{-12}\ \frac{\mathrm{J}}{\mathrm{nucleon}}$$

The energy released in the formation of a nucleus equals the *negative* of the binding energy. Therefore, the energy released in the formation of ^{56}Fe is 1.41×10^{-12} J/nucleon and in the formation of ^{235}U is 1.22×10^{-12} J/nucleon. Since ^{56}Fe releases more energy per nucleon, it is a more stable nucleus than ^{235}U.

Since Ernest Rutherford's α-scattering experiments in the early twentieth century, scientists have been intrigued by the fact that protons with the same positive charge are confined to a small volume of a nucleus and still remain stable.

An early theory that has helped us to understand this phenomenon is the **π-meson theory**. It states that a third nucleon, a π-meson, maintains nuclear stability by a series of nuclear reactions occurring at nearly the speed of light: a rapid interconversion of nucleons between protons and neutrons via π-mesons. The π-mesons may have a positive charge, a negative charge, or no charge.

$$n^0 \rightleftharpoons p^+ + \pi^-$$

$$p^+ \rightleftharpoons n^0 + \pi^+$$

$$p^+ \rightleftharpoons p^+ + \pi^0$$

$$n^0 \rightleftharpoons n^0 + \pi^0$$

At any given instant, two nucleons share the same π-meson. This sharing, analogous to the sharing of electrons by atoms in a bond, is the "glue" that holds nucleons together in nuclei.

SECTION 19.5 NUCLEAR REACTIONS

Objectives

- To predict the possible radioactive decay modes of an unstable nucleus which has a neutron to proton ratio outside of the stability band.
- To determine the product nucleus from a nuclear transmutation reaction.
- To describe the process of fission, its release of tremendous amounts of energy, and its associated environmental problems.
- To describe how the breeder reactor resembles the conventional nuclear reactor.
- To describe the fusion process and some of its technological problems.
- To describe how the sun releases its energy.

Focus

This lengthy section about nuclear reactions is conveniently broken down into the following subsections:

- Radioactivity
- Artificial Transmutations

- Fission
- Breeder Reactors
- Fusion
- Stellar Energy

Radioactivity

A stable nucleus has a neutron to proton (n/p) ratio that lies within a narrow **stability band**. The n/p ratio varies from about 1 for the lighter elements to about 1.54 in $^{208}_{82}Pb$. Thus we deduce that apparently more neutrons are needed to maintain a stable nucleus in heavier elements.

A nucleus with a n/p ratio to the *right* of the stability band (text, Figure 19.6) has a n/p ratio *too low* for stability. As a result, the nucleus transforms, by one of the following modes of decay, so as to increase that ratio.

(a) $^{0}_{+1}\beta$ (positron) emission: $^{1}_{1}p \rightarrow {}^{1}_{0}n + {}^{0}_{+1}\beta$, resulting in 1 less proton and 1 more neutron.

(b) $^{1}_{1}H$ (proton) emission, resulting in 1 less proton.

(c) Electron capture (EC) of a K-shell electron: $^{1}_{1}p + e^{-} \rightarrow {}^{1}_{0}n$, resulting in 1 less proton and 1 more neutron.

(d) $^{4}_{2}He$ (α) emission, resulting in a n/p ratio reduction of one.

Notice that in each case the n/p ratio in the product nucleus is larger than that for the radioactive nucleus.

Example 19.5 The following nuclei are radioactive. The decay mode for each is given. Calculate the n/p ratio of the nuclide before and after decay.

(a) $^{30}_{15}P \rightarrow {}^{30}_{14}Si + {}^{0}_{+1}\beta$
(b) $^{28}_{13}Al \rightarrow {}^{27}_{12}Mg + {}^{1}_{1}H$
(c) $^{71}_{32}Ge + {}^{0}_{-1}e \rightarrow {}^{71}_{31}Ga$
(d) $^{235}_{92}U \rightarrow {}^{231}_{90}Th + {}^{4}_{2}He$

Answer

		n/p		n/p
(a)	$^{30}_{15}P$	$\frac{15}{15} = 1.000$	$^{30}_{14}Si$	$\frac{16}{14} = 1.14$
(b)	$^{28}_{13}Al$	$\frac{15}{13} = 1.154$	$^{27}_{12}Mg$	$\frac{15}{12} = 1.250$
(c)	$^{71}_{32}Ge$	$\frac{39}{32} = 1.269$	$^{71}_{31}Ga$	$\frac{40}{31} = 1.290$
(d)	$^{235}_{92}U$	$\frac{143}{92} = 1.554$	$^{231}_{90}Th$	$\frac{141}{90} = 1.567$

In each case, decay results in an increase in the n/p ratio.

A nucleus with a n/p ratio to the *left* of the stability band has a n/p ratio *too high* for stability. The unstable nucleus undergoes decay, decreasing the number of neutrons relative to the number of protons by:

1. ${}_{-1}^{0}\beta$ (electron) emission: ${}_{0}^{1}n \rightarrow {}_{1}^{1}p + {}_{-1}^{0}\beta$, resulting in 1 less neutron and 1 more proton.
2. ${}_{0}^{1}n$ (neutron) emission, resulting in 1 less neutron.

Example 19.6 The following nuclei have a *n/p* ratio to the left of the stability band and therefore undergo decay. Calculate the *n/p* ratio before and after decay.

(a) ${}_{90}^{234}Th \rightarrow {}_{91}^{234}Pa + {}_{-1}^{0}\beta$
(b) ${}_{53}^{137}I \rightarrow {}_{53}^{136}I + {}_{0}^{1}n$

Answer

		n/p		*n/p*
(a)	${}_{90}^{234}Th$	$\frac{144}{90} = 1.600$	${}_{91}^{234}Pa$	$\frac{143}{91} = 1.571$
(b)	${}_{53}^{137}I$	$\frac{84}{53} = 1.585$	${}_{53}^{136}I$	$\frac{83}{53} = 1.566$

In both cases, decay results in a decrease in the *n/p* ratio.

Artificial Transmutations

Bombarding a nucleus with particles such as ${}_{1}^{1}H$, ${}_{0}^{1}n$, ${}_{2}^{4}He$ can produce a nucleus of another element. The second nucleus is generally radioactive. For example, when ${}_{17}^{35}Cl$ is bombarded with neutrons, the product is ${}_{17}^{36}Cl$. But ${}_{17}^{36}Cl$ is unstable (a high *n/p* ratio) and undergoes proton decay to form ${}_{16}^{35}S$:

$$ {}_{17}^{35}Cl + {}_{0}^{1}n \rightarrow [{}_{17}^{36}Cl]^{*} \rightarrow {}_{16}^{35}S + {}_{1}^{1}H $$

Example 19.7 What nuclei are formed in the following nuclear transmutations?

(a) ${}_{7}^{14}N + {}_{1}^{1}H \rightarrow [\ \ ]^{*} \rightarrow \ ? + {}_{0}^{0}\gamma$
(b) ${}_{20}^{43}Ca + {}_{2}^{4}He \rightarrow [\ \ ]^{*} \rightarrow \ ? + {}_{1}^{1}H$
(c) ${}_{98}^{249}Cf + {}_{6}^{12}C \rightarrow [\ \ ]^{*}\ ?\ \ ? + 4\,{}_{0}^{1}n$
(d) ${}_{92}^{235}U + {}_{1}^{2}H \rightarrow [\ \ ]^{*} \rightarrow \ ? + 6\,{}_{0}^{1}n$

Answer

(a) ${}_{7}^{14}N + {}_{1}^{1}H \rightarrow [{}_{8}^{15}O]^{*} \rightarrow {}_{8}^{15}O + {}_{0}^{0}\gamma$
(b) ${}_{20}^{43}Ca + {}_{2}^{4}He \rightarrow [{}_{22}^{47}Ti]^{*} \rightarrow {}_{21}^{46}Sc + {}_{1}^{1}H$
(c) ${}_{98}^{249}Cf + {}_{6}^{12}C \rightarrow [{}_{104}^{261}Ha]^{*} \rightarrow {}_{104}^{257}Ha + 4\,{}_{0}^{1}n$
(d) ${}_{92}^{235}U + {}_{1}^{2}H \rightarrow [{}_{93}^{237}Np]^{*} \rightarrow {}_{93}^{231}Np + 6\,{}_{0}^{1}n$

Fission

For some heavier unstable nuclides, fission is the mode of radioactive decay. In this process, a nucleus divides (or fissions), producing two nuclei that have higher B.E./nucleon—they are more stable.

From the text, Figure 19.9, when a heavy nucleus fissions, nuclei are produced with a greater B.E./nucleon. In Example 19.4, Study Guide, the energy released in the formation of ${}^{56}_{26}Fe$ is 1.41×10^{-12} J/nucleon, whereas for ${}^{235}_{92}U$ it is only 1.22×10^{-12} J/nucleon. The difference is the amount of energy released per nucleon if ${}^{235}_{92}U$ were to fission and form the ${}^{56}_{26}Fe$ nucleus as one of its products.

Nuclear fissions generally produce several neutrons that *another* fissionable nucleus can absorb. In turn, several neutrons are produced when the nucleus fissions. This causes a chain reaction and, as a result, tremendous amounts of energy are released in a very short period of time. The insertion of "chain breakers"—Cd and/or B rods which are good neutron absorbers—can control or stop the chain-reaction.

Example 19.8 The following nuclear fission reaction is typical of the fission process occurring in a nuclear reactor:

$${}^{235}_{92}U + {}^{1}_{0}n \rightarrow [{}^{236}_{92}U]^* \rightarrow {}^{142}_{56}Ba + {}^{91}_{36}Kr + 3{}^{1}_{0}n$$

$$\Delta E = -1.9 \times 10^{10} \text{ kJ/mol}$$

What mass of water in kg is heated from 0°C to 100°C from the fission of 1 mol ${}^{235}_{92}U$? The specific heat for H_2O is 4.184 J/g°C.

Answer 4.5×10^4 kg

Solution Given quantity: $+1.9 \times 10^{10}$ kJ

$\Delta T = 100°C$

Desired quantity: kg H_2O

Conversion factors: $\dfrac{4.184 \text{ J}}{\text{g °C}}$

$$\text{mass } H_2O = 1.9 \times 10^{10} \cancel{\text{J}} \times \frac{\cancel{\text{g}}\ \cancel{°\text{C}}}{4.184 \cancel{\text{J}}} \times \frac{\text{kg}}{10^3 \cancel{\text{g}}} \times \frac{1}{100 \cancel{°\text{C}}}$$

$= 4.5 \times 10^4$ kg = 45 metric tons H_2O from 0°C to 100°C with only 1 mole ${}^{235}_{92}U$!! That is a lot of energy from 235 g (about ½ pound) or uranium!

Breeder Reactor

A breeder reactor is designed to manufacture, or *breed*, fissionable nuclei. The absorption of a neutron by ${}^{238}_{92}U$ produces ${}^{239}_{94}Pu$:

$${}^{238}_{92}U + {}^{1}_{0}n(\text{fast}) \rightarrow [{}^{239}_{92}U]^* \xrightarrow{-{}^{0}_{-1}\beta} [{}^{239}_{93}Np]^* \xrightarrow{-{}^{0}_{-1}\beta} {}^{239}_{94}Pu$$

${}^{239}_{94}Pu$ is a fissionable nucleus which decays further:

$${}^{239}_{94}Pu + {}^{1}_{0}n(\text{fast}) \rightarrow [{}^{240}_{94}Pu]^* \rightarrow \text{fission products} + \text{energy}$$

Therefore, two fast neutrons are needed to sustain a breeder reaction—one for the production of ${}^{239}_{94}Pu$ and the other for the fission of ${}^{239}_{94}Pu$. Compare this to the fission of ${}^{235}_{94}U$ in a conventional nuclear reactor where only one neutron must be effective.

Much of the controversy over the expansion of nuclear power stems from the storage and disposal of radioactive wastes. Review some of the references listed in the text before you form an opinion on its use for the generation of electric power. As a scientist you should review both sides of the controversy.

Fusion

Figure 19.9 shows that the B.E./nucleon for lightweight nuclei is small compared to heavier nuclei. Therefore, the fusion of two lightweight nuclei into a heavier nucleus increases the B.E./nucleon with the difference released as energy.

The main technical difficulty of controlled fusion reactions is the tremendous temperatures, 10^6 K to 10^7 K, needed to overcome the electrostatic repulsions between the nuclei. The technology necessary to generate these temperatures has not yet been perfected.

In the fusion reaction,

$$^{2}_{1}H + ^{3}_{1}H \rightarrow ^{4}_{2}He + ^{1}_{0}n$$

notice that a He atom and a neutron are the only products. The energy released, $\Delta E = -1.8 \times 10^9$ kJ, for the fusion of 2 g of $^{2}_{1}H$ (deuterium) and 3 g of $^{3}_{1}H$ (tritium) is comparable to the fission of 235 g $^{235}_{92}U$, $\Delta E = -1.9 \times 10^{10}$ kJ. In addition, the fission process produces many undesirable radioactive products; fusion does not.

In the future, fusion reactors could be very beneficial from an energy and an environmental viewpoint; however, a reliable fusion technology needs to be developed.

Stellar Energy

The sun's energy is generated from a series of fusion reactions occurring at about 10^6 K. Although heavier nuclei, such as $^{12}_{6}C$, $^{14}_{7}N$, $^{15}_{8}O$, and $^{15}_{7}N$, are generated in the fusion process, the overall reaction is

$$4\,^{1}_{1}H \rightarrow ^{4}_{2}He + 2\,^{0}_{+1}\beta \qquad \Delta E = -2.38 \times 10^9 \text{ kJ}.$$

SECTION 19.6 RATE OF RADIOACTIVE DECAY

Objectives

- To explain the relationship between half-life and nuclear stability.
- To calculate the time for a radioisotope to decay.
- To calculate the quantity of radioisotope remaining at the end of the time measurement.

Focus

All radioactive decay reactions are *first order*—the decay rate is directly proportional to the concentration or number of radioactive nuclei.

$$\text{rate} = k \times \text{no. of radioactive nuclei}$$

The **half-life** is the time for one-half of the given number of radioactive nuclei to decay. Nuclei with a shorter half-life decay more rapidly and therefore are less stable than those with longer half-lives. Each radioisotope has its own half-life.

For radioactive decay calculations, memorize the equation derived from the definition of a half-life:

$$q_t = q_0 \times (0.5)^{t/t_{\frac{1}{2}}}$$

where q_t = quantity (grams, number of atoms, decays per second, percent of sample) after time, t, has elapsed

q_0 = quantity (same units as q_t) at time zero

$t_{1/2}$ = half-life of the radioisotope

Example 19.9 The half-life of $^{59}_{26}Fe$ is 45 days. How long does it take for 85% of a $^{59}_{26}Fe$ sample to disappear? (Or more specifically, if a piece of sheet metal contains only 15% of the $^{59}_{26}Fe$ it had when the sheet metal was made at the foundry, how long ago was the sheet metal manufactured given that the half-life of $^{59}_{26}Fe$ is 45 days?)

Answer 123 days

Solution Assume at time zero 100% was present. After time, t, elapses 85% disappears, or 15% remains. Remember q_t is the quantity that *remains*. Therefore,

$$q_t = 15\% \text{ or } 0.15$$

$$q_0 = 100\% \text{ or } 1.00$$

$$t_{1/2} = 45 \text{ days}$$

$$q_t = q_0 \times (0.5)^{t/t_{\frac{1}{2}}}$$

$$0.15 = 1.00 \times (0.5)^{t/45\text{d}}$$

$$\log \frac{0.15}{1.00} = \frac{t}{45 \text{ d}} \log 0.5$$

$$-0.824 = \frac{t}{45 \text{ d}} (-0.301)$$

$$t = 45 \text{ d} \left(\frac{-0.824}{-0.301}\right) = 123 \text{ days}$$

Example 19.10 The half-life of $^{198}_{79}Au$ is 64.8 hours. Starting with a 4.00 mg sample, how much $^{198}_{79}Au$ remains (in mg) after 7.00 days? After 30.00 days? At what point in time does the amount of $^{198}_{79}Au$ reach 0.400 mg?

Answer 7 days: 0.66 mg
30 days: 1.82×10^{-3} mg = 1.8 μg
Time for reduction to 0.400 mg: 215 hr or 8.96 days

Solution First convert 7 days to hours:

$$t = 7 \text{ days} \times \frac{24 \text{ hr}}{\text{day}} = 168 \text{ hr}$$

Then convert 30 days to hours:

$$t = 30 \text{ days} \times \frac{24 \text{ hr}}{\text{day}} = 720 \text{ hr}$$

$$q_0 = 4.00 \text{ mg}$$
$$t_{1/2} = 64.8 \text{ hr}$$

For 7 days,

$$\begin{aligned} q_t &= q_0 \times (0.5)^{t/t_{\frac{1}{2}}} = 4.00 \text{ mg} \times (0.5)^{168 \text{ hr}/64.8 \text{ hr}} \\ &= 4.00 \text{ mg} \times (0.5)^{2.59} = 4.00 \text{ mg} \times 0.166 \\ &= 0.66 \text{ mg} \end{aligned}$$

For 30 days,

$$\begin{aligned} q_t &= 4.00 \text{ mg} \times (0.5)^{720 \text{ hr}/64.8 \text{ hr}} = 4.00 \text{ mg} \times (0.5)^{11.1} \\ &= 4.00 \text{ mg} \times 4.6 \times 10^{-4} \\ &= 1.8 \times 10^{-3} \text{ mg, or } 1.8\ \mu\text{g} \end{aligned}$$

Time for 0.400 mg: $q_t = 0.400$.

$$\begin{aligned} 0.400 \text{ mg} &= 4.00 \text{ mg} \times (0.5)^{t/64.8 \text{ hr}} \\ \log \frac{0.400}{4.00} &= \frac{t}{64.8 \text{ hr}} \log 0.5 \\ -1.0 &= \frac{t}{64.8 \text{ hr}} (-0.301) \\ t &= 64.8 \text{ hr} \left(\frac{-1.0}{-0.301} \right) = 215 \text{ hr} = 8.96 \text{ days} \end{aligned}$$

SECTION 19.7 RADIOCHEMISTRY

Objectives

- To calculate the age of materials using radiochemical dating.
- To explain how radioisotopes are used to follow the rate and mechanism of a chemical reaction.
- To identify and explain how some radioisotopes are used in medicine and in industry.
- To explain the principles of neutron activation analysis.

Focus

This section introduces the variety of radioisotopes currently used in chemical research, medicine, and industry.

Radiochemical Dating

The age of materials and rocks can be dated using a radioisotope with an appropriate half-life. $^{14}_{6}C$, which has a half-life of 5.73×10^3 years, is commonly used for dating artifacts (plants and animals) alive in ancient times, and ^{238}U, ^{40}K, and ^{87}Rb is used for dating fossils or rocks. A scintillation counter measures the age of a substance by recording the **activity** of the sample—the number of disintegrations of a weighed sample per unit time.

Example 19.11 What is the $^{14}_{6}C$ activity of a wooden soup ladle discovered in an archeological dig at a village known to exist in 1000 B.C.?

Answer 9.8 d/(min g)

Solution $^{14}_{6}C$ has activity in living plants and animals of 14 d/(min g) and $t_{\frac{1}{2}} = 5.73 \times 10^3$ years. We assume that 3000 years have elapsed since the village was in existence. 3000 years is less than one half-life; thus the activity is greater than ½(14 d/(min g)) or 7.0 d/(min g).

$$q_t = q_0 \times (0.5)^{t/t_{\frac{1}{2}}}$$
$$q_t = 14\ \text{d/(min g)} \times 0.5^{3000\ \text{yr}/5730\ \text{yr}}$$
$$= 14\ \text{d/(min g)} \times 0.70$$
$$= 9.8\ \text{d/(min g)}$$

Tracer Elements

The decay of a radioisotope is independent of its physical environment (temperature, pressure, state) and chemical environment (the oxidation state, bond strength, bond length, a gas, a liquid, a solid, and so on).

Because radioisotopes have the same chemistry as the stable isotopes of the same element they can by used to follow (trace) the rate and mechanism of chemical reactions. The activity of a radioisotope is determined in the reactant before reaction and in the product afterward (at equilibrium).

Medical Applications

The text has recognized a few radioisotopes currently used in medicine. Some additional radioisotopes in medicine are:

- $^{24}_{11}Na$—as $^{24}NaCl$ to monitor blood circulation in the body.
- $^{99}_{43}Tc$—as TcO_4^- to generate brain scans.
- $^{3}_{1}H$—as (^{3}HHO) to measure total body water.
- $^{45}_{20}Ca$—to detect bone tumors.
- $^{32}_{15}P$—to detect brain tumors.
- $^{42}_{19}K$—to monitor the metabolism and distribution of potassium.

Industrial Applications

Radioisotopes are used in industry

1. For gauging the thickness of metal foils and plastic sheets.
2. For determining the effectiveness of lubricants.
3. For indicating the boundary between products moving through a pipeline.

Neutron Activation Analysis

Neutron activation analysis is a relatively new analytical technique. A sample containing an unknown element is exposed to neutrons. The neutrons are absorbed by the nuclei of the element, resulting in a higher n/p ratio. This causes the nuclei to be unstable and undergo decay. The activity of these radioisotopes identifies the element in the sample and its concentration. This analytical technique is used on milligram samples with some element concentrations of only one part per million.

QUESTIONS

Section 19.1

19.1 Write a balanced equation for the nuclear reactions:

(a) $^{78}_{32}Ge \rightarrow$ ______ $+ \ ^{0}_{-1}\beta$ (β-decay)

(b) $^{122}_{53}I \rightarrow$ ______ $+ \ ^{0}_{+1}\beta$

(c) ______ $+ \ ^{1}_{0}n \rightarrow \ ^{56}_{25}Mn + \ ^{4}_{2}He$

(d) $^{58}_{28}Ni + \ ^{1}_{1}H \rightarrow$ ______ $+ \ ^{4}_{2}He$

(e) $^{235}_{92}U + \ ^{1}_{0}n \rightarrow$ ______ $^{97}_{36}Kr + 4^{1}_{0}n$

(f) $^{95}_{42}Mo +$ ______ $\rightarrow \ ^{95}_{43}Tc + \ ^{1}_{0}n$

(g) $^{200}_{79}Au +$ ______ $\rightarrow \ ^{200}_{78}Pt$

Section 19.2

19.2 Calculate the energy change for the nuclear reactions:

(a) $^{11}_{6}C \rightarrow \ ^{11}_{5}B + \ ^{0}_{+1}\beta$

molar masses, g/mol

11.01143 11.00931 0.000549

(b) $^{234}_{92}U \rightarrow \ ^{230}_{90}Th + \ ^{4}_{2}He$

molar masses, g/mol

234.0409 230.0331 4.0026

(c) $^{213}_{85}At \rightarrow \ ^{209}_{83}Bi + \ ^{4}_{2}He$

molar masses, g/mol

212.9931 208.9804 4.0026

(d) $^{27}_{13}Al + \ ^{2}_{1}H \rightarrow \ ^{25}_{12}Mg + \ ^{4}_{2}He$

molar masses, g/mol

26.9815 2.0141 24.9859 4.0026

(e) $^{59}_{27}Co + \ ^{1}_{0}n \rightarrow \ ^{60}_{27}Co + \ ^{0}_{0}\gamma$

molar masses, g/mol

58.93319 1.00867 59.93355 0

Section 19.3

19.3 Which of the following nuclei possess a magnetic moment? Identify the number of unpaired protons and neutrons in the nucleus.
(a) $^{14}_{6}C$; (b) $^{31}_{15}P$; (c) $^{11}_{5}B$; (d) $^{56}_{26}Fe$.

19.4 Does a proton pair with a neutron in nuclear energy levels? Explain.

Section 19.4

19.5 Calculate the B.E./nucleon for

(a) $^{122}_{50}Sn$; molar mass: 121.9034 g/mol

(b) $^{208}_{82}Pb$; molar mass: 207.9766 g/mol

(c) $^{7}_{3}Li$; molar mass: 7.01600 g/mol

Compare each of these with the B.E./nucleon for $^{56}_{26}Fe$ in Example 19.4. Which nucleus is most stable and which is least stable?

Section 19.5

19.6 Determine which of the following are probably radioactive. For those that are radioactive, is the *n/p* ratio too high or too low?

(a) $^{30}_{14}Si$ (d) $^{18}_{9}F$ (h) $^{37}_{17}Cl$
(b) $^{25}_{11}Na$ (e) $^{32}_{16}S$ (i) $^{127}_{53}I$
(c) $^{37}_{18}Ar$ (g) $^{213}_{85}At$ (j) $^{40}_{20}Ca$

19.7 Which of the following increase the *n/p* ratio for $^{62}_{29}Cu$?

(a) positron emission (d) electron capture
(b) $_{-1}^{0}\beta$ emission (e) neutron emission
(c) proton emission (f) alpha emission

19.8 (a) In a $_{+1}^{0}\beta$ emission the atomic number ________ by ________.
(b) In α-decay the atomic mass ________ by ________.
(c) The atomic number ________ by ________ for an electron capture decay.
(d) In $_{-1}^{0}\beta$ emission the number of neutrons ________ by ________ and the number of protons ________ by ________.

19.9 What is the nucleus formed in each of the following nuclear transmutations?

(a) $^{10}_{5}B + ^{1}_{0}n \rightarrow [\ \ ]^* \rightarrow$ ________ $+ ^{4}_{2}He$
(b) $^{51}_{23}V + ^{2}_{1}H \rightarrow [\ \ ]^* \rightarrow$ ________ $+ 2^{1}_{0}n$
(c) $^{24}_{12}Mg + ^{1}_{0}n \rightarrow [\ \ ]^* \rightarrow$ ________ $+ ^{1}_{1}H$

19.10 (a) Calculate ΔE for the reaction

$$^{1}_{1}H \quad + \quad ^{2}_{1}H \quad \rightarrow \quad ^{3}_{2}He$$

molar masses, g/mol

1.0078 2.0141 3.0160

(b) What is the quantity of heat released per gram of $^{3}_{2}He$?

19.11 (a) Calculate ΔE for the reaction

$$^{235}_{92}U \quad + \quad ^{1}_{0}n \quad \rightarrow \quad ^{94}_{38}Sr \quad + \quad ^{139}_{54}Xe \quad + \quad 3^{1}_{0}n$$

molar mass, g/mol

235.0439 1.0087 93.9154 138.9179 1.0087

(b) What is the quantity of heat released per gram of $^{235}_{92}U$? Compare this to the answer in Question 19.10b.

Section 19.6

19.12 $^{60}_{27}Co$, used for cancer therapy, has a half-life of 5.26 years. If a hospital orders a 100 mg sample, how much remains after 15 years?

19.13 $^{90}_{38}Sr$, one of the feared radioactive wastes from a nuclear power plant because of its ability to displace calcium from the bone structure, has a half-life of 27.6 years. If the maximum dosage of $^{90}_{38}Sr$ in an adult human is that amount which undergoes 3.70×10^4 emissions (or nuclear disintegrations) per second, how many emissions per second would remain after 50.0 years (the normal adult life)?

Section 19.7

19.14 An ancient wood carving found in archeological dig shows a ^{14}C activity of 2.77 disintegrations per minute per gram. Assuming $^{14}_{6}C$ has a normal activity of 14.0 d/(min g) and $t_{1/2} = 5.73 \times 10^3$ years, how old is the wood carving?

19.15 Identify a medical application for each radioisotope:

(a) ^{59}Fe (f) ^{125}I
(b) ^{60}Co (g) ^{111}In
(c) ^{14}C (h) ^{64}Cu
(d) ^{131}I (i) ^{67}Ga
(e) ^{201}Tl

ANSWERS

19.1 (a) $^{78}_{33}As$ (e) $^{135}_{56}Ba$
(b) $^{122}_{52}Te$ (f) $^{1}_{1}H$
(c) $^{59}_{27}Co$ (g) $_{-1}^{0}e$
(d) $^{55}_{27}Co$

19.2 (a) -1.41×10^{11} J/mol
(b) -4.68×10^{11} J/mol
(c) -9.09×10^{11} J/mol
(d) -6.48×10^{11} J/mol
(e) -7.48×10^{11} J/mol

19.3 (b) $^{31}_{15}P$; one unpaired proton; (c) $^{11}_{5}B$; one unpaired proton

19.4 No, the protons and neutrons have their own set of energy levels.

19.5 (a) -1.36×10^{-12} J/nucleon
(b) -1.26×10^{-12} J/nucleon
(c) -9.01×10^{-13} J/nucleon
For Fe: -1.37×10^{-12} J/nucleon
$^{122}_{50}Sn$ is most stable and $^{7}_{3}Li$ is least stable.

19.6 (a) $^{30}_{14}Si$: *n/p* is too high
(b) $^{25}_{11}Na$: *n/p* is too high
(c) $^{37}_{18}Ar$: *n/p* is too low
(g) $^{213}_{85}At$: *n/p* is too high

19.7 (a) positron emission
(b) proton emission
(d) electron capture
(f) alpha emission

19.8 (a) decreases by one; (b) decreases by four; (c) decreases by one; (d) decreases by one; increases by one.

19.9 (a) $[^{11}_{5}B]^* \rightarrow {}^{7}_{3}Li$

(b) $[^{53}_{24}Cr]^* \rightarrow {}^{51}_{24}Cr$

(c) $[^{25}_{12}Mg]^* \rightarrow {}^{24}_{11}Na$

19.10 (a) $\Delta E = -5.31 \times 10^{11}$ J/mol

(b) $\Delta E = -1.76 \times 10^{11}$ J/g ${}^{3}_{2}He$

19.11 (a) $\Delta E = -1.74 \times 10^{13}$ J/mol

(b) $\Delta E = -7.40 \times 10^{10}$ J/g ${}^{235}_{92}U$

More energy is released per gram in the fusion than in the fission process.

19.12 13.9 mg

19.13 1.05×10^4 emissions/sec

19.14 1.34×10^4 yr

19.15 (a) measurement of red blood cell production

(b) treatment of cancer

(c) volume of blood

(d) diagnose thyroid disorders

(e) appraise the condition of the heart

(f) monitor blood circulation and locate blood clots.

(g) locate brain tumors

(h) locate brain tumors

(i) locate lesions resulting from Hodgkin's disease.

Chapter 20

CONCEPTS OF COVALENT BONDING

SECTION 20.1 OVERLAP OF ATOMIC ORBITALS

Objectives

- To define and describe the formation of a sigma (σ) bond and a pi (π) bond.
- To relate atomic orbital (AO) overlap to bond strength.

Focus

A **σ bond** results from the head-to-head overlap of two atomic orbitals, producing a region of high electron density along the internuclear axis between two bonded atoms. A σ bond can form from three types of overlap:

1. Two s atomic orbitals, called a σ_s bond.
2. An *s* and a *p* atomic orbital, called a σ_{sp} bond.
3. Two *p* atomic orbitals, called a σ_p bond.

All σ bonds have an electron density that has cylindrical symmetry about the internuclear axis. Therefore, free rotation of the atoms along a σ bond can occur. (See text, Figure 20.1.)

A **π bond** forms from the parallel (side-by-side) overlap of two *p* atomic orbitals, producing a region of high electron density above and below the internuclear axis. The overlap of two p_y atomic orbitals is designated as a π_y bond, and two p_z atomic orbitals is a π_z bond. The overlap of two p_x AO's forms a σ_p bond.

The electron density for a π_y bond lies mainly in the region above and below the *xz*-plane; for a π_z bond the electron density is in front of and behind the *xy* plane. (See Figures 20.1, 20.2, and 20.3). There is no free rotation between atoms held together by a π bond (text, Figures 20.3 and 20.4).

Two atoms may be bonded in several ways:

1. By a single bond, the σ bond.
2. By a double bond, the σ bond and a π bond.
3. By a triple bond, the σ bond plus a π_y and a π_z bond.

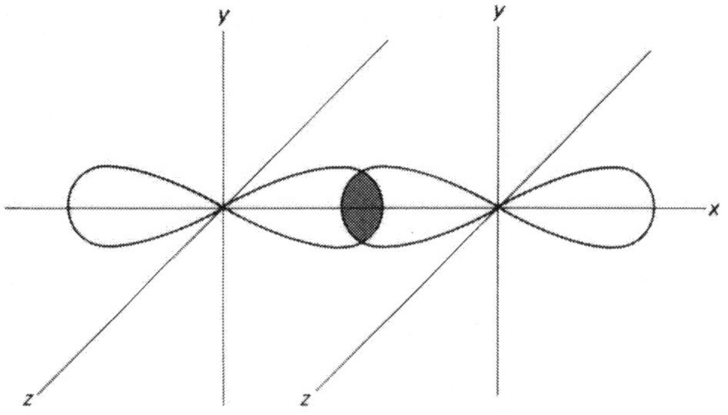

Figure 20.1 σ bond.

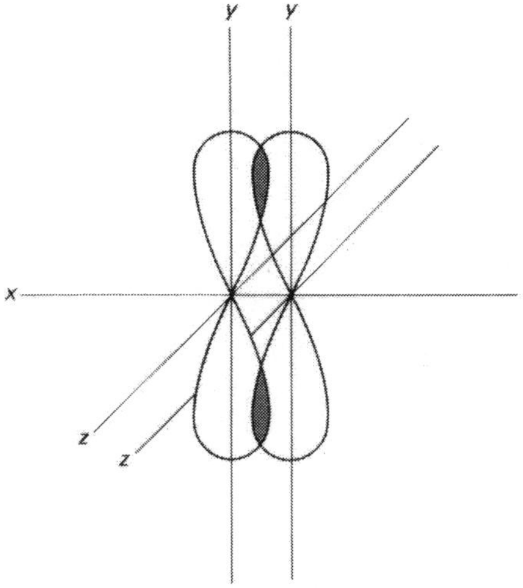

Figure 20.2 π_y bond.

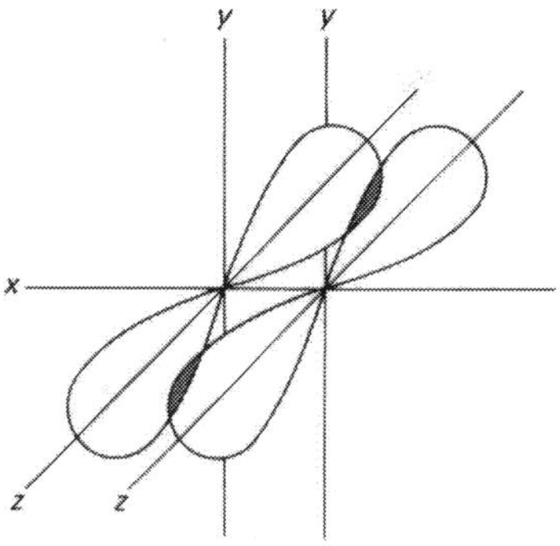

Figure 20.3 π_z bond.

In both the σ and the π bonds, the *overlap* is important. The greater the overlap of the AO's, the shorter is the internuclear distance (shorter bond length), and the greater is the energy necessary to break apart the bonded atoms (greater bond strength).

Example 20.1 Sketch the σ bond that forms between a H atom and an I atom.

Answer See Figure 20.4.

Solution First, write the energy-level diagram for valence shell electrons in each atom.

H 1*s* ↑ I 5*s* ↑↓ 5*p* ↑↓ ↑↓ ↑

The spherical *s* AO on the H atom overlaps with the dumbbell-shaped *p* AO on the I atom to produce a σ bond. The two electrons forming the bond are paired and have a high probability of cylindrical symmetry along the internuclear axis between the H atom and the I atom (as shown in Figure 20.4).

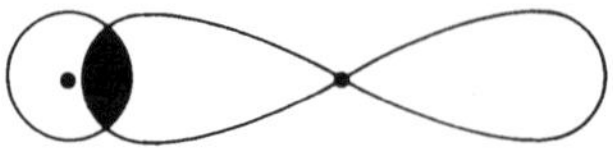

Figure 20.4 Head to head overlap produces the σ bond, a σ_{sp} bond.

SECTION 20.2 SINGLY BONDED POLYATOMIC MOLECULES

Objective

- To note that the AO's on the central atom of a polyatomic molecule are arranged differently from those in the nonbonded atom.

Focus

The valence shell AO's are generally the only orbitals used in covalent bonding. The AO's on the central atom of a polyatomic molecule often shift or reorient in space, thereby increasing the bond angles and the AO overlap with other atoms. A more detailed explanation is found in Section 20.3.

SECTION 20.3 HYBRIDIZATION OF ATOMIC ORBITALS

Objective

- To explain the formation of a set of equivalent hybridized atomic orbitals used for bond formation.

Focus

Remember this discussion is theory, a rational explanation of observed data.

An atom tends to bond to other atoms, so that the molecule has a minimum of energy. The theory holds that the valence shell orbitals on the central atom are oriented and redefined so that a set of lowest energy AO's are available for bonding.

First our theory says that the AO's in the valence shell on the *isolated* (nonbonded) atom are not necessarily the same as those on the bonded atom. Instead, the AO's blend (hybridize) to form an equivalent number of new orbitals that can overlap better when a bond is formed. This overlap provides for a stronger, stabler, lower energy bond between the atoms. The *s* AO, having the lowest energy in the valence shell, is always a part of the set of hybrid orbitals.

For *two* equivalent orbitals and therefore for two equivalent bonds, the central atom of a polyatomic molecule blends an *s* AO and a *p* AO to form two *sp* hybrid orbitals; two *p* orbitals remain unhybridized:

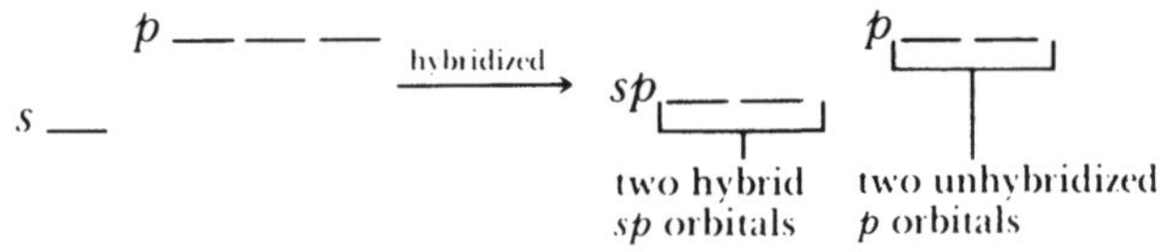

For *three* equivalent orbitals on a bonded central atom, the *s* AO and two *p* AO's blend to form three sp^2 hybrid orbitals. In the valence shell, one *p* orbital remains unhybridized:

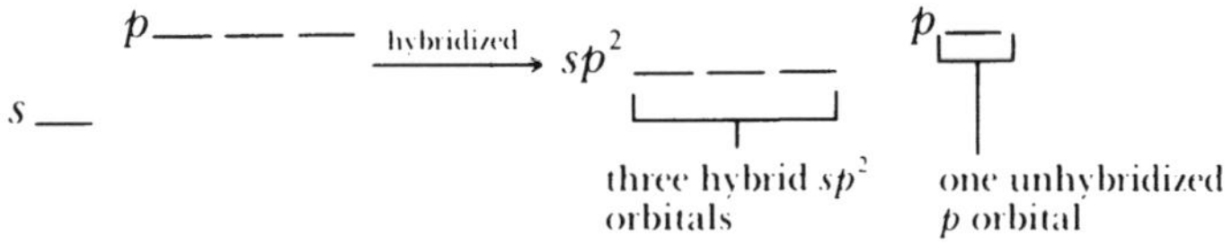

For *four* equivalent orbitals, the *s* AO and three *p* AO's blend to form four sp^3 hybrid orbitals.

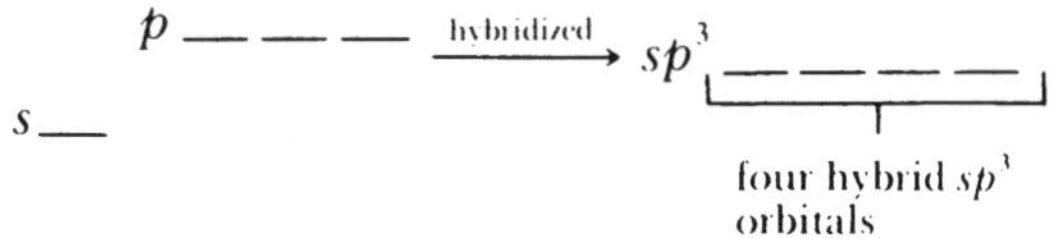

Become familiar with Table 20.2 (text), the common types of AO hybridization, bond angles, geometric arrangements, and fractions of *s* and *p* character of a hybrid AO. Remember that the more *s* character in a hybrid AO, the more stable is the resulting σ bond.

Example 20.2 Describe the AO's used by Al in the formation of $AlBr_3$. All bonds are equal in length and strength.

Answer Three sp^2 orbitals

Solution Al has one 3*s* AO and three 3*p* AO's available for bonding in the valence shell. Al has 3 valence electrons: $3s^23p^1$.

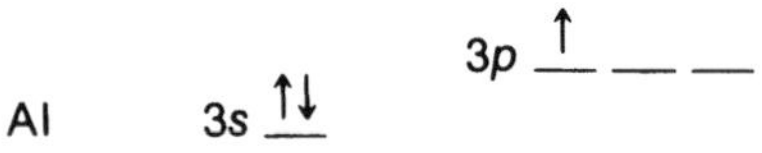

Br (7 valence electrons) has a single electron in the 4*p* AO available for sharing with an Al electron,

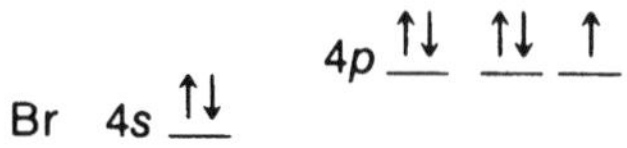

To form three equivalent bonds, Al hybridizes its valence shell AO's to provide three equivalent sp^2 orbitals, each having an electron available for sharing with a Br 4*p* electron.

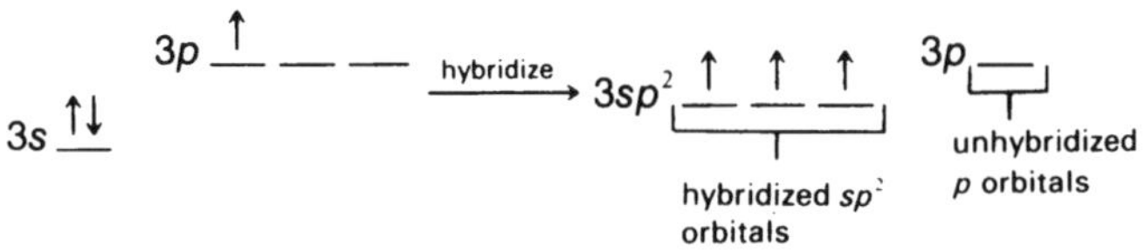

Three σ bonds form because of the head-to-head overlap of AO's (see Figure 20.5). One empty unhybridized *p* AO remains on the Al atom.

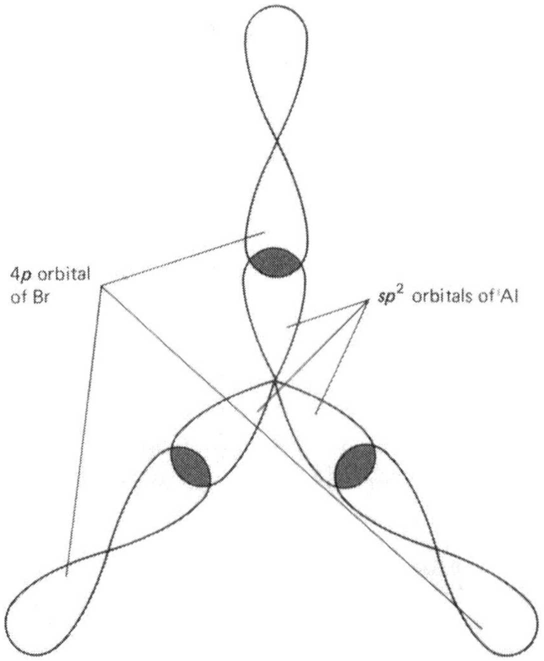

Figure 20.5 Head-to-head overlap of AO's to form three σ bonds in $AlBr_3$.

SECTION 20.4 MORE ON SHAPE AND HYBRIDIZATION

Objectives

- To predict the type of hybrid AO's used for bonding based upon the experimental bond angle.
- To explain the effect of an unshared pair of electrons on bond angle.

- To explain the use of hybrid AO's in the covalent bonding of second period elements.

Focus

We can use the experimental bond angle to determine which hybrid AO's the central atom uses for bonding. H_2O has an H—O—H bond angle of 105°; we assume sp^3 hybridization because this angle is closest to the 109.5° of an sp^3 hybridized central O atom: H_2S has an H—S—H bond angle of 92°; we assume no hybridization because this angle is closest to the 90° orientation of unhybridized *p* orbitals on the central S atom.

An unshared electron pair in the valence shell of the central atom frequently causes distortion of the bond angle from the pure hybridized bond angle. Whereas two nuclei attract a bonded pair of electrons, only one nucleus (that of the central atom) attracts an unshared pair of electrons. The unshared pair, being closer to the nucleus of the central atom, repels adjacent valence shell bonding pairs. The bond angle, measured by the position of the bonded nuclei in a molecule, is therefore reduced.

Around the central atom, two unshared pairs of electrons will have the greatest repulsion, an unshared pair and a bonded pair will have less repulsion, and two bonded pairs will repel one another the least.

Study these two sketches in Figure 20.6.

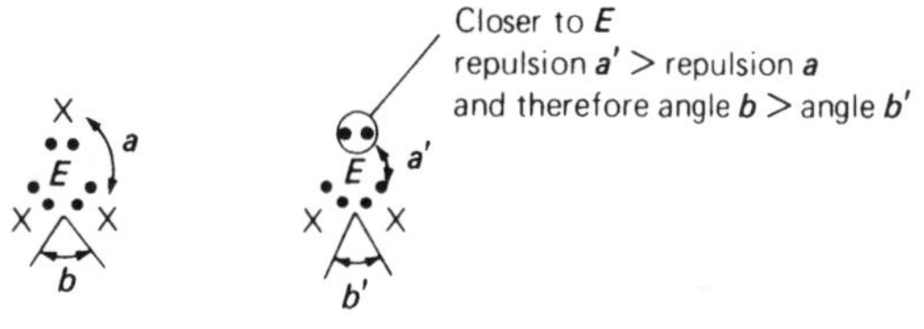

Figure 20.6 Effect of electron pair repulsion.

Second period elements, those with very small atoms, normally hybridize their valence shell AO's in forming σ bonds. Thus larger bond angles and stronger σ bonds, resulting from greater AO overlap, form to produce a more stable molecule. This also reduces the electron pair repulsion around the central atom.

Example 20.3 Which has the larger bond angle, $:NH_3$ or $H_2\ddot{O}:$? Explain.

Answer NH_3

Solution N and O are second period elements and tend to hybridize. Each requires four AO's; $:NH_3$ needs three AO's for bonding and one AO for its unshared electron pair. $H_2\ddot{O}:$ needs two AO's for bonding and two AO's for its two unshared electron pairs.

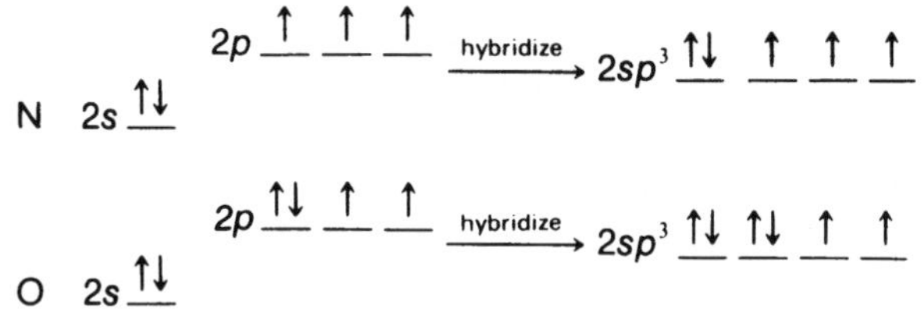

Because $H_2\ddot{O}$: has two unshared electron pairs closer to the oxygen nucleus (the central atom), the bonding electron pairs are pushed closer, resulting in a smaller bond angle.

$$\angle H_2O = 105° \qquad \angle NH_3 = 107°$$

Remember that we are trying to explain why a molecule is as it is; we cannot tell the molecule how it should act. It forms a configuration most suitable to itself, one having a configuration of the lowest possible energy and the most stable form (see text, Box 20.2, Section 20.4).

SECTION 20.5 HYBRIDIZATION IN MULTIPLY-BONDED MOLECULES

Objectives

- To predict from Lewis formulas the type of hybrid AO's in molecules having multiple and single bonds.
- To predict molecular structure based upon an assumed hybridization of AO's on the central atom.

Focus

First review the difference between a σ bond and a π bond. Remember these guidelines in predicting the type of hybrid AO's used in bonding.

1. Every single bond is a σ bond.
2. Every multiple bond has one σ bond and one or more π bonds.
3. For the second period elements, we can use the Lewis formula to determine the extent of hybridization of an atom by counting the number of σ bonds formed and the number of unshared electron pairs on the atom. For example, in $H_2\ddot{O}$:, the O atom is sigma bonded to *two* H atoms and has *two* unshared pairs of electrons (total = four) and therefore needs *four* hybridized AO's for bonding. O is sp^3 hybridized. Odd electrons and π bonded electrons (electrons in unhybridized p orbitals) are not counted when determining the number of hybrid AO's of the central atom. This guideline does not necessarily apply to 3rd, 4th, 5th, and 6th period elements.
4. π bonds result from the overlap of *un*hybridized, parallel p orbitals and/or d orbitals on adjacent atoms. Remember that the electron density of a π bond is above and below the internuclear axial plane, not all around the axis (see text, Figure 20.13c).
5. Hybrid orbitals are for unshared electron pairs and for σ bonding, never for π bonding.

Example 20.4 (a) Determine the hybridization on the C atom in $:CH_2$. (b) From Table 20.3, predict its shape. (c) Determine the approximate H—C—H bond angle.

Answer (a) sp^2; (b) triangular planar; (c) < 120°

Solution (a) A Lewis formula for $:CH_2$ is

$$H:\ddot{C}:H$$

C forms two σ bonds and has an unshared electron pair. This involves hybridization of three AO's on the C atom, resulting in sp^2 hybridization with an "empty p orbital" remaining unhybridized.

$$\begin{array}{l} 2p\ \underline{\uparrow}\ \underline{\uparrow}\ \underline{\ \ } \\ 2s\ \underline{\uparrow\downarrow} \end{array} \xrightarrow{\text{hybridize}} 2sp^2\ \underline{\uparrow\downarrow}\ \underline{\uparrow}\ \underline{\uparrow} \quad 2p\ \underline{\ \ }$$

(b) sp^2 hybridization produces a triangular planar structure (see Figure 20.7).

(c) sp^2 hybridization indicates a H—C—H bond angle of 120°. However, the repulsion between one unshared electron pair and the two bonding pairs causes the bond angle to be slightly less than 120°.

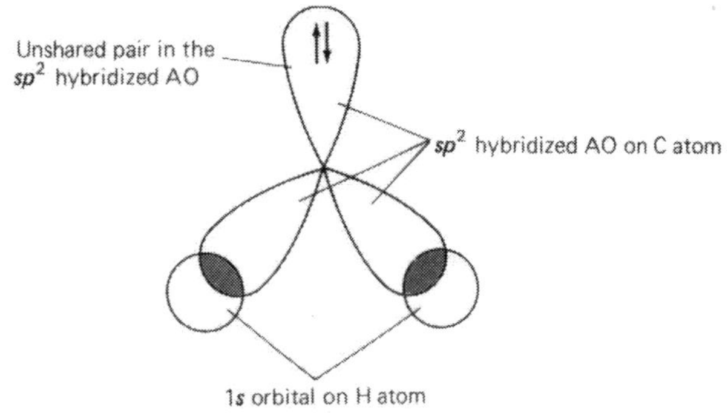

Figure 20.7 sp^2 hybridization in CH_2.

Example 20.5 Predict the hybridization of each atom in the hyponitrite ion, $N_2O_2^{2-}$, having the Lewis formula

$$\left[:\ddot{\underset{\cdot\cdot}{O}}—\ddot{N}=\ddot{N}—\ddot{\underset{\cdot\cdot}{O}}:\right]^{2-}$$

Answer Each O atom is sp^3, and each N atom is sp^2 hybridized.

Solution Both N and O are second period elements and therefore do hybridize for more stable bond formation.

Each O atom forms one σ bond and has three unshared pairs for a total of four hybridized AO's; therefore, each O atom is sp^3 hybridized.

Each N atom forms two σ bonds and has one unshared pair; therefore each is sp^2 hybridized.

$$\begin{array}{l} 2p\ \underline{\uparrow\downarrow}\ \underline{\uparrow}\ \underline{\uparrow} \\ 2s\ \underline{\uparrow\downarrow} \end{array} \xrightarrow{\text{hybridize}} 2sp^2\ \underline{\uparrow\downarrow}\ \underline{\uparrow}\ \underline{\uparrow} \quad 2p\ \underline{\uparrow}$$

The two unhybridized p orbitals on the adjacent N atoms overlap to form the π bond, the second bond of the double bond between the N atoms.

SECTION 20.6 HYBRIDIZATION WITH *d* ORBITALS

Objective

• To predict the hybridization for five and six pairs of electrons on the central atom by using *d* atomic orbitals.

Focus

For atoms, in which the number of σ bonds plus the number of unshared electron pairs exceed four (sp^3 hybridization), hybridization of *d* orbitals is necessary on the central atom. Obviously, elements in the second period cannot hybridize any *d* orbitals because there are no *d* orbitals in the second energy level.

Molecules with five electron pairs on the central atom hybridize five atomic orbitals—one *s*, three *p*, and one *d*—forming five sp^3d hybrid orbitals of equal energy and a trigonal bipyramidal structure.

Molecules with six pairs of electrons have six sp^3d^2 hybrid orbitals of equal energy with an octahedral structure.

Example 20.6 Predict the hybridization of Se in $:SeCl_4$.

Answer sp^3d

Solution Se forms four σ bonds and has one unshared electron pair. Thus Se hybridization forms five equivalent $4sp^3d$ AO's with four $4d$ orbitals remaining unhybridized.

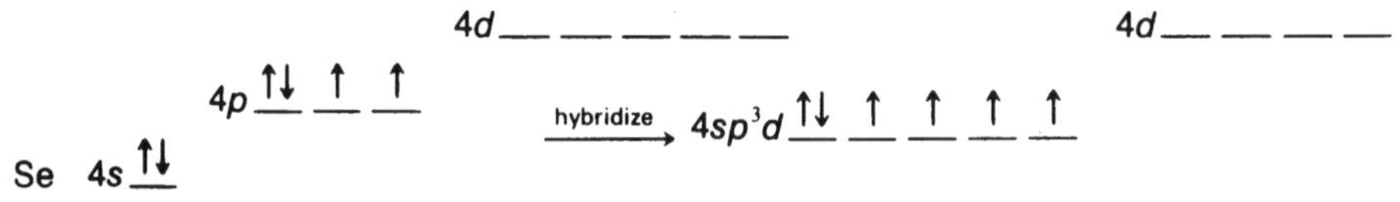

SECTION 20.7 EXTENDED π BONDING

Objective

• To describe the delocalization of electrons in π bonds for a molecule or an ion having Lewis formula resonance structures.

Focus

In Chapter 9, resonance structures were necessary for writing the Lewis formulas of various molecules and ions. The π bonded system explains why it was necessary.

Example 20.7 Predict the hybridization and show the extended π system that exists for CO_3^{2-}.

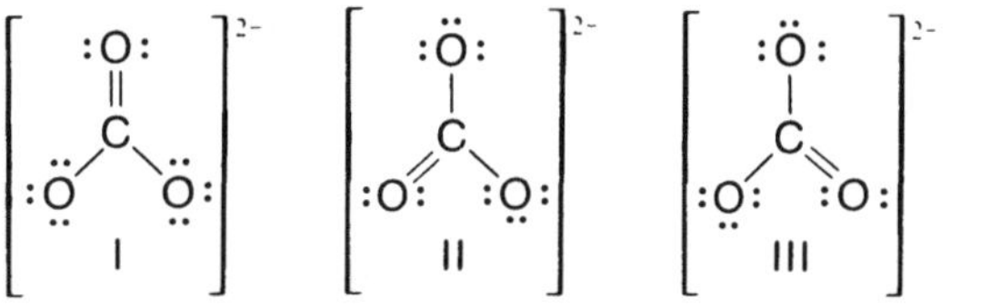

Answer All C and O atoms are sp^2 hybridized. The ion is resonance stabilized by the four electrons that are delocalized in four parallel, unhybridized p orbitals.

Solution C forms three σ bonds and has no unshared pairs, regardless of the Lewis formula resonance structure. C is sp^2 hybridized, forming a triangular planar structure, with one electron in an unhybridized P orbital, an electron available for π bonding.

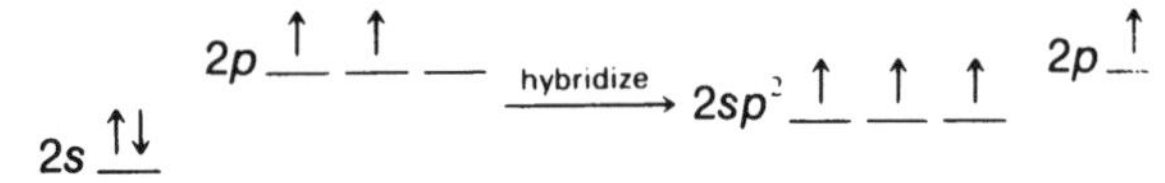

The π bonded O atom in structure I is also sp^2 hybridized (one σ bond, two unshared pairs).

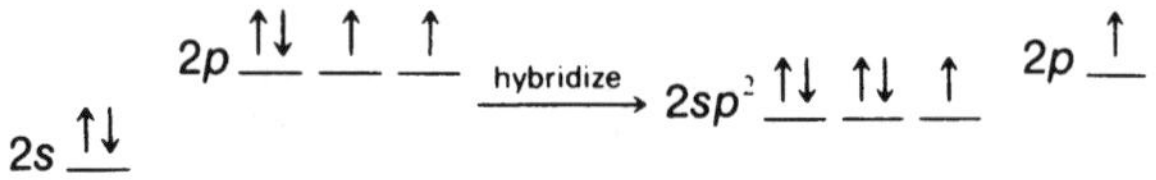

The one electron in an unhybridized p orbital is available for π bonding. Similarly, in structures II and III, each π bonded oxygen atom is sp^2 hybridized. Each has one electron in an unhybridized p orbital available for π bonding to the central C atom.

Therefore, the bonding in CO_3^{2-} is interpreted as having three σ bonds and a total of four electrons in four parallel, unhybridized p orbitals forming a delocalized system of π electrons over the CO_3^{2-} ion (see Figure 20.8). A more stable system results.

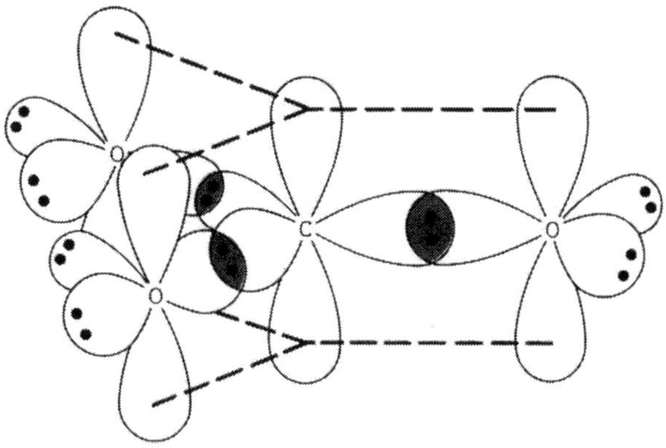

Figure 20.8 Delocalized π bonding system in CO_3^{2-}.

SECTION 20.8 MOLECULAR ORBITAL (MO) THEORY

Objectives

- To interpret the energy states in molecules in terms of molecular orbitals (MO's) in the same manner that energy states in atoms are interpreted by AO's.
- To describe a bonding MO and an antibonding MO.

Focus

Molecular orbital (MO) **theory** is to molecules what atomic orbital (AO) theory is to atoms. For atoms, electrons can have only quantized energy states. Atomic theory is based on the fact that atomic spectra consist of discrete wavelengths of light corresponding to photons with an energy, $E = hc/\lambda$. These photons originate from electron transitions from higher to lower energy states.

Excited molecules also produce line spectra, interpreted as originating from electronic transitions between energy states in molecules. The electrons in the atomic energy states are in AO's; therefore electrons in molecular energy states are in MO's.

MO's are derived from a combination of AO's; two AO's form two MO's, one of lower energy (bonding molecular orbital, MO) and one of higher energy (antibonding molecular orbital, MO*) than the AO's. Two AO's that form a σ bond result in a σ bonding MO and a σ^* antibonding MO*; two AO's that form a π bond result in a π bonding MO and a π^* antibonding MO*. In the text, Figures 20.23 and 20.24 show these various combinations. Study them closely. Remember that a combination of two *s* AO's, an *s* AO and a *p* AO, or two *p* AO's forms a σ bond. In MO theory these form σ_s and σ_s^*, σ_{sp} and σ_{sp}^*, and σ_p and σ_p^* MO's, respectively.

SECTION 20.9 MOLECULAR ORBITAL ENERGY LEVELS

Objective

- To write the electronic configuration for a molecule from the energy-level diagram.
- To predict the relative strength, the ionization energy, and the electron affinity of molecules.

Focus

The electronic configuration for a molecule or molecular ion is written in the same manner as for atoms. The energy-level diagram for molecules (text) can be filled with electrons, starting with the lowest energy σ_{1s} bonding MO. This is like starting with the 1*s* AO when filling the energy-level diagram for atoms. Hund's rule and the Pauli exclusion principle also apply for the MO energy-level diagram.

To construct an MO energy diagram, remember the following:

1. The number of MO's equals the sum of the number of atomic orbitals (orbitals are conserved!).
2. The antibonding MO* formed from lower energy AO's (for example, 1*s* AO's) has lower energy than the bonding MO formed from next higher energy AO's (for example, 2*s* AO's).
3. The greater the atomic orbital overlap, the lower the energy of the bonding MO, and the higher the energy of the antibonding MO*. In a completed MO energy-level diagram (text, Figure 20.25), note the following:

   (a) In order for a molecule to be stable, there must be more electrons in bonding MO's than in antibonding MO*'s.

   (b) The more electrons in bonding MO's than in antibonding MO*'s, the stronger the bond.

   (c) Electrons in antibonding MO*'s are easier to remove (ionize) than those in bonding MO's.

Example 20.8 (a) Write the electronic configurations for CN^- and F_2^+. Use text, Figure 20.25.
(b) Which ion forms the stronger bond?
(c) Which ion has the higher ionization energy?
(d) Which ion is paramagnetic?

Answer (a) See solution
(b) CN^-
(c) CN^-
(d) Fe^+

Solution (a) Use Figure 20.25 as a guide for filling MO's. CN^- has a total of 6 + 7 + 1 = 14 electrons; the electronic configuration is

$$CN^- \quad (\sigma_{1s})^2 (\sigma_{1s}^*)^2 (\sigma_{2s})^2 (\sigma_{2s}^*)^2 (\pi_{2y})^2 (\pi_z)^2 (\sigma_{2p})^2$$

For F_2^+, which has a total of 9 + 9 − 1 = 17 electrons, the electronic configuration is

$$F_2^+ \quad (\sigma_{1s})^2 (\sigma^*_{1s})^2 (\sigma_{2s})^2 (\sigma^*_{2s})^2 (\pi_y)^2 (\pi_z)^2 (\sigma_{2p})^2 (\pi^*_y)^2 (\pi^*_z)^1$$

(b) CN^- has the stronger bond with 10 electrons in bonding MO's and 4 in antibonding MO*'s. F_2^+ also has 10 electrons in bonding MO's but 7 in antibonding MO*'s.

(c) CN^- has the higher ionization energy because an electron would be removed from a bonding MO; in F_2^+ it would be removed from an antibonding MO*, a higher energy MO. The higher the energy of the MO, the easier it is to remove an electron.

SECTION 20.10 METALLIC BONDING

Objective

- To use MO theory to differentiate between electron and conduction bands.

Focus

As the number of MO's in a metallic atom increases, the energy difference between successive bonding MO's and antibonding MO*'s diminishes to the extent that the energy levels are continuous. Electrons in these MO's can therefore move from an MO to an MO* and from one atom to the next with virtually no change in energy; the electrons are thus "free to roam" the metal's entire structure.

A group of closely spaced bonding and antibonding MO's derived from like AO's in the metal (that is, the 1*s* AO on two atoms form σ_{1s} and σ^*_{1s} MO's) form an **electron band**. If electrons only partially fill the electron band, the electrons can readily move from one atom to the next. This is the metal's **conduction band** and accounts for its conductivity.

SECTION 20.11 METALLIC PROPERTIES IN TERMS OF MO BAND THEORY

Objective

- To explain the electrical and thermal conductivity, the metallic luster, and the malleability of metals.

Focus

Electrons in partially filled MO's roam the entire metallic structure. Thus, metals conduct electric current and thermal energy. However, a temperature increase may reduce the metal's conductivity because the additional vibration of the positive metallic ions may interfere with electron motion.

Electrons may absorb photons and be excited to a higher energy state, that is, a higher energy MO. When the electrons deexcite, photons are emitted, accounting for some of the shiny appearance of metals.

SECTION 20.12 INSULATORS AND SEMICONDUCTORS

Objective

- To differentiate between an insulator, a conductor, and a semiconductor in terms of MO's, and electron and conduction bands.

Focus

In an **insulator** there is a large energy separation between the MO's of the filled electron band and the MO's of an empty conduction band. The gap is so large that conventional energy sources cannot excite electrons from the electron band to the conduction band and therefore insulators are nonconductors.

In a **semiconductor**, a narrow energy gap separates the two bands, a gap that approximates thermal energies. With heat, the conductivity of semiconductors increases because electrons are excited into the empty MO's of the conduction band.

Insulators can be "doped" with elements or compounds having a conduction band at a slightly lower energy than the conduction band of the insulator. Thermal energy promotes the electrons of the doping element into the conduction band of the insulator, thus converting an insulator into a semiconductor.

Example 20.9 "Doped" semiconductors are classified as *n* type and *p* type. *n* types are those in which the doping element has one more electron than the insulator, and *p* types are those in which the doping element has one less electron than the insulator. Explain how each conducts current in terms of electron and conduction bands.

Answer The *n* type is electron rich. This increases the probability for an electron from the doping element to be excited into the conduction band of the insulator.

The *p* type semiconductor is electron poor. An electron from the insulator is easily excited into the conduction band of the doping element, producing "holes" in the electron band MO. The electrons in the metal's electron band MO's are now free to move and thereby conduct current.

QUESTIONS

Section 20.1

20.1 List three ways in which a σ bond forms.

20.2 How is a π bond formed? What is the difference between a π_y bond and a π_z bond?

Section 20.2

20.3 What is the angle between atomic *p* orbitals?

Section 20.3

20.4 What is a hybrid AO? Why is such an orbital proposed?

20.5 How many sp^2 hybrid AO's are in a central atom's valence shell? What is the bond angle in a molecule that has an sp^2 hybridized central atom? How much *s* character is in each sp^2 hybrid AO?

20.6 Give the type of hybrid AO's used by the central atom. Show the electron distribution in the hybrid AO's of the central atom in forming equivalent σ bonds.

(a)
```
   Br
Br C Br
   Br
```

(b)
```
[   Cl   ]2−
[ Cl Be Cl ]
[   Cl   ]
```

(c)
```
[   H   ]+
[ H P H ]
[   H   ]
```

(d)
```
F   F
  Ge
  F
```

(e) $[\text{I}\ \ \text{I}\ \ \text{I}]^-$

Section 20.4

20.7 Predict the type of AO's used for bonding in

(a) PH_3 (bond angle = 97.7°)
(b) NF_3 (bond angle = 102.1°)
(c) NH_3 (bond angle = 107°)
(d) NO_2^+ (bond angle = 180°)
(e) NH_4^+ (bond angle = 109°)
(f) H_2Te (bond angle = 89.5°)
(g) Cl_2O (bond angle = 110.8°)
(h) PF_3 (bond angle = 97.8°)

20.8 Which has the smaller F—P—F bond angle, PF_4^+ or PF_4^-? (*Hint*: Write the Lewis formula for each.)

Section 20.5

20.9 Predict the hybrid AO's used by the underlined atom to accommodate its unshared electron pair(s) and for the formation of its σ bonds.

(a) $F_2\underline{C}{=}\underline{\ddot{O}}$
(b) $\underline{N}O_2^+$
(c) $[:\underline{C}{\equiv}\underline{N}{-}\underline{\ddot{O}}:]^-$
(d) $H_2\underline{C}{=}\underline{\ddot{N}}H$
(e) $H_2\underline{C}{=}\underline{N}{=}\underline{\ddot{N}}:$
(f) $H{-}\underline{\ddot{O}}{-}Cl$
(g) $:\underline{\ddot{N}}H_2^-$
(h) $\underline{N}H_3^{2+}$

20.10 Draw the Lewis formula and predict the type of hybrid AO's on the underlined central atom and their shape.

(a) $H_3\underline{O}^+$
(b) $\underline{N}Br_2^-$
(c) $Cl_2\underline{O}$
(d) $O\underline{C}Cl_2$ (O C Cl Cl)
(e) H $\underline{N}$ C
(f) $O\underline{S}Cl_2$ (O S Cl Cl)
(g) $O_2\underline{S}Cl_2$ (O O S Cl Cl)
(h) $\underline{C}S_2$
(i) $\underline{N}_2H_4$ (H H N N H H)

20.11 Predict the C—S—C bond angle in $H_3C{-}S{-}CH_3$ assuming that (a) there is and (b) there is not hybridization of the S atom.

Section 20.6

20.12 Write the Lewis formulas and predict the hybrid AO's used for the unshared pairs and σ bonds.

(a) $TeBr_4$
(b) $TlBr_4^-$
(c) BiI_4^-
(d) IF_5
(e) AsF_3
(f) XeF_4

Section 20.7

20.13 Predict the type of hybrid AO's for the central atom. Over what atoms in the molecule or ion is there an extended or delocalized π system?

(a) O_2NF $\left(\begin{matrix} O & & \\ & N & F \\ O & & \end{matrix}\right)$

(b) CHO_2^- $\left(\begin{matrix} H & C & O \\ & O & \end{matrix}\right)^-$

(c) O_3 (O O O)

(d) CN_2^{2-} $(N\ \ C\ \ N)^{2-}$

(e) $HNNO_2^-$ $\left(\begin{matrix} & & O \\ H\ \ N\ \ N & & \\ & & O \end{matrix}\right)^-$

(f) SO_2 $\left(\begin{matrix} & S & \\ O & & O \end{matrix}\right)$

Section 20.8

20.14 The head-to-head overlap of an s AO and a p AO produces a σ_{sp} MO and a σ^*_{sp} MO. What MO's does the side-by-side (parallel) overlap of two p_y orbitals produce?

Section 20.9

20.15 Complete the following table:

	MO Electronic Configuration	Stable? yes or no	Paramagnetic? yes or no
HeH			
LiB			
BeH			
BeH^+			
NO^-			
HeH^-			
HF			
F_2^{2-}			
Be_2			

20.16 Which molecular ion has the higher electron affinity, CO^+ or NO^+? Write the electronic configuration for each.

20.17 Which has a more stable bond?

(a) BN or CN

(b) Ne_2 or Ne_2^+

(c) O_2^+ or O_2^{2+}

20.18 Which has the higher ionization energy?

(a) H_2^- or H_2

(b) O_2 or O_2^+

Section 20.12

20.19 What MO's make up the conduction band in K and Ca?

20.20 Which has the largest energy between the filled electron and the conduction bands: a conductor, a semiconductor, or an insulator?

20.21 Explain why a temperature rise decreases the conductivity of conductors and increases that of semiconductors.

ANSWERS

20.1 Overlap of two *s* atomic orbitals; overlap of two *p* atomic orbitals; overlap of an *s* atomic orbital with a *p* orbital

20.2 π bond is formed by the side-by-side overlap of *p* orbitals. A π_y bond is formed by two *p* orbitals in the *xy*-plane, a π_z by two *p* orbitals in the *xz*-plane.

20.3 90°

20.4 A hybrid AO is a blend of two or more AO's. The hybrid AO provides a better overlap for bond formation.

20.5 Three sp^2 hybrid AO's. Bond angle = 120°. 1/3 *s* character in each sp^2 hybrid AO.

20.6 (a)

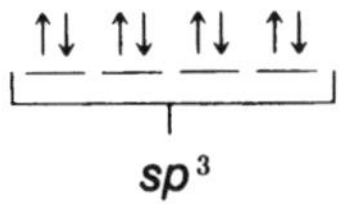

(b)

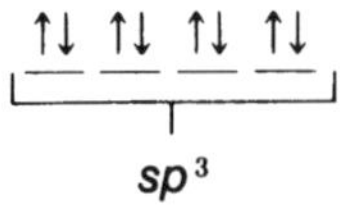

(c)

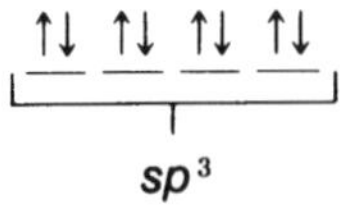

(d)

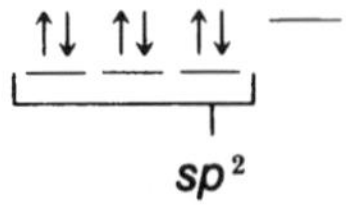

(e) 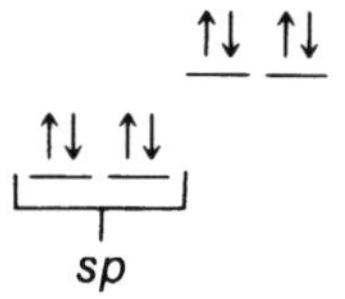

20.7 (a) *p* AO's
(b) sp^3 AO's
(c) sp^3 AO's
(d) sp^2 AO's
(e) sp^3 AO's
(f) *p* AO's
(g) sp^3 AO's
(h) *p* AO's

20.8 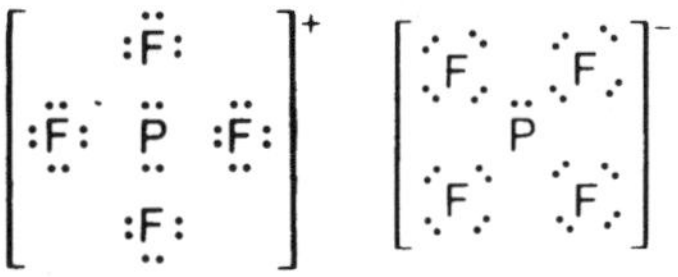

PF_4^+ uses sp^3 hybrid AO's, 109.5°. PF_4^- has smaller bond angle.

20.9 (a) C, sp^2; O, sp^2
(b) N, *sp*
(c) C, *sp*; N, *sp*; O, sp^3
(d) C, sp^2; N, sp^2
(e) C, sp^2; N, *sp*; N, sp^2
(f) O, sp^3
(g) N, sp^3
(h) N, sp^2

20.10

	Lewis formula	Hybrid AO's	Shape
(a)	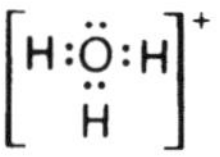	sp^3	Trigonal pyramid
(b)	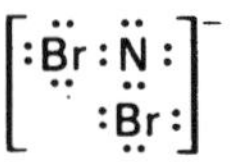	sp^3	Bent
(c)	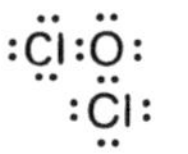	sp^3	Bent
(d)	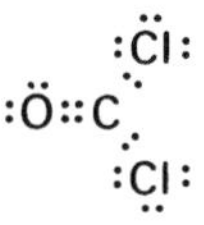	sp^2	Triangular planar
(e)	H: N::: C:	*sp*	Linear
(f)	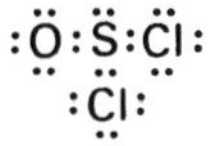	sp^3	Trigonal pyramid

(Continued)

20.10 (Continued)

	Lewis formula	Hybrid AO's	Shape
(g)	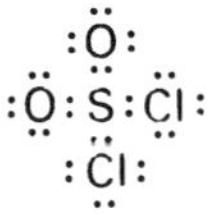	sp^3	Tetrahedral
(h)	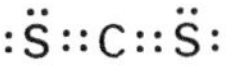	sp	Linear
(i)	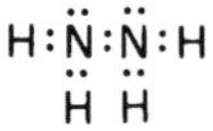	sp^3	Trigonal pyramid

20.11 (a) Hybridization: sp^3, 109.5°
(b) No hybridization: p orbital overlap, 90°

20.12

	Lewis formula	Hybrid AO's
(a)	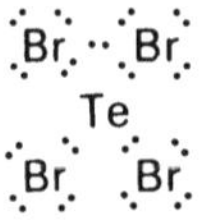	sp^3d
(b)	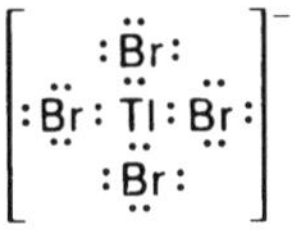	sp^3
(c)	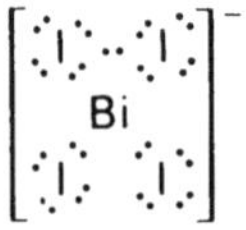	sp^3d
(d)	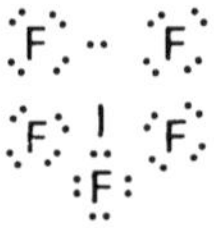	sp^3d^2
(e)	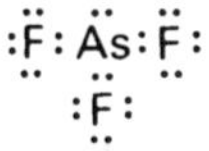	sp^3
(f)	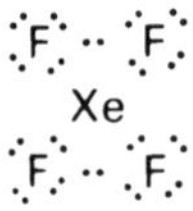	sp^3d^2

20.13

	Lewis formula	Hybrid AO's	Atoms in π System
(a)	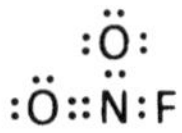	sp^2	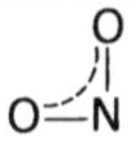

(b)	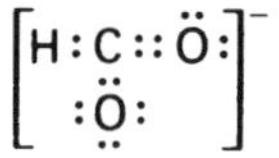	sp^2	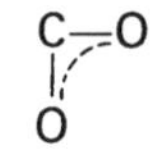
(c)	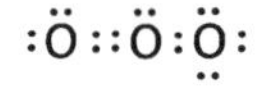	sp^2	
(d)	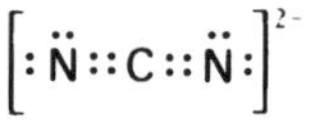	sp	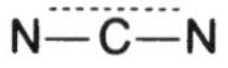
(e)	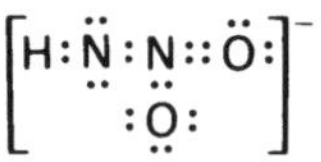	sp^2	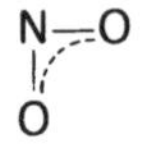
(f)	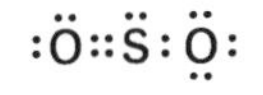	sp^2	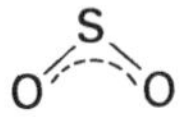

20.14 π_y and π_y^*

20.15

	MO Electron Configuration	Stable	Paramagnetic
HeH	$(\sigma_{1s})^2 (\sigma^*_{1s})^1$	Yes	Yes (1)
LiB	$(\sigma_{1s})^2 (\sigma^*_{1s})^2 (\sigma_{2s})^2 (\sigma^*_{2s})^2$	No	No
BeH	$(\sigma_{1s})^2 (\sigma^*_{1s})^2 (\sigma_{2s})^1$	Yes	Yes (1)
BeH^+	$(\sigma_{1s})^2 (\sigma^*_{1s})^2$	No	No
NO^-	$(\sigma_{1s})^2 (\sigma^*_{1s})^2 (\sigma_{2s})^2 (\sigma^*_{2s})^2 (\pi_y)^2 (\pi_z)^2$ $(\sigma_{2p})^2 (\pi^*_y)^1 (\pi^*_z)^1$	Yes	Yes (2)
HeH^-	$(\sigma_{1s})^2 (\sigma^*_{1s})^2$	No	No
HF	$(\sigma_{1s})^2 (\sigma^*{}_{1s})^2 (\sigma_{2s})^2 (\sigma^*_{2s})^2 (\pi_y)^1 (\pi_z)^1$	Yes	Yes (2)
F_2^{2-}	$(\sigma_{1s})^2 (\sigma^*_{1s})^2 (\sigma_{2s})^2 (\sigma^*_{2s})^2 (\pi_y)^2 (\pi_z)^2$ $(\sigma_{2p})^2 (\pi^*_y)^2 (\pi^*_z)^2 (\sigma^*_{2p})^2$	No	No
Be_2	$(\sigma_{1s})^2 (\sigma^*_{1s})^2 (\sigma_{2s})^2 (\sigma^*_{2s})^2$	No	No

20.16 CO^+, because the electron enters a bonding MO.
CO^+ $(\sigma_{1s})^2 (\sigma^*_{1s})^2 (\sigma_{2s})^2 (\sigma^*_{2s})^2 (\pi_y)^2 (\pi_z)^2 (\sigma_{2p})^1$
NO^+ $(\sigma_{1s})^2 (\sigma^*_{1s})^2 (\sigma_{2s})^2 (\sigma^*_{2s})^2 (\pi_y)^2 (\pi_z)^2 (\sigma_{2p})^2$

20.17 (a) CN; (b) Ne^{2+}; (c) O_2^{2+}.

20.18 (a) H_2; (b) O_2^+

20.19 K σ_{4s} MO's
Ca σ_{4p} MO's

20.20 an insulator

20.21 The conductivity of conductors decreases with temperature because the added heat increases the vibration of the positive ions in the crystal lattice producing resistance to electron flow. The conductivity of semiconductors increases with temperature because the added heat easily promotes electrons into the conduction band.

Chapter 21

TRANSITION ELEMENTS; COORDINATION COMPOUNDS

SECTION 21.1 INTRODUCTION

Objective

- To identify some of the properties of transition elements.

Focus

The transition elements are listed under Groups IIIB through IB in the periodic table. The transition elements have partially filled *d* or *f* orbitals. The Group IIB elements, with filled *d* orbitals, are not considered transition elements.

All transition elements are metals and have their expected properties: they are hard, can conduct heat and electricity, have high melting points, and are lustrous. Additional properties of the transition metals include:

1. Most compounds are colored.
2. Most compounds are paramagnetic (unpaired electrons).
3. Most metals have positive reduction potentials, except Cr, Mn, Fe, Co, and Ni.
4. The Pt and Pd triads, Ag, and Au are chemically inert to most acids but dissolve in **aqua regia**, a 3:1 mixture of $HCl:HNO_3$.
5. Most metals have variable positive oxidation states, with the highest being that of the group number. The higher oxidation states are more common in the 2nd and 3rd row transition elements than in the 1st row elements; metals with a higher oxidation state form bonds that are more covalent; the +2 and +3 oxidation states, which are most common, tend to form ionic bonds.

SECTION 21.2 COMPLEX FORMATION

Objectives

- To define a complex and a coordination compound.
- To define a ligand.
- To determine the oxidation state of the central metal atom in a complex.

Focus

A **complex** is a transition metal ion surrounded by anions and/or molecules. These anions and/or molecules, called **ligands**, are Lewis bases (electron-pair donors). The metal ions are electron-pair acceptors; therefore, the formation of a complex is the reaction between a Lewis acid (the metal ion) and a Lewis base (the ligand). A neutral complex or a compound containing a complex ion is a **coordination compound**.

The charge on the complex equals the sum of the metal ion's oxidation number and the cumulative charge on the ligands.

Example 21.1 Identify the ligands and determine the oxidation state of the metal ion in the following complexes: $K_2[ZnCl_4]$, $[Co(NH_3)_4(CN)_2]^+$, $[Pt(NH_3)_4]^{2+}$, $[Cr(NH_3)_4ClBr]^+$, $[Ni(SCN)_2(NO)_2]$.

Answer

Complex ion	Ligand(s)	Oxidation State of Metal Ion
$K_2[ZnCl_4]$	Cl^-	$2(K^+) + Zn + 4(Cl^-) = 0$ $2(+1) + Zn + 4(-1) = 0$ $Zn = +2$
$[Co(NH_3)_4(CN)_2]^+$	NH_3, CN^-	$Co + 4(NH_3) + 2(CN^-) = +1$ $Co + 4(0) + 2(-1) = +1$ $Co = +3$
$[Pt(NH_3)_4]^{2+}$	NH_3	$Pt + 4(NH_3) = 0$ $Pt + 4(0) = +2$ $Pt = +2$
$[Cr(NH_3)_4ClBr]^+$	NH_3, Cl^-, Br^-	$Cr + 4(NH_3) + 1(Cl^-) + (Br^-) = +1$ $Cr + 4(0) + (-1) + (-1) = +1$ $Cr = +3$
$[Ni(SCN)_2(NO)_2]$	SCN^-, NO	$Ni + 2(SCN^-) + 2(NO) = 0$ $Ni + 2(-1) + 2(0) = 0$ $Ni = +2$

In the absence of other ligands, transition metal ions form a complex with H_2O molecules (a Lewis base) in aqueous solution. For example, the Fe^{2+} ion in water is the $[Fe(H_2O)_6]^{2+}$ complex.

SECTION 21.3 COORDINATION NUMBER

Objectives

- To identify the coordination number of a complex.
- To distinguish between mono-, bi- and multidentate ligands.

Focus

The number of atoms bonded to the metal ion is called the **coordination number** of the complex. In addition, some ligands can form more than one bond with the metal; a ligand having only one bonding site is a monodentate ligand; one with two bonding sites, a bidentate ligand; one with three, a tridentate ligand, and so on. Monodentate ligands are the most common.

Example 21.2 Designate the coordination number in the following complex ions: $[Pd(en)_2]^{2+}$, $[Fe(en)_2(CN)_2]^+$, $[ZnCl_4]^{2-}$, $[Co(NH_3)_3(NO_2)_3]$, $[Cr(DAS)_2(CO)_2I]^+$.
Both en = ethylenediammine and DAS = diarsine are bidentate liquids.

Answer

Complex Ion	Coordination Number
$[Pd(en)_2]^{2+}$	2 en's, each a bidentate = 4
$[Fe(en)_2(CN)_2]^+$	2 en's, each a bidentate (=4) and 2 CN^-'s, each a monodentate (=2) = 6
$[ZnCl_4]^{2-}$	4 Cl^-'s, each a monodentate = 4
$[Co(NH_3)_3(NO_2)_3]$	3 NH_3's and 3 NO_2^-'s, each a monodentate = 6
$[Cr(DAS)_2(CO)_2I]^+$	2 DAS's, each a bidentate (= 4) and 2 CO's, each a monodentate (= 2) and I^-, a monodentate (= 1) = 7

SECTION 21.4 WERNER'S COORDINATION THEORY

Objectives

- To distinguish between the oxidation number (Werner's principal valence) and coordination number (Werner's auxiliary valence) for a complex.
- To recognize the chemical reactivity of complexes as determined by whether the ligand is bonded to a metal ion.

Focus

Werner explained the chemical behavior of coordination compounds in terms of the metal ion's **principal valence** and its **auxiliary valence**. We now interpret these terms to mean:

Principal valence = oxidation number.

Auxiliary valence = coordination number.

The chemical properties of a coordination compound depends on whether the ligand is bonded to the central atom as part of the complex or is merely in solution as the anion of a salt. For example, the two isomers $[Co(NH_3)_5Br]SO_4$ and $[Co(NH_3)_5SO_4]Br$ undergo the following reactions:

$$[Co(NH_3)_5Br]SO_4 + Ba(NO_3)_2 \rightarrow [Co(NH_3)_5Br](NO_3)_2 + BaSO_4(c)$$

$$[Co(NH_3)_5Br]SO_4 + AgNO_3 \rightarrow \text{no precipitate formed}$$

In $[Co(NH_3)_5Br]SO_4$, the Br^- is bonded to Co and is not free Br^- in solution. AgBr does not precipitate. The SO_4^{2-}, the anion of the salt, is in solution and not bonded to Co; it precipitates with the Ba^{2+} ion but not with the silver ion.

$$[Co(NH_3)_5SO_4]Br + Ba(NO_3)_2 \rightarrow \text{no precipitate formed}$$

$$[Co(NH_3)_5SO_4]Br + AgNO_3 \rightarrow [Co(NH_3)_5SO_4]NO_3 + AgBr(c)$$

In $[Co(NH_3)_5SO_4]Br$, the SO_4^{2-} is bonded to Co and does not precipitate with Ba^{2+}. The Br^-, which is not a part of the complex, is in solution and precipitates as AgBr. $BaBr_2$ is soluble and therefore does not precipitate.

Example 21.3 A coordination compound has the empirical formula, $FeCl_2Br(NH_3)_3(H_2O)_3$. When $AgNO_3$ is added to a solution containing 1 mole of the compound, only 1 mole of AgCl precipitates. When 1 mole of the salt is heated strongly, 2 moles of H_2O are released. The addition of HCl does not produce NH_4^+. What is the formula of the compound?

Answer $[Fe(NH_3)_3BrClH_2O]Cl(H_2O)_2$

Solution Since only 1 mol AgCl precipitates, then one Cl^- is *not* bonded to the Fe; the Br^- and one Cl^- must be ligands. Since heat easily removes 2 mol H_2O, then only 1 mol H_2O is bonded directly to the Fe. The NH_3 molecules must also be bonded to Fe because no reaction occurs with HCl.

SECTION 21.5 COORDINATION NUMBER AND SHAPE

Objectives

- To recognize the shapes of complexes having common coordination numbers.
- To identify *cis-* and *trans-*isomers.

Focus

The common coordination numbers for metal ions are 2, 4, and 6. These result in ligand arrangements about the central atom as linear —coordination number = 2

tetrahedral or square planar —coordination number = 4

octahedral —coordination number = 6

For square planar molecules (coordination number of 4) having two identical ligands, two arrangements—resulting in two isomers (geometric stereoisomers)—are possible:

1. Identical ligands can be adjacent to each other (the *cis*-isomer).
2. Identical ligands can be diagonally opposite to each other (the *trans*-isomer).

Cis-trans-isomerism also occurs for octahedral complexes.

Example 21.4 Draw the *cis*- and *trans*-isomers for the octahedral complex $[Co(NH_3)_4(CN)_2]^+$

Answer

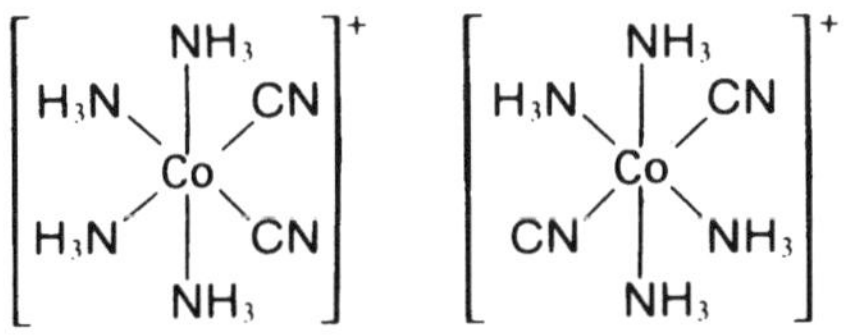

cis-isomer *trans*-isomer

Solution Since the coordination number is 6, the shape of the complex is octahedral. The *cis*-isomer has two CN^- ligands on adjacent bonding sites and the *trans*-isomer has two CN^- ligands at opposite bonding sites to the Co.

SECTION 21.6 METAL IONS IN BIOCHEMICALS: PRACTICAL APPLICATIONS

Objective

• To realize that transition metal complexes are involved in many biochemical and industrial processes.

Focus

Refer to the text for a discussion of several ways transition metal complexes are important in biochemical systems and industrial processes.

Ligands affect the reduction potentials of metal ions. Some examples are:

$$[Ni(H_2O)_6]^{2+} + 2e^- \rightleftharpoons Ni \qquad \mathcal{E}^0 = -0.250\text{ V}$$
$$[Ni(NH_3)_6]^{2+} + 2e^- \rightleftharpoons Ni \qquad \mathcal{E}^0 = -0.476\text{ V}$$

$$Fe(H_2O)_6{}^{3+} + e^- \rightleftharpoons Fe(H_2O)_6{}^{2+} \qquad \mathcal{E}^0 = +0.771\text{ V}$$
$$[Fe(CN)_6]^{3-} + e^- \rightleftharpoons [Fe(CN)_6]^{4-} \qquad \mathcal{E}^0 = +0.36\text{ V}$$

Other examples of complex ion formation are found in any qualitative analysis scheme used for cation identification. The insoluble salts $AgCl(c)$ and $Hg_2Cl_2(c)$ are separated by dissolving AgCl in an aqueous NH_3 solution. The soluble complex $[Ag(NH_3)_2]^+$ forms

$$AgCl(c) + 2NH_3 \rightleftharpoons [Ag(NH_3)_2]^+ + Cl^-$$

whereas Hg_2Cl_2 forms the complex $HgNH_2^+$, which is insoluble as the Cl^-.

$$HgCl_2(c) + 2NH_3 \rightleftharpoons Hg(\ell) + [HgNH_2]Cl(c) + NH_4^+ + Cl^-$$

SECTION 21.7 NOMENCLATURE OF COORDINATION COMPOUNDS

Objective

- To name coordination compounds.

Focus

Review the rules in the text for the names of

1. anionic ligands
2. neutral ligands
3. cationic complexes
4. anionic complexes

and then practice the following. Remember that ligands are named alphabetically.

Example 21.5 Complete the following table.

	Formula	Name
(a)		Potassium hexafluorocobaltate(III)
(b)	$[Co(NH_3)_6]F_3$	
(c)		Ammonium diaquatetracyanoferrate(III)
(d)	$[PtBr_2(en)_2](ClO_4)_2$	
(e)		Diamminediaquabromochloro rhodium(III) phosphate
(f)	$K_3[Fe(CN)_5CO]$	
(g)		Diamidotetraamminecobalt(III) oxide
(h)	$K[PtCl_3(C_2H_4)]$	

Answer

(a) $K_3[CoF_6]$
(b) hexaamminecobalt(III) fluoride
(c) $NH_4[Fe(CN)_4(H_2O)_2]$

(d) dibromo"bis"ethylenediamine platinum(IV) perchlorate
(e) $[RhBrCl(H_2O)_2(NH_3)_2]_3PO_4$
(f) potassium carbonylpentacyanoferrate(II)
(g) $[Co(NH_3)_4(NH_2)_2]_2O$
(h) potassium trichloro(ethylene)platinate(IV)

SECTION 21.8 BONDING IN COORDINATION COMPOUNDS

Objectives

- To explain crystal field theory (CFT) bonding in complexes.
- To account for the splitting of d orbitals in an octahedral complex.
- To describe the relationship between the crystal field splitting energy, Δ, and the color of the complex ion.
- To account for the magnetism of a complex ion in terms of "high spin" and "low spin" complexes.

Focus

Crystal field theory assumes that the bonding between the metal ion and the ligand is ionic. Three d orbitals (d_{xy}, d_{xz}, and d_{yz}) of a transition metal ion are directed in *between* x-, y-, and z-axes, and two d orbitals ($d_{x^2-y^2}$ and d_{z^2}) are directed *along the* x-, y-, and z-axes.

The six ligands in an octahedral complex bonded to a transition metal ion are all located along the x-, y-, and z-axes. Remember that ligands are Lewis bases; they have at least one unshared electron pair. Electrons also occupy d orbitals on the central metal ion. The electrons on the ligands directed along the axes are more strongly repelled by the electrons in the $d_{x^2-y^2}$ and d_{z^2} orbitals than those in the d_{xy}, d_{yz}, and d_{xz} orbitals. Therefore in an octahedral complex, an energy difference between the two sets of d orbitals occurs. The energy difference between the two sets of d orbitals is the **crystal field splitting energy**, Δ.

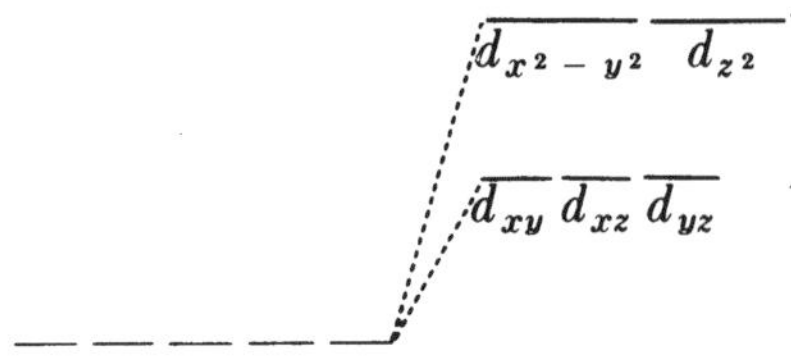

$$\Delta = \Delta E = h\nu = h\frac{c}{\lambda}$$

The splitting energy is about the same as the energy of photons in the visible region of the spectrum. Thus, the complex absorbs and emits visible light when d orbital $\rightarrow$ d orbital electron transitions occur. This electron excitation and deexcitation can occur only if the metal ion has *partially filled d* orbitals—only metal ions with partially filled d orbitals show color. Therefore, the representative ions, Na^+, F^-, or Ba^{2+}, are colorless because their d orbitals are either filled or empty and no electron transitions can occur. Transition metal ions with empty *d* orbitals, Sc^{3+} or Ti^{4+}, and with filled *d* orbitals, Zn^{2+}, Ag^{+}, or Cu^{+}, are colorless for the same reason.

Recall that white light consists of all wavelengths in the visible range; black light is the absence of any wavelengths of visible light. Therefore, if a substance reflects all wavelengths in the visible region, it appears white (or colorless); if it absorbs all wavelengths, it appears black.

The visible radiation that an electron does *not* absorb when excited from a lower to a higher energy *d* orbital is transmitted through the sample and detected as color. Therefore, the color that is seen is the light *not* absorbed. If a *d* electron absorbs high energy (short wavelength) visible light, such as blue, because of a large Δ, the complementary low energy (long wavelength) visible light, such as orange, is transmitted and observed as the visible color of the sample (see text, Color Plate IV).

Example 21.6 Salts of $[Co(NH_3)_5Cl]^+$ absorb light of wavelength 430 nm. What is the characteristic color of these salts and the value of Δ? 430 nm corresponds to purple light in the visible spectrum (see text, Color Plate IV).

Answer green; $\Delta = 3.77 \times 10^{-19}$ J

Solution If purple light is absorbed, its complementary color, green (Color plate), is transmitted. The crystal field splitting energy is $\Delta = \dfrac{hc}{\lambda}$.

$$h = \text{Planck's constant} = 6.63 \times 10^{-34} \text{ J sec/photon}$$

$$c = \text{speed of light} = 3.00 \times 10^{8} \text{ m/sec}$$

$$\lambda = 430 \text{ nm} = 430 \cancel{\text{nm}} \times \frac{10^{-9} \text{ m}}{\cancel{\text{nm}}} = 430 \times 10^{-9} \text{ m}$$

$$\Delta = \frac{hc}{\lambda} = \frac{6.63 \times 10^{-34} (\text{J} \cancel{\text{sec}}/\text{photon})(3.00 \times 10^{8} \cancel{\text{m}/\text{sec}})}{430 \times 10^{-9} \cancel{\text{m}}}$$

$$= 4.63 \times 10^{-19} \text{ J/photon}$$

The stronger the bond between the metal ion and ligand, the larger the Δ; the larger the Δ, the higher is the energy of the photon absorbed from the visible spectrum; the higher the energy of the photon absorbed, the shorter is the wavelength absorbed; the shorter the wavelength absorbed, the longer are the wavelengths of light transmitted. Thus complexes with strong field ligands absorb in the short wavelength (violet) region and transmit the long wavelengths (yellow).

The relative Δ's for various ligands are given in the text. CN^- is a strong field ligand, causing a large Δ; I^- is a weak field ligand.

Example 21.7 Which of the complexes has a greater crystal field splitting energy, Δ, $[CrCl_6]^{3-}$ or $[Cr(CN)_6]^{3-}$?

Answer $[Cr(CN)_6]^{3-}$

Solution CN^- is a strong field ligand. Accordingly, these six Lewis base ligands cause a greater interaction with the *d* orbitals on the Cr^{3+} ion directed along the *x*-, *y*-, and *z*-axes. This causes a larger Δ. We can picture the splitting as follows: The Cr^{3+} ion has a $3d^3$ electron configuration.

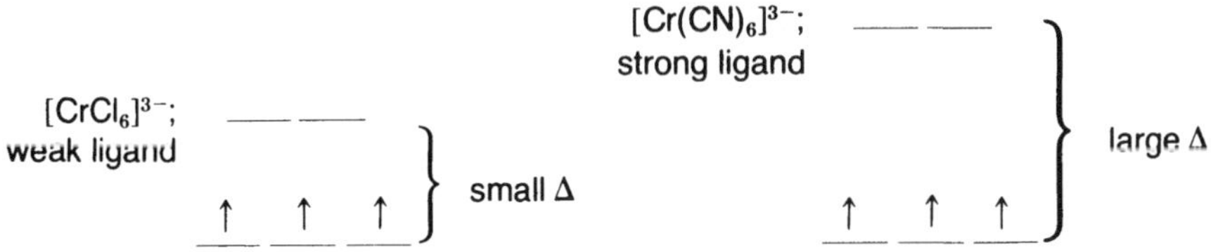

For a large Δ, short wavelength light is absorbed, allowing the complementary longer wavelength light to be transmitted.

Weak field ligands produce a sufficiently small Δ so that the *d* electron arrangement on the transition metal ion is not affected. This produces a **high spin complex**, one with a maximum number of unpaired electrons (remember that an atom or an ion with one or more unpaired electrons is paramagnetic). Strong field ligands produce a large Δ; the *d* electrons rearrange and occupy the lowest available *d* orbital energy state, one with a minimum number of unpaired electrons (an atom or ion with all electrons paired is diamagnetic). The result is a **low spin complex**. For Co^{2+}, with a $3d^7$ electron configuration, we can have two configurations:

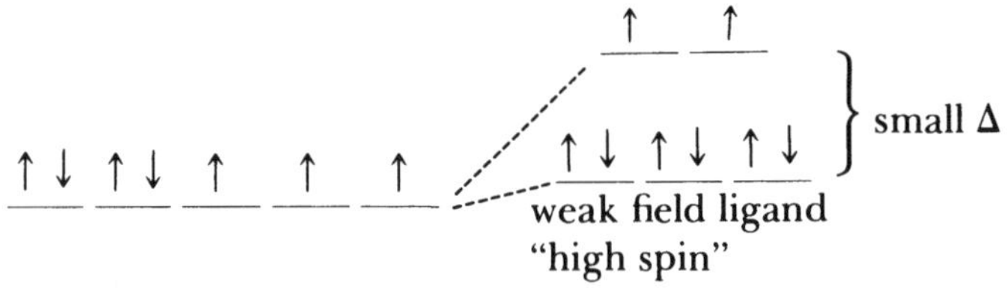

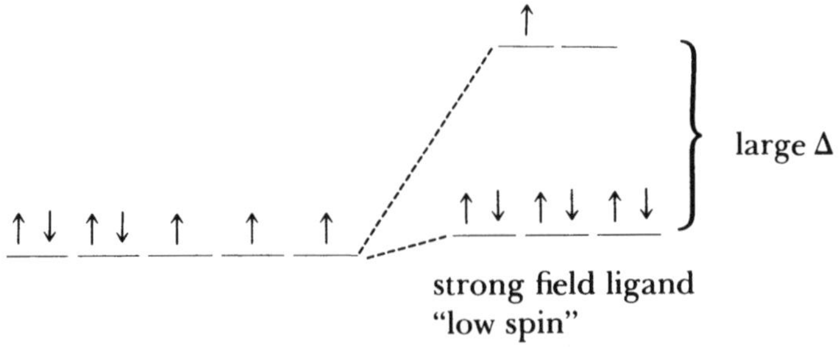

High spin complexes tend to be more paramagnetic than low spin complexes.

Example 21.8 $[Co(NH_3)_6]^{3+}$ is diamagnetic, $[CoF_6]^{3-}$ is paramagnetic. Co has an oxidation state of +3 in both complex ions. Explain the magnetism in terms of crystal field theory.

Answer Co^{3+} has a $3d^6$ electron configuration. NH_3 is a strong field ligand; F^- is a weak field ligand. In $[Co(NH_3)_6]^{3+}$ the six *d* electrons are paired as expected for a low spin complex with a large Δ. The *d* electrons remain unpaired in $[CoF_6]^{3-}$, resulting in a smaller Δ.

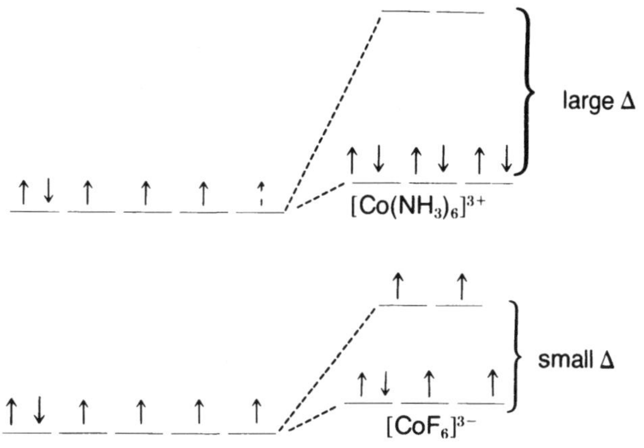

$[Co(NH_3)_6]^{3+}$ is a low spin complex with no unpaired electrons (diamagnetic); $[CoF_6]^{3-}$ is a high spin complex with four unpaired electrons (paramagnetic).

SECTION 21.9 THE STABILITY OF COMPLEX IONS

Objectives

- To distinguish between labile and inert complexes.
- To distinguish between kinetic stability and thermodynamic stability in complexes.

Focus

A **liabile complex** is one in which there is a free exchange of ligands with the metal ion in solution. An **inert complex** exchanges ligands very slowly.

We can make a distinction between **thermodynamically stable** and **kinetically stable** complexes by comparing the values for the free energy of the reaction, ΔG^0, versus that for activation, $\Delta G\ddagger$.

$\Delta G^0 = G^0_{\text{products}} - G^0_{\text{reactants}}$: If $G^0_{\text{products}} > G^0_{\text{reactants}}$, then ΔG^0 is positive and the reaction is thermodynamically stable. For this value we are not concerned about the "how" of the reaction, only the before and after.

For $\Delta G\ddagger$, we are interested in the "how," that is, the mechanism and transition state of the reaction

$$\Delta G\ddagger = G\ddagger_{\substack{\text{transition state}\\ \text{intermediate}}} - G^0_{\text{reactants}}$$

$\Delta G\ddagger$ is always a positive value; the higher the $\Delta G\ddagger$ value, the more kinetically stable is the complex.

A comparison of ΔG^0 and $\Delta G\ddagger$ gives the following conditions (remember $\Delta G\ddagger$ is always positive; ΔG^0 may be positive or negative).

1. If a positive $\Delta G^0 < \Delta G\ddagger$, and $G\ddagger \simeq G^0_{\text{products}}$, then the reaction is thermodynamically stable but kinetically unstable (see Figure 21.1a).

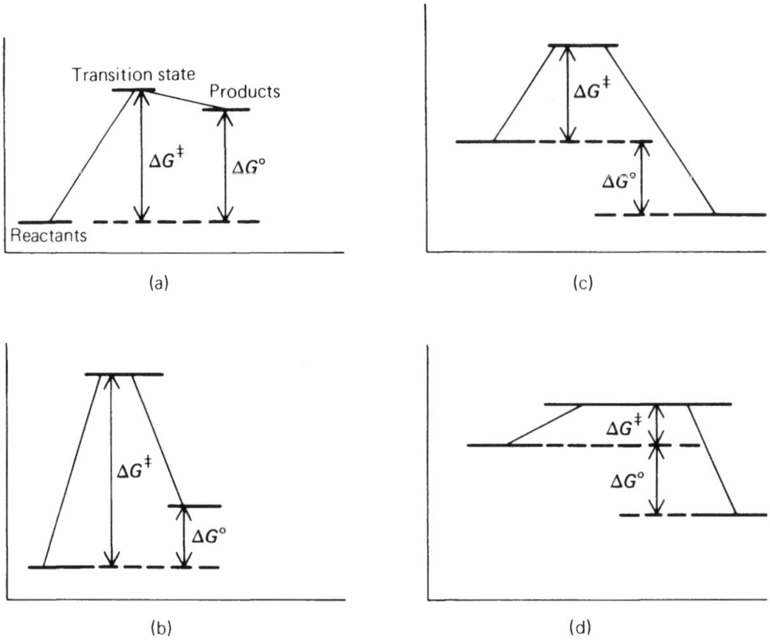

Figure 21.1 Gibbs energy diagrams illustrating reactions that are (a) thermodynamically stable but kinetically unstable, (b) thermodynamically and kinetically stable, (c) thermodynamically unstable but kinetically stable, and (d) thermodynamically and kinetically unstable.

2. If a positive $\Delta G^0 <<< \Delta G\ddagger$, then the reaction is thermodynamically and kinetically stable (see Figure 21.1b).
3. If a negative $\Delta G^0 << \Delta G\ddagger$, then the reaction is thermodynamically unstable but kinetically stable (see Figure 21.1c).
4. If a negative $\Delta G^0 < \Delta G\ddagger$ where $G\ddagger \simeq G^0_{\text{reactants}}$, then the reaction is thermodynamically and kinetically unstable (see Figure 21.1 d).

Inert complexes are kinetically stable, a large $\Delta G\ddagger$ and small rate constant, k, but may be either thermodynamically stable or unstable. These are conditions (2) and (3).

Labile complexes, however, are kinetically unstable, a small $\Delta G\ddagger$ and large k, but may be either thermodynamically stable or unstable. These are conditions (1) and (4).

The equilibrium constant for a complex ion equilibrium is only a *thermodynamic* measure of the complex's stability. Recall that

$$\Delta G^0 = -2.303RT \log K$$

K is a ratio of the rate constant for the forward reaction, k_f, divided by the rate constant for the back reaction, k_b, or

$$K = \frac{k_f}{k_b}$$

K is neither a measure of the rate nor a mechanism of the forward and back reactions, nor is it a measure of $\Delta G\ddagger$.

It is thus possible for K to be small, while k_f and k_b are either small or large. It is also possible for K to be large, while k_f and k_b are either small or large:

$K = \dfrac{10^9}{10^{15}} = 10^{-6}$	$K = \dfrac{10^{-15}}{10^{-9}} = 10^{-6}$
thermodynamically stable kinetically unstable labile complex	thermodynamically stable kinetically stable inert complex
$K = \dfrac{10^{15}}{10^9} = 10^6$	$K = \dfrac{10^{-9}}{10^{-15}} = 10^6$
thermodynamically unstable kinetically unstable labile complex	thermodynamically unstable kinetically stable inert complex

QUESTIONS

Section 21.1

21.1 Select the more appropriate choice of each of the following.

(a)	has a higher reduction potential	Au	Fe
(b)	+5 is the more stable oxidation state	Nb	V
(c)	greater bond covalency	$TiCl_2$	$TiCl_4$
(d)	the highest oxidation state for Mn	5+	7+
(e)	releases H_2 from an acidic solution	Cr	Cu
(f)	has a $4s^2 3d^6$ electron configuration	Co	Fe
(g)	has partially filled *f* orbitals	Eu	W
(h)	is not considered a transition metal	Hg	Sc
(i)	the number of valence electrons in Mn, Fe, Co, Ni	2	3
(j)	dissolves only in aqua regia	Rh	Fe

Section 21.2

21.2 What is the oxidation number of the metal ion in the following complexes?

(a) $[Ag(CN)_2]^-$

(b) $[Ni(NH_3)_2I_2]$

(c) $[Cr(NH_3)_4(SCN)_2]Cl$

(d) $K_4[Fe(CN)_6]$

(e) $K_3[Fe(CN)_6]$

Section 21.3

21.3 Designate the coordination number of the metal ion for each complex ion in Question 21.2.

Section 21.4

21.4 Calculate the mass (grams) of AgCl that precipitates from a solution containing 200 grams of (a) $[Cr(NH_3)_4Cl_2]Cl$ and (b) $[Cr(NH_3)_5Cl]Cl_2$.

Section 21.5

21.5 Draw the *cis*- and *trans*-isomers for the square planar complex $[Au(CN)_2Cl_2]$.

Section 21.7

21.6 Complete the following table.

Formula	Name
(a) $[Co(NH_3)_5Cl]Cl_2$	
(b)	Chloro"bis"ethylenediamminenitro cobalt(III) ion
(c) $K[Au(CN)_4]$	
(d)	Sodium dicyanoargentate(I)
(e) $Fe(CO)_5$	
(f)	Calcium tetrabromomercurate(II)
(g) $[Zn(NH_3)_4][ZnCl_4]$	
(h)	Diaquatetrahydroxoaluminate(III) ion

Section 21.8

21.7 $[RhCl_6]^{3-}$ absorbs violet light at a wavelength of 439 nm. $[RhBr_6]^{3-}$ absorbs green light at a wavelength of 526 nm. What is the color of each complex ion? Which has a larger crystal field splitting energy? Which ligand is stronger?

21.8 Predict the number of unpaired *d* electrons for two Co^{2+} complex ions, one that is a high spin octahedral complex, and the other a low spin complex. Which has the larger crystal field splitting energy? Which complex ion shows a warmer color, for example, yellow, orange, or red?

21.9 Draw the electron configurations for a small *and* a large Δ in an octahedral complex of (a) Mn(II); (b) Co(III); (c) Ti(III). Which configuration is the more paramagnetic for each? Which configuration in each case absorbs the higher energy photons?

20.10 List the following complexes in order of an increasing energy of photons absorbed: (increasing Δ): $[FeCl_6]^{3-}$, $[Fe(CN)_6]^{3-}$, $[Fe(H_2O)_6]^{3+}$, $[FeCl_4(H_2O)_2]^-$, $[Fe(NO_2)_6]^{3-}$, $[Fe(H_2O)_3(OH)_3]$. Which one absorbs the longest wavelength?

ANSWERS

21.1 (a) Au (b) Nb (c) $TiCl_4$ (d) 7^+ (e) Cr (f) Fe (g) Eu (h) Hg (i) 2 (j) Rh

21.2 (a) 1+ (b) 2+ (c) 3+ (d) 2+ (e) 3+

21.3 (a) 2 (b) 4 (c) 6 (d) 6 (e) 6

21.4 (a) 126 g AgCl; (b) 234 g AgCl

21.5

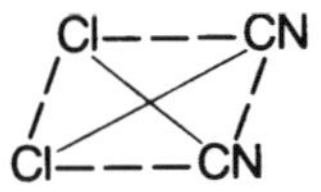

cis-$[Au(CN)_2Cl_2]$

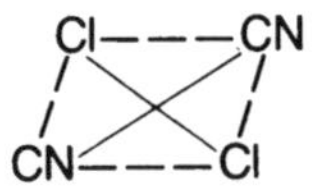

trans-$[Au(CN)_2Cl_2]$

21.6
(a) pentaamminechlorocobalt(III) chloride
(b) $[Co(en)_2ClNO_2]^+$
(c) potassium tetracyanoaurate(III)
(d) $Na[Ag(CN)_2]$
(e) pentacarbonyl iron(0)
(f) $Ca[HgBr_4]$
(g) tetraamminezinc(II) tetrachorozincate(II)
(h) $[Al(OH)_4(H_2O)_2]^-$

21.7 $[RhCl_6]^{3-}$ appears yellow; $[RhBr_6]^{3-}$ appears purple. $[RhCl_6]^{3-}$ has a larger crystal field splitting energy. Cl^- is a stronger ligand.

21.8 high spin complex: 3
low spin complex: 1
The low spin complex (larger Δ) has a warmer color.

21.9 (a)

small Δ
more paramagnetic

large Δ
absorbs higher energy photon

(b)

small Δ
more paramagnetic

large Δ
absorbs higher energy photons

(c)

small Δ	large Δ
no difference in paramagnetism	absorbs higher energy photons

21.10 $[FeCl_6]^{3-}$—smallest Δ
$[FeCl_4(H_2O)_2]^-$
$[Fe(H_2O)_3(OH)_3]$
$[Fe(H_2O)_6]^{3+}$
$[Fe(NO_2)_6]^{3-}$
$[Fe(CN)_6]^{3-}$—largest Δ
$[FeCl_6]^{3-}$ absorbs longest wavelength

Chapter 22

ORGANIC CHEMISTRY—STRUCTURE AND FUNCTIONALITY

SECTION 22.1 WHAT MAKES CARBON UNIQUE?

Objective

- To state the versatility of carbon in forming stable compounds.

Focus

Carbon is unique in that carbon atoms bond to one another to form extended stable structures. Although carbon forms bonds to nearly every other element, the carbon-to-carbon bond energy is higher than most.

Three kinds of carbon-to-carbon bonds form: the C—C, C═C, and C≡C bonds. In addition, the carbon atoms can be linked to form straight chain, branch chain, cross-linked chain, and cyclic compounds. Of course carbon also forms covalent bonds to almost all nonmetals. The nonmetals most frequently bonded to carbon are H, O, N, S, the halogens, P, and Si.

This diversity of bond formation with itself and with other elements makes carbon an element that has prompted many extensive research programs and the consequent development of plastics, rubber, drugs, fuels, synthetic fibers, and insecticides.

SECTION 22.2 ALKANE HYDROCARBONS

Objective

- To define an alkane, a homologous series, and an alkyl group.

Focus

Hydrocarbons are compounds that consist of only C and H. The **alkanes** are a class of hydrocarbons that have the general formula C_nH_{2n+2}. All bonds are single σ bonds. All C atoms are sp^3 hybridized and all C atoms bond to four other atoms. Continuous (unbranched) C—C chains of more than 3 carbon atoms are called *n*-alkanes.

Learn the names of the first 10 alkanes because we will use the stems of these names in naming compounds in other classes of organic compounds. Removing a H atom from an alkane produces an **alkyl group** with a covalency requirement of one.

Example 22.1 The alkanes represent a homologous series with the formula C_nH_{2n+2}. What group of atoms separates one member of the series from the next?

Answer $—CH_2—$

Solution Let's write the first four "normal" or straight-chained members of the alkane series.

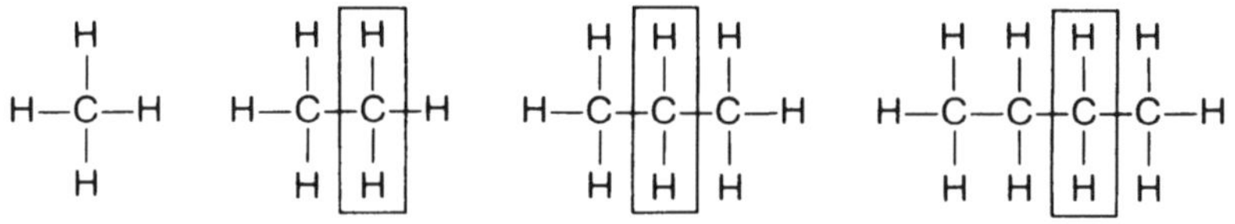

The only difference is the insertion of a $—CH_2—$ group shown in the rectangle.

SECTION 22.3 NAMING ALKANES

Objectives

- To name an alkane given its structural formula.
- To write the structural formula for an alkane given its name.

Focus

In naming alkanes learn and use the five rules in the text plus the rule for choosing the longest chain. Also, study the rules for punctuation and alphabetization.

Example 22.2 Name the following alkane.

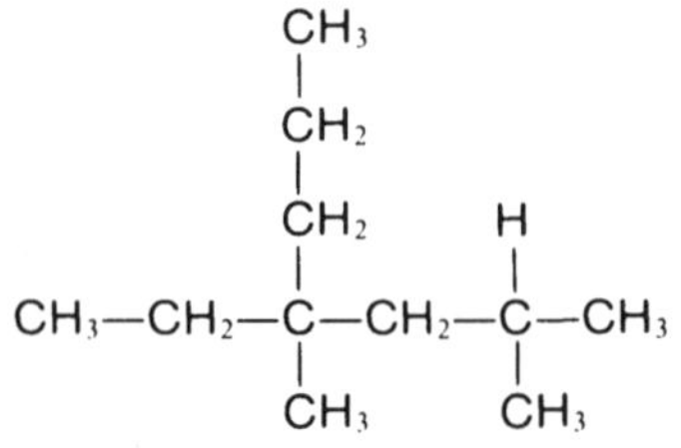

Answer 2,4-dimethyl-4-ethylheptane

Solution

Rule (a). The longest chain is 7 carbons long, a heptane.

Rule (b). We start our numbering with the carbon atom at the right because we want the branching groups on low-numbered C atoms. If the numbering had started at the "top" (C^7), the first branching chain would have been on C^4 instead of on C^2.

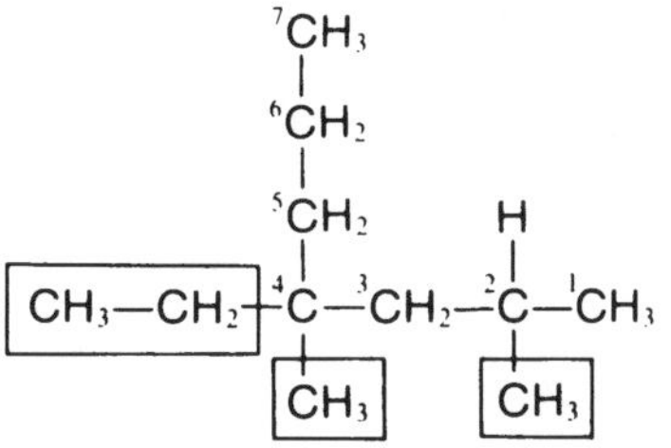

Rule (c). Three branched chains are the two $-CH_3$'s (methyls), one at C^2, another at C^4, and one $-CH_2-CH_3$ (ethyl) at C^4.

Rule (d). The names of the branched chains are 2,4-dimethyl and 4-ethyl.

Rule (e). Alphabetize the branched chains ("d" in dimethyl comes before "e" in ethyl) to get 2,4-dimethyl-4-ethyl and attach to heptane to get

2,4-dimethyl-4-ethylheptane

Note the use of the dashes to separate the numbers from the words and the commas to separate the numbers. The words are connected in the name for the compound; there is no space between ethyl and heptane.

Example 22.3 Write the formula for 3-chloro-2,2-dimethyl-4-fluoropentane.

Solution Work backward through the name starting with pentane. Write the skeleton structure for pentane and number the C atoms.

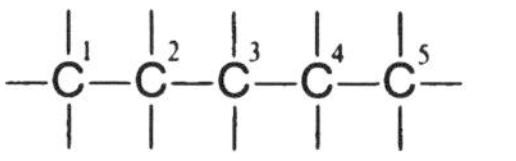

In sequence place $-F$ on C^4, 2 $-CH_3$'s on C^2, and $-Cl$ on C^3. Fill out the remaining bonds with $-H$.

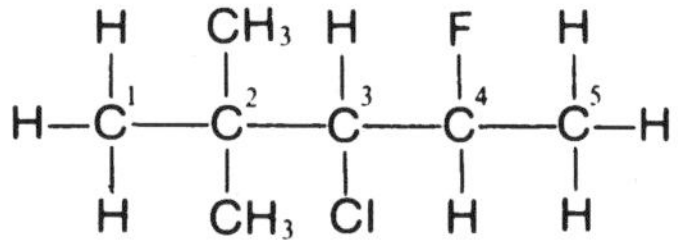

SECTION 22.4 CONFORMATIONS

Objective

- To define a conformational structure of an alkane.

Focus

In alkanes all C atoms are sp^3 hybridized, and all bonds are σ bonds. Rotation can occur along a C—C σ bond. Therefore, more than one orientation of atoms is possible. The different arrangements are termed **conformational structures**. These structures exist for any molecule which has three or more successive σ bonds, A—X—Y—B, where A, X, Y, and B may be the same or different atoms.

Example 22.4 How many conformations exist for the compound

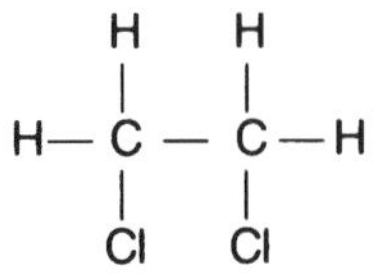

Answer four

Solution The σ-bonded sequence is Cl—C—C—Cl or Cl—C—C—H or H—C—C—H. According to the text, Figure 22.3, both *eclipsed* and *staggered* forms exist for ethane. Draw all the conformational structures of this compound using the same view of the molecule.

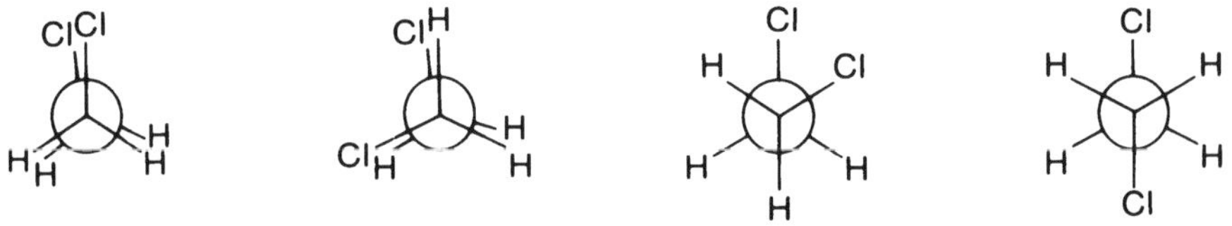

eclipsed, Cl's aligned — eclipsed, H and Cl aligned — Staggered, Cl's are adjacent — staggered, Cl's are opposite

SECTION 22.5 MULTIPLE BONDS; *CIS-TRANS*-ISOMERISM

Objectives

- To define and name alkene and alkyne hydrocarbons.
- To recognize *cis*- and *trans*-stereoisomers in alkenes.

Focus

An **alkene** is a hydrocarbon having at least one C=C. The general formulas for the alkene homologous series containing *only one* C=C is C_nH_{2n}. **Alkynes** have at least one C≡C; the general formula for alkynes containing only one C≡C is C_nH_{2n-2}.

In naming alkenes and alkynes, begin with the longest chain containing the double or triple bond. Number the carbon chain so that the first C atom having C=C or C≡C has the smallest possible number. Name the longest chain as an -ene (or diene, triene, and so on) or -yne (or diyne, triyne, and so on), and name the branches as you did for alkanes.

Cis- and *trans*-isomers are possible for alkenes because of the restricted rotation around the C=C. *Cis*- and *trans*-isomerism does not exist for alkane and alkyne molecules.

Example 22.5 How many isomers are there for chloropropene? Write and name each isomer.

Answer four

Solution First, let's determine all the structural isomers of chloropropene.

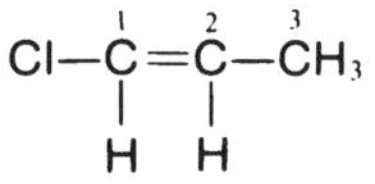

1-chloropropene (we do not have to name the compound 1-chloro-1-propene because in propene, the C=C must be on the first C atom)

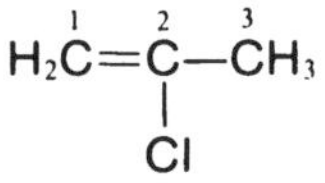

2-chloropropene

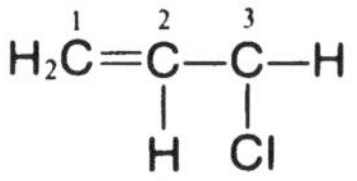

3-chloropropene

Second, let's determine which of these have *cis*- and *trans*-isomers:

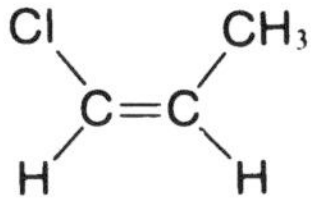

cis-1-chloropropene (H atoms are on the same side of the C=C)

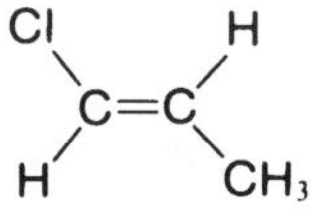

trans-1-chloropropene (H atoms are on opposite sides of the C=C)

No *cis*- and *trans*-isomers exist for 2-chloropropene or 3-chloropropene.

SECTION 22.6 CYCLOALKANES

Objective

- To recognize and name some simple cycloalkanes.

Focus

In cycloalkanes, each C atom bonds to four other atoms by means of σ bonds only. Because of its cyclic nature, no free rotation occurs about the C—C bond although ring twisting and puckering does occur. Cycloalkanes are isomers of open-chained alkenes having only one C=C bond. Both have the general formula C_nH_{2n}.

The cycloalkanes are named according to the rules used to name alkanes, except the prefix "cyclo" is attached to the name of the chain.

Example 22.6 Name each of the following compounds:

(a) 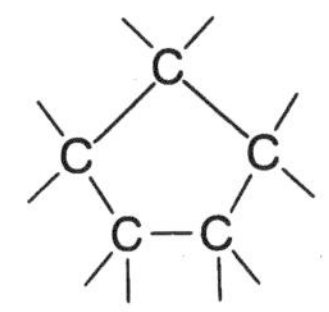

(b) 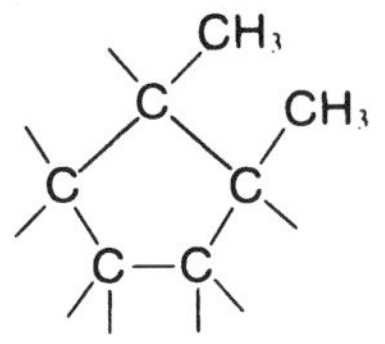

(c) 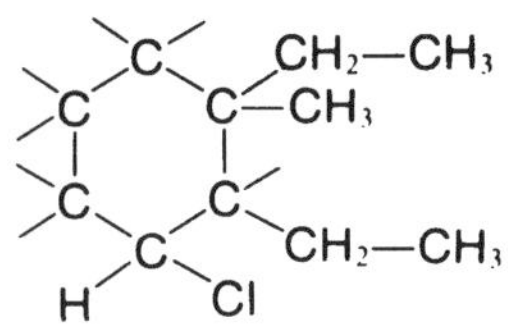

Answer (a) cyclopentane

(b) 1,2-dimethylcyclopentane

(c) 3-chloro-1,2-diethyl-1-methylcyclohexane

Solution (a) All C—C bonds, cyclopentane.

(b) Number the C atoms, using the smallest possible number on the C atoms where the branches are attached. The following figure shows how to number the ring.

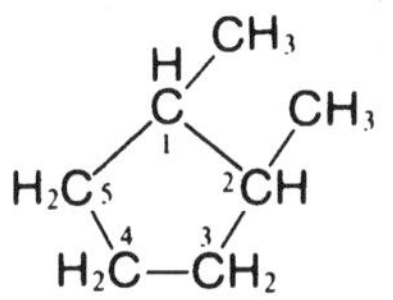

The $—CH_3$'s on C^1 and C^2 are named 1,2-dimethyl. Attach this to the name of the cyclic chain to get 1,2-dimethylcyclopentane.

(c) Again, number the C atoms keeping the numbering of the branches low. The branches are named accordingly.

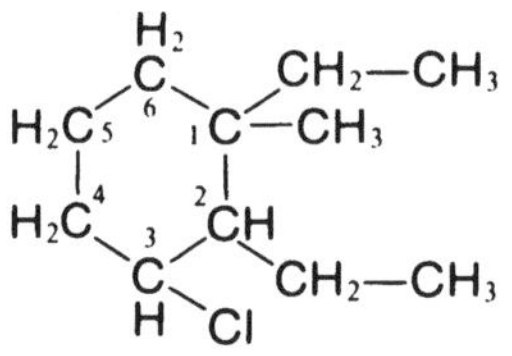

The $—CH_2—CH_3$'s on C^1 and C^2; 1,2-diethyl-. One $—CH_3$ on C^1; 1-methyl-. One —Cl on C^3; 3-chloro-.

Alphabetize the branches and add this to the name of the cyclic chain (cyclohexane). We have 3-chloro-1,2-diethyl-1-methylcyclohexane.

SECTION 22.7 BENZENE AND AROMATIC COMPOUNDS

Objective

- To recognize and name aromatic hydrocarbons.

Focus

Benzene is a cyclic structure having the molecular formula C_6H_6 and is represented by the structure

Benzene and compounds having a benzenelike structure are **aromatic compounds**. Benzene rings may be fused together to form high molecular weight aromatic compounds, such as naphthalene.

The naming of aromatic compounds is like other hydrocarbons. The C atoms on the benzene ring are numbered like cycloalkanes or they can be designated by the prefixes *ortho*- (for 1,2 disubstituted benzenes), *meta*- (for 1,3 disubstituted benzenes), and *para*- (for 1,4 disubstituted benzenes):

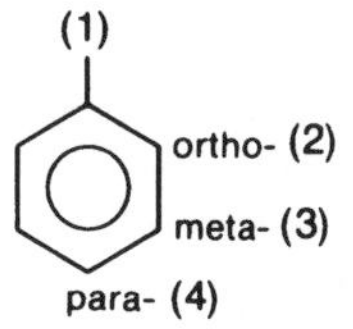

Example 22.7 If

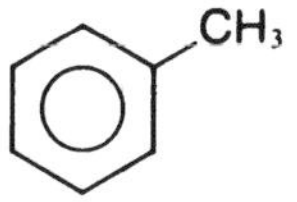

is the formula for toluene, then what is the formula for *p*-chlorotoluene?

Answer

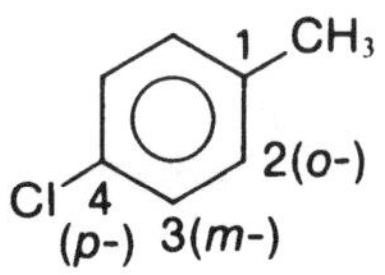

Solution *p*- (or *para*-) is in the fourth C position from the —CH_3 group in toluene.

SECTION 22.8 STRUCTURE OF BENZENE

Objective

- To explain the structure and stability of benzene.

Focus

Each C atom in benzene is sp^2 hybridized, forming a σ bond to each of two adjacent C atoms and a σ bond to a H atom. The remaining electron on each of the six C atoms is in the unhybridized p orbital.

These six p electrons in the unhybridized p orbitals are free to interact with each other to form a delocalized cyclic π bonding system over benzene's entire molecular structure. Because these six p electrons are "free to roam," they are lower in energy (and therefore more stable) than when localized to one or two double bonds. This lowering of energy resulting from the delocalization of electrons in a π system is called **delocalization energy**. The lower the energy (or the greater the delocalization of electrons), the more stable is the molecule.

Example 22.8 The heat of formation (ΔH_f°) of 1,2-butadiene is +162 kJ/mol, whereas that of 1,3-butadiene is +109 kJ/mol. Based upon delocalization, explain the difference in stability.

Answer First draw the structures of the two isomers.

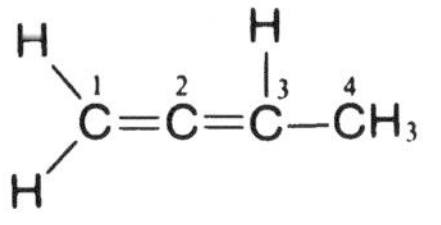

1,2-butadiene

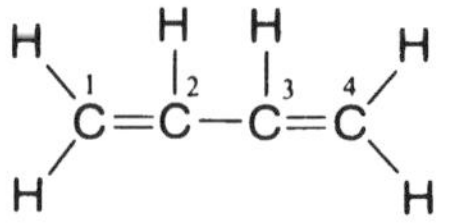

1,3-butadiene

In the 1,2-butadiene molecule, C^2 is *sp* hybridized, forming a π bond to C^1 *perpendicular* to a π bond formed with C^3. Because the unhybridized *p* orbitals on C^2 in 1,2-butadiene are not parallel but are perpendicular to each other, they cannot overlap with each other. The *p* electrons remain localized between the C atoms.

In the 1,3-butadiene molecule, all C atoms are sp^2 hybridized. The unhybridized *p* orbitals on the C atoms are parallel and are able to overlap. The *p* electrons roam freely over the molecule's entire length, thus producing a more stable structure. Generally, delocalization occurs only if *every other* C atom is double bonded (alternating single and double bonds).

SECTION 22.9 CHEMISTRY OF HYDROCARBONS

Objective

• To illustrate some basic reactions of the alkanes, alkenes, alkynes, and benzene.

Focus

When compared to other hydrocarbons, alkanes are relatively inert. Alkenes and alkynes readily react by means of an **addition reaction** in which atoms or groups of atoms are added to the multiple (π) bond. Alkanes and benzene commonly react by means of a **substitution reaction**, substituting an H atom with other atoms (or groups) such as a halogen or a $-NO_2$ group.

Example 22.9 Write equations for the following:

(a) $H_3C-CH_3 + Br_2 \xrightarrow{uv}$
(b) $H_2C=CH_2 + Br_2 \rightarrow$
(c) $H_3C-CH=CH-CH_3 + HBr \rightarrow$
(d) $H_3C-CH=CH-CH_3 + HOSO_2OH(H_2SO_4) \rightarrow$

(e)  $+ Cl_2 \xrightarrow{FeCl_3}$

Answer
(a) $H_3C-CH_3 + Br_2 \xrightarrow{uv} H_3C-CH_2Br + HBr$
(substitution)

(b) $H_2C{=}CH_2 + Br_2 \rightarrow BrH_2C{-}CH_2Br$
(addition)

(c) $H_3C{-}CH{=}CH{-}CH_3 + HBr \rightarrow H_3C{-}CHBr{-}CH_2{-}CH_3$
(addition)

(d) $H_3C{-}CH{=}CH{-}CH_3 + HOSO_2OH \rightarrow H_3C{-}CH(OSO_2OH){-}CH_2{-}CH_3$
(addition)

(e) 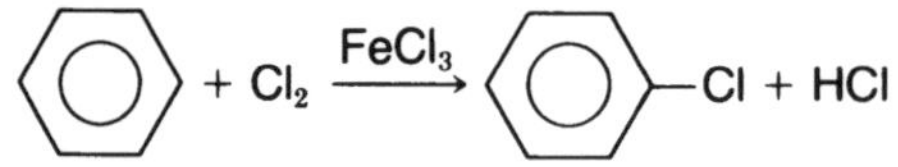

(substitution)

SECTION 22.10 STEREOISOMERISM

Objectives

- To identify stereoisomers that are enantiomers.
- To recognize a chiral center in a molecule.

Focus

Stereoisomers are two molecules having the same bonding sequence but a different spatial arrangement of atoms. Stereoisomers that are not superimposable mirror images are **enantiomers**.

If a molecule has a **chiral center**, an atom to which four *different groups* are attached, then the molecule has an enantiomer. Enantiomers are said to be chiral or asymmetric. In organic compounds, the chiral center is most often a C atom, one bonded to four different atoms or groups of atoms. For example, each of the following compounds has a chiral center, marked by an asterisk, and therefore each molecule has an enantiomer.

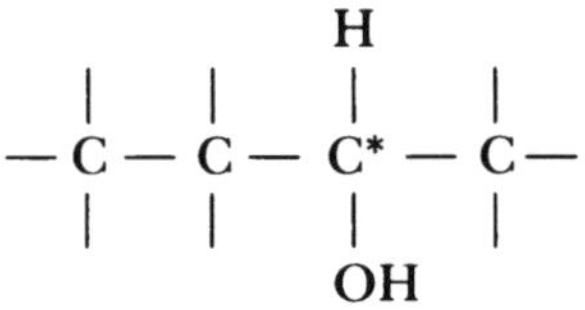

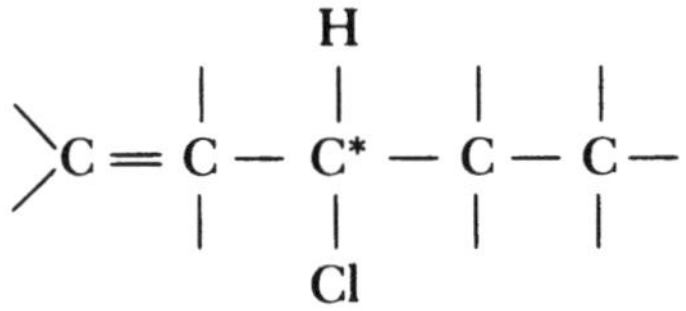

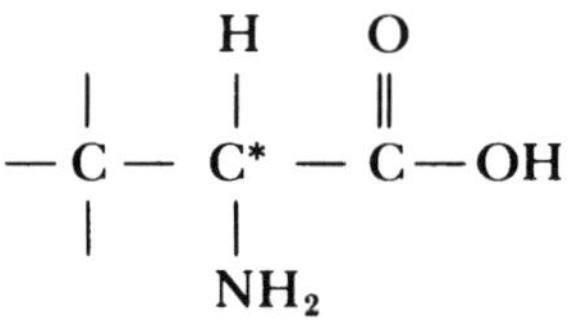

and most other α-amino acids

Many biochemicals are enantiomers because of the complexity of the molecules. Somewhere in the C—C chain at least one C atom is asymmetric.

Example 22.10 What is the simplest hydrocarbon that can have an asymmetric C atom? Name it.

Answer

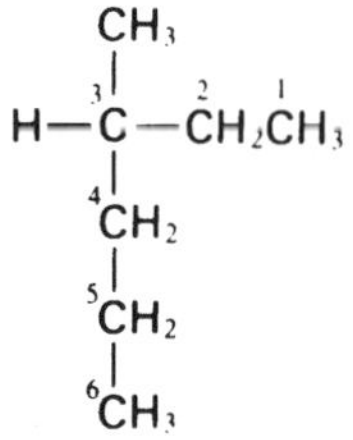

3 methylhexane

Solution Four different groups must be attached to an asymmetric C atom: one of which can be —H, another —CH_3, the next —CH_2CH_3, and the fourth —$CH_2CH_2CH_3$. C^3 is the chiral center.

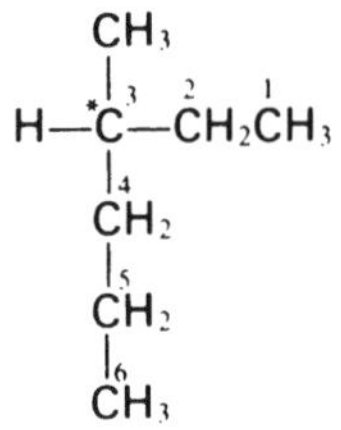

The longest chain of C atoms is six, a hexane with substitution of a —CH_3 group on C^3, or 3-methylhexane.

SECTION 22.11 THE FUNCTIONAL GROUPS

Objectives

- To recognize the functional groups: the alkyl halides, alcohols, ethers, aldehydes, ketones, carboxylic acids, esters, amines, and amides.
- To state some common chemical reactions for each functional group.

Focus

Alkyl halides: R—X, where X = F, Cl, Br, or I
Alkyl halides commonly undergo **elimination reactions**, where alcoholic KOH or $NaNH_2$ eliminates an H—X molecule from adjacent C atoms yielding a multiply bonded product, or **displacement reactions** where a base stronger than $:\ddot{\underset{..}{X}}:^-$ displaces —X.

Example 22.11 Complete the following reactions for *n*-propyl bromide.

(a) $H_3C-CH_2-CH_2-Br + NaNH_2 \xrightarrow{\text{alcohol}}$
(b) $H_3C-CH_2-CH_2-Br + OH^- \rightarrow$

Answer

(a) $H_3C-C(H)=CH_2 + HBr$
(b) $H_3C-CH_2-CH_2-OH + Br^-$

Alcohols: R—OH

Alcohols are structures similar to H_2O; an alkyl group replaces −H. Therefore, alcohols have chemical properties similar to H_2O:

$$ROH + Na \rightarrow RO^-Na^+ + \tfrac{1}{2}H_2$$

$$ROH + HCl \rightarrow ROH_2^+ + Cl^-$$

Methanol, CH_3OH, wood alcohol, makes ethanol unfit for drinking. A 5% CH_3OH-95% C_2H_5OH solution is called denatured alcohol. Methanol is also used as a fuel for racing cars such as those in the Indianapolis 500.

Ethanol, C_2H_5OH, grain alcohol or "drinking alcohol," is made from the natural fermentation of grains such as barley, corn, potatoes, and rice. Absolute ethanol, 100% C_2H_5OH, is 200 proof alcohol.

Alcohols react with HX acids to produce alkyl halides and with concentrated H_2SO_4 to eliminate H—OH from adjacent C atoms forming an alkene. The —OH group readily oxidizes to a carbonyl, C=O, group.

Example 22.12 Complete the following reactions for ethanol, CH_3CH_2OH.

(a) $H_3C{-}CH_2{-}OH + HCl \rightarrow$

(b) $H_3C{-}CH_2{-}OH \xrightarrow{\text{conc. } H_2SO_4}$

(c) $H_3C{-}CH_2{-}OH + [O] \rightarrow$

Answer

(a) $H_3C{-}CH_2{-}Cl + H_2O$

(b) $H_2C{=}CH_2$

(c) $\begin{array}{l} H_3C{-}C{=}O + H_2O \\ \phantom{H_3C{-}}| \\ \phantom{H_3C{-}}H \end{array}$

Replacing an H atom on the benzene ring with an —OH group yields a phenol. Phenol is a weak acid.

Ethers: $R{-}\ddot{\underset{..}{O}}{-}R$ or $R{-}\ddot{\underset{..}{O}}{-}R'$

An ether is similar to H_2O, except that alkyl groups replace both H atoms. Like H_2O, ethers are polar molecules and serve as B/L and Lewis bases because of the availability of the unshared electron pairs on the oxygen atom. Ethers can be prepared by the dehydration of an alcohol.

Example 22.13 Write the formula for the ether formed from the following alcohols.

(a) $2\,H_3CCH_2OH \xrightarrow[\text{heat}]{H_2SO_4}$

(b) $\begin{array}{c} CH_3 \\ | \\ 2\,H{-}C{-}OH \\ | \\ CH_3 \end{array} \xrightarrow[\text{heat}]{H_2SO_4}$

(c) $2\,H_3CCH_2CH(OH)CH_3 \xrightarrow[\text{heat}]{H_2SO_4}$

Answer

(a) $H_3CCH_2—O—CH_2CH_3$

(b) $H—C(CH_3)_2—O—C(CH_3)_2—H$

(c) $H_3CCH_2—CH(CH_3)—O—CH(CH_3)—CH_2—CH_3$

Aldehydes, $R—C(H)=O$, **and Ketones,** $R—C(R)=O$ or $R—C(R')=O$

These two classes of organic compound contain the **carbonyl group** C=O. When one alkyl group is attached to a carbonyl group, it is an aldehyde; when two are attached, it is a ketone.

H_2 adds across the C=O, producing $—C(H)—OH$.

Mild oxidizing agents oxidize aldehydes to carboxylic acids.

Example 22.14 Write the formula for the product of the reaction.

(a) $H_3CCH_2\,C(=O)—H + H_2 \xrightarrow{Pt}$

(b) $H_3CCH_2C(=O)—H + [O] \rightarrow$

(c) $H_3CC(=O)CH_3 + H_2 \rightarrow$

Answer

(a) $H_3CCH_2C(OH)(H)—H$

(b) $H_3CCH_2\overset{O}{\overset{\|}{C}}-OH$

(c) $H_3C\overset{OH}{\overset{|}{\underset{\underset{H}{|}}{C}}}CH_3$

Carboxylic Acids: $R-\overset{O}{\overset{\|}{C}}-OH$

Carboxylic acids are weak organic acids with $K_a \simeq 10^{-5}$ (text, Table 15.3). Soaps are prepared by the addition of a strong base to a high molecular weight, naturally occurring carboxylic acid. This is called *saponification*.

$$R-\overset{O}{\overset{\|}{C}}-OH + NaOH \rightarrow \underset{\text{soap}}{R-\overset{O}{\overset{\|}{C}}-O^-Na^+} + H_2O$$

$$R-\overset{O}{\overset{\|}{C}}-OH + Ca(OH)_2 \rightarrow \underset{\text{soap}}{(R-\overset{O}{\overset{\|}{C}}-O^-)_2Ca^{2+}} + 2H_2O$$

Some carboxylic acids are known for their foul odor:

- Butyric acid— odor of rancid butter
- Valeric acid— stale perspiration
- Caproic acid— odor of limburger cheese
- Lactic acid— odor of sour milk and formed in muscles during exercise

Esters: $R-\overset{O}{\overset{\|}{C}}-O-R$ or $R-\overset{O}{\overset{\|}{C}}-O-R'$

Carboxylic acids react with alcohols in a reaction catalyzed by strong acid to produce esters. This is called *esterification*.

$$R-\overset{O}{\overset{\|}{C}}-OH + HOR' \xrightarrow{H^+} R-\overset{O}{\overset{\|}{C}}-O-R' + H_2O$$

The reverse reaction, hydrolysis, occurs in dilute aqueous acid solutions.

Esters hydrolyze in a solution containing OH^- to produce an alcohol and the anion of the organic acid. This is a saponification (soap-forming) reaction.

$$R-\overset{O}{\overset{\|}{C}}-OR' + Na^+OH^- \rightarrow R-\overset{O}{\overset{\|}{C}}-O^-Na^+ + R'OH$$

Animal *fats*, such as lard, butter, chicken fat, and beef tallow, are esters of saturated fatty acids. Vegetable *oils*, such as corn oil, soybean oil, peanut oil, and safflower oil, are esters of *un*saturated fatty acids.

Esters are known for their natural pleasant odors. Natural flavors and scents of many flowers and fruits are usually mixtures of one or more esters.

$$C_3H_7\overset{\overset{O}{\|}}{C}-O-CH_3 \qquad \text{pineapple}$$

$$C_3H_7\overset{\overset{O}{\|}}{C}-O-C_2H_5 \qquad \text{apricot}$$

$$CH_3\overset{\overset{O}{\|}}{C}-O-C_8H_{15} \qquad \text{orange}$$

$$CH_3\overset{\overset{O}{\|}}{C}-O-C_5H_{11} \qquad \text{banana}$$

$$H_2NC_6H_4-\overset{\overset{O}{\|}}{C}-O-CH_3 \qquad \text{grape}$$

Amines: RNH_2, R_2NH, or R_3N

Just as alcohols have a structure similar to water (an alkyl group substitutes for a H atom), amines are structurally similar to ammonia, NH_3 (at least one alkyl group substitutes for a H atom). Amines are weak organic bases and, as B/L proton acceptors, form cations resembling the ammonium ion, NH_4^+.

$$RNH_3^+,\ R_2NH_2^+, \quad \text{or} \quad R_3NH^+$$

The K_b for amines range from 10^{-4} to 10^{-8} (see text, Table 15.3).

Low molecular weight amines have strong, unpleasant odors, like that of dead fish. Amphetamines, drugs that act as nervous system stimulants, are amines.

Amides: $R-\overset{\overset{O}{\|}}{C}-NH_2$, $R-\overset{\overset{O}{\|}}{C}-NHR$, or $R-\overset{\overset{O}{\|}}{C}-NR_2$

Amines and ammonia, being bases, react with carboxylic acids at high temperatures to produce amides, just as alcohols react with carboxylic acids to produce esters.

$$R-\overset{\overset{O}{\|}}{C}-OH + R'NH_2 \xrightarrow{\Delta} R-\overset{\overset{O}{\|}}{C}-NHR' + H_2O$$

SECTION 22.12 POLYFUNCTIONAL COMPOUNDS

Objectives

- To identify the functional groups in amino acids and carbohydrates
- To write the reaction for the preparation of a polypeptide.

Focus

Amino acids have the $-NH_2$ and $-\overset{\overset{O}{\|}}{C}-OH$ functional groups as part of the molecule. The α-amino acids have the $-NH_2$ on the $\alpha-C$, the C adjacent to the $-\overset{\overset{O}{\|}}{C}-OH$ group. These amino acids are the building blocks for proteins.

α-amino acids form dipolar ions or Zwitterions,

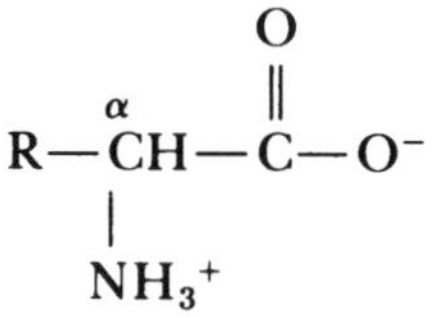

Zwitterions have strong I.M. forces causing amino acids to be crystalline solids.

α-amino acids also react with one another to produce a peptide. A number of amino acids linked together is a polypeptide:

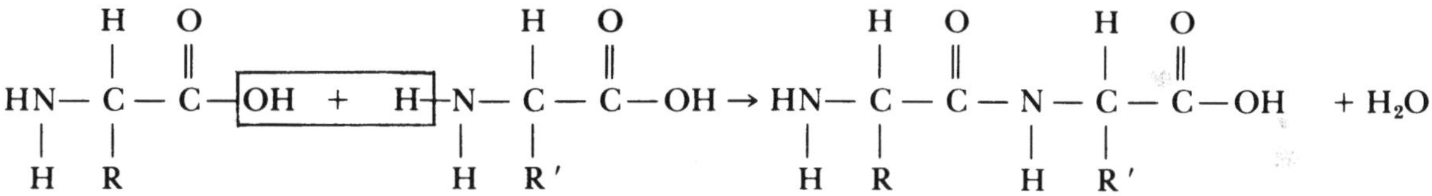

The product, still an amino acid, reacts with another amino acid molecule by eliminating a H_2O molecule:

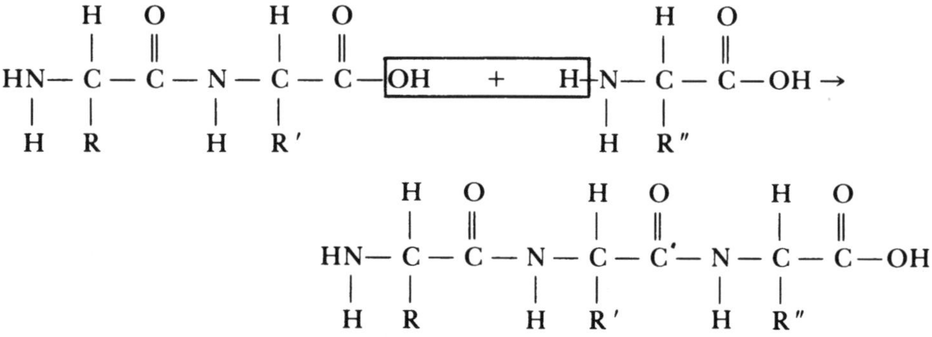

and so on.

Example 22.15 Using Table 22.4, write the two dipeptides that form alanine and valine.

Answer

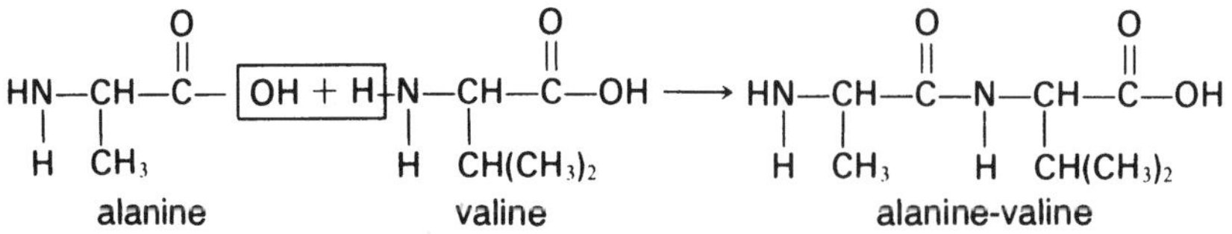

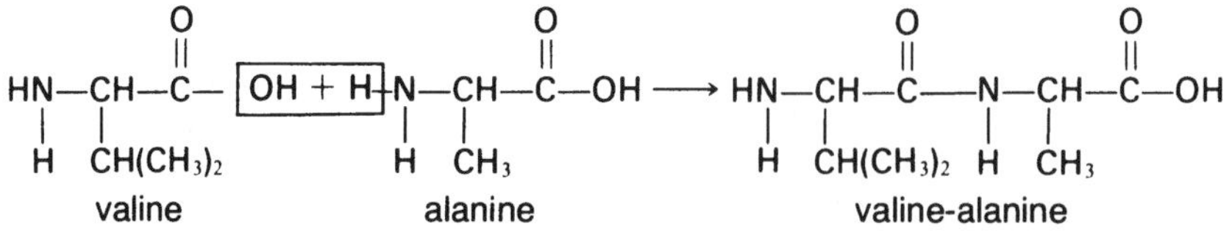

Carbohydrates, such as sugars, starches, and cellulose, have the —OH and >C=O functional groups as part of the molecule. The most common sugar is glucose, $C_6H_{12}O_6$, which when chemically bonded to fructose, $C_5H_{10}O_5$, forms sucrose or common table sugar.

Glucose forms a red precipitate in the Fehling's test, a positive test indicating the presence of a —C(H)=O group.

QUESTIONS

Section 22.1

22.1 Using Table 6.2, list the bond energies for

(a) C—C (e) C—O
(b) C=C (f) C—N
(c) C≡C (g) C—S
(d) C—H

Section 22.2

22.2 Circle the compounds that are alkanes.

C_4H_{10}, C_6H_{12}, C_2H_6, $C_{12}H_{26}$, $C_{10}H_{20}$

22.3 Name each of the following alkyl groups.

(a) CH_3 ________
(b) CH_3CH_2 ________
(c) $CH_3CH_2CH_2$ ________

22.4 How many butyl groups are possible? Write the formula for each.

Section 22.3

22.5 Name the following alkanes.

(a) 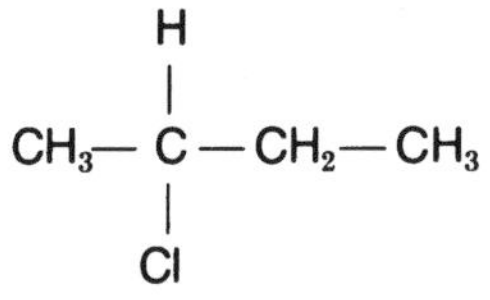

(b) 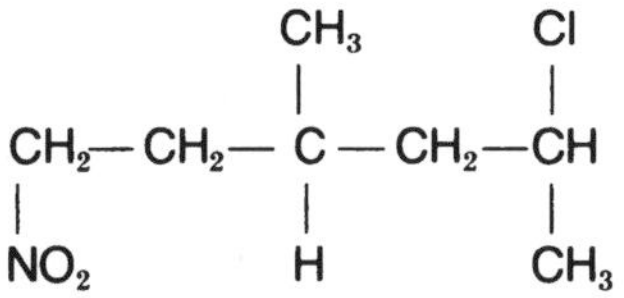

(c) $CF_3-CF_2-CF_3$

(d) 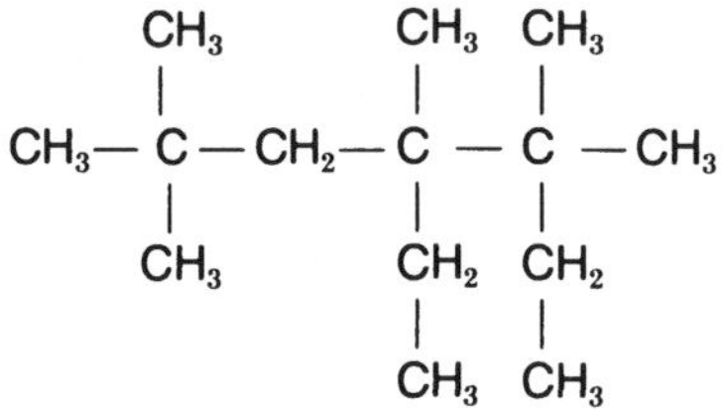

22.6 Write the formula for the following compounds.

(a) 1-nitro-3,3,4-trimethylpentane

(b) 1,4-diaminobutane

(c) 2,2-dimethyl-3,3,4-triethylheptane

Section 22.5

22.7 Write and name the isomers of dichlorobutene, $C_4H_6Cl_2$.

Section 22.6

22.8 Draw the structural formula for

(a) 1,2-dimethyl-4-nitrocyclohexane

(b) 3,4-dichloro-1-ethyl-1-cyclohexene

(c) 1,3-cyclohexadiene

Section 22.7

22.9 Write the formula for the following aromatic compounds.

(a) 1,2-dichorobenzene

(b) 1-ethyl-4-bromobenzene

(c) *o*-nitrotoluene

(d) *m*-iodotoluene

22.10 Name the following compounds.

(a) 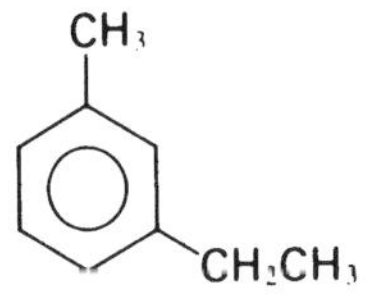

(b) 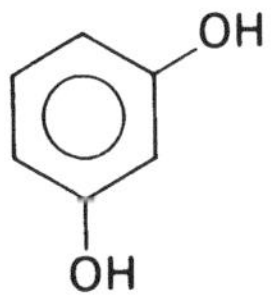

(c) 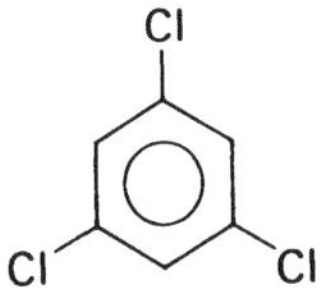

(d) 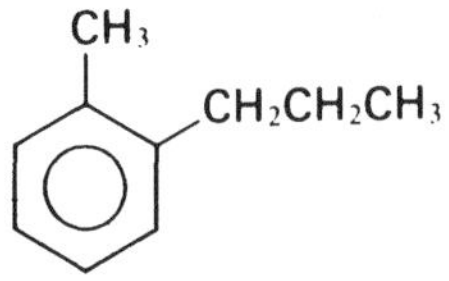

Section 22.8

22.11 Which compound is stabler?

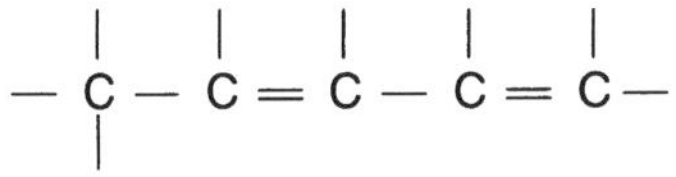

or 

Section 22.9

22.12 Distinguish between a saturated and an unsaturated hydrocarbon.

23.13 Write equations for the following.

(a) The conversion of ethane to chloroethane

(b) The conversion of propene to propane

(c) The conversion of benzene to nitrobenzene

(d) The conversion of 2-butene to 2-iodobutane

(e) The conversion of 2-butyne to 2-bromobutene

(f) 2-butyne + 2 Br_2 →

(g) ethyne + 2 HBr →

(h) 2-butene + HOCl →

(i)  + $HOSO_2OH$(conc.) →

Section 22.10

22.14 Identify the molecules having a chiral center.

(a) 1-pentanol

(b) 2-pentanol

(c) 3-pentanol

(d) 2-methyl-1-pentanol

(e) 2-methyl-2-pentanol

(f) 2-methyl-3-pentanol

(g) 2-bromo-2-chlorobutane

(h) 2-chloro-1-propanol

(i) $\left[CH_3-\underset{\underset{Cl}{|}}{\overset{\overset{H}{|}}{N}}-CH_2-CH_3\right]^+$

Section 22.11

22.15 Identify the major product(s) in the following.

(a) $CH_3-\overset{\overset{O}{\|}}{C}-OH + C_2H_5OH \xrightarrow{H^+}$

(b) $CH_3CH_2CH_2OH + [O] \rightarrow$

(c) $C_3H_7\overset{\overset{O}{\|}}{C}-OH + C_2H_5NH_2 \xrightarrow{\Delta}$

(d) $CH_3CH_2CH_2Cl \xrightarrow[KOH]{alcoholic}$

(e) $CH_3CH_2CH_2OH + HCl \rightarrow$

(f) $CH_3CH_2\overset{\overset{H}{|}}{C}=O + [O] \rightarrow$

(g) $C_4H_9\overset{\overset{O}{\|}}{C}-O-C_2H_5 + H_2O \xrightarrow{dilute\ H^+}$

(h) $CH_3(CH_2)_{16}\overset{\overset{O}{\|}}{C}-OH + NaOH \rightarrow$

(i) $CH_3NH_2 + HCl \rightarrow$

(j) $CH_3CH_2\underset{\underset{OH}{|}}{C}HCH_3 \xrightarrow[H_2SO_4]{conc.}$

(k) $CH_3CH_2CH_2Br + NH_3 \rightarrow$

(l) 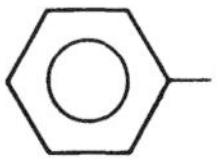$-CH_2CH_2OH + HI \rightarrow$

(m) $2C_3H_7OH \xrightarrow{H_2SO_4}$

(n) $CH_3CH_2\overset{\overset{O}{\|}}{C}CH_3 + H_2 \xrightarrow{Pt}$

(o) $C_3H_7{-}O{-}C_2H_5 + HBr \rightarrow$

(p) $C_3H_7{-}\overset{\overset{\displaystyle O}{\|}}{C}{-}OH + NaOH \rightarrow$

22.16 Match the functional group with the name.

(a) ______	$-\overset{\overset{\displaystyle O}{\|}}{C}-OR$	1. ether
(b) ______	$-X(X = F, Cl, Br, I)$	2. alcohol
(c) ______	$-\overset{\overset{\displaystyle O}{\|}}{C}-OH$	3. amide
(d) ______	$-OH$	4. amine
(e) ______	$-NH_2$	5. carboxylic acid
(f) ______	$-\overset{\overset{\displaystyle H}{\vert}}{C}{=}O$	6. alkyl halide
(g) ______	$-\overset{\overset{\displaystyle R'}{\vert}}{C}{=}O$	7. ester
(h) ______	$-\overset{\overset{\displaystyle O}{\|}}{C}-NH_2$	8. ketone
(i) ______	$-O-$	9. aldehyde

Section 22.12

22.17 Using Table 22.4, write the two dipeptides that form from glycine and leucine.

ANSWERS

22.1
(a) 339 kJ
(b) 586 kJ
(c) 809 kJ
(d) 416 kJ
(e) 336 kJ
(f) 285 kJ
(g) 237 kJ

22.2 C_4H_{10}, C_2H_6, $C_{12}H_{26}$

22.3
(a) methyl
(b) ethyl
(c) propyl

22.4 four:

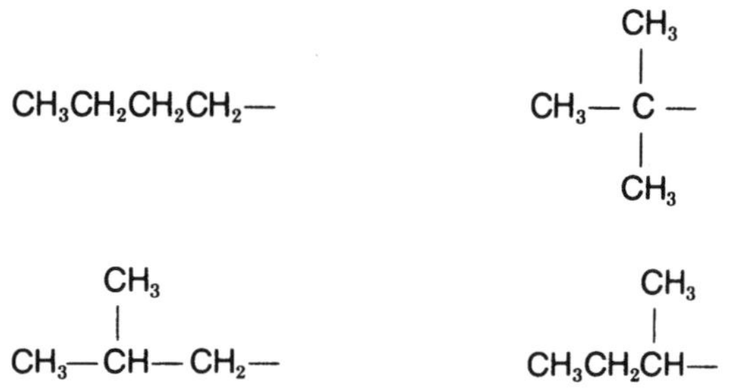

22.5 (a) 2-chlorobutane

(b) 5-chloro-3-methyl-1-nitrohexane

(c) 1,1,1,2,2,3,3,3-octafluoropropane (perfluoropropane)

(d) 4-ethyl-2,2,4,5,5-pentamethylheptane

22.6

(a) 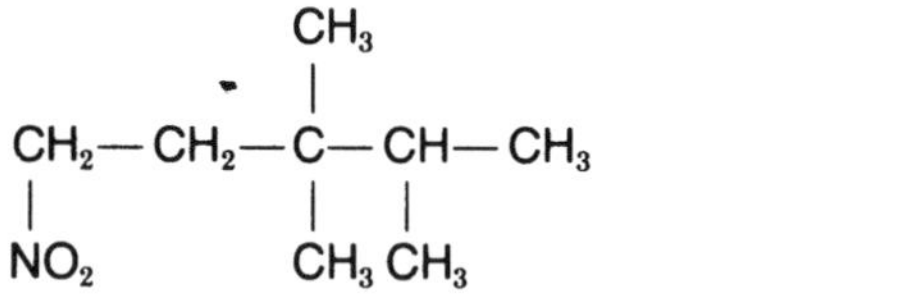

(b) 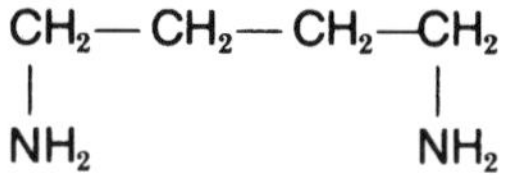

(c) 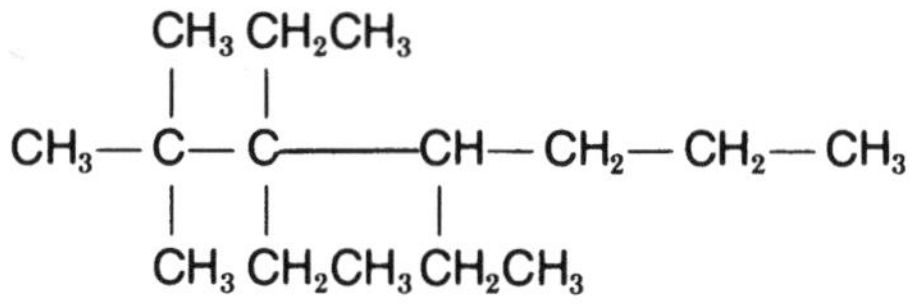

22.7

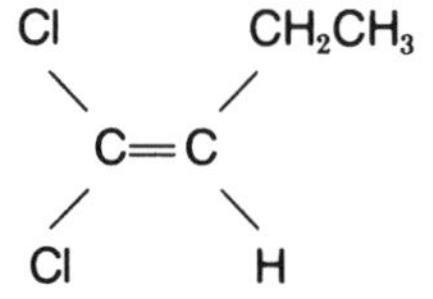

1,1-dichloro-1-butene

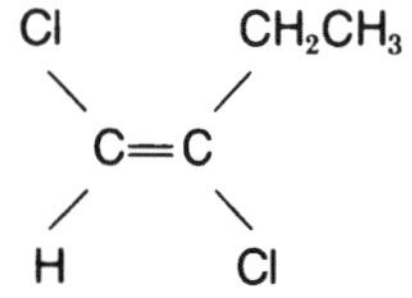

trans-1,2-dichloro-1-butene

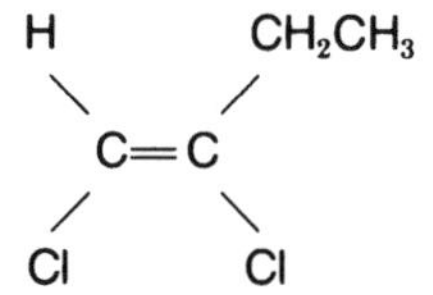

cis-1,2-dichloro-1-butene

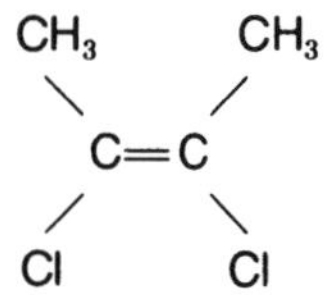

cis-2,3-dichloro-2-butene

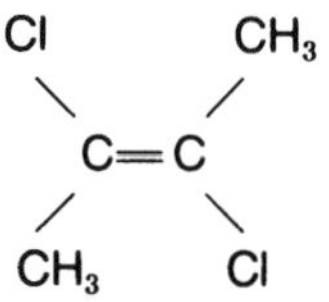

trans-2,3-dichloro-2-butene

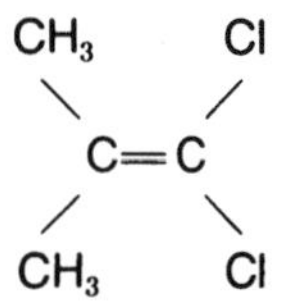

1,2-dichloro-2-methylpropene

22.8 (a)

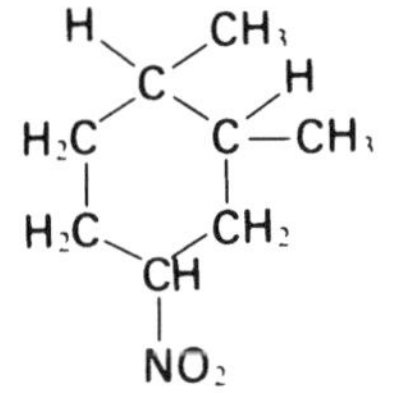

(c)

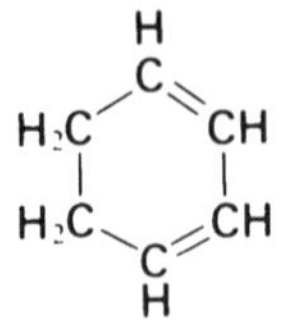

(b)

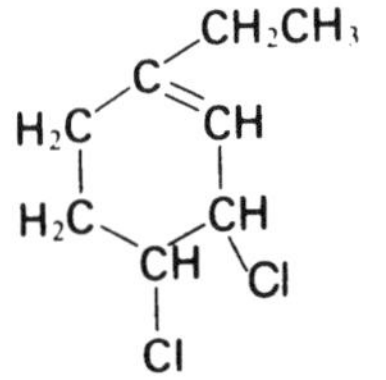

22.9 (a)

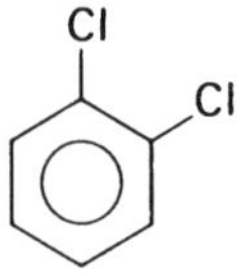

(c)

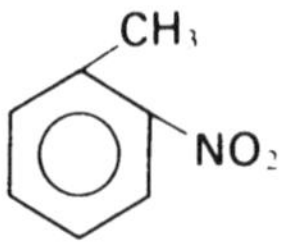

(b)

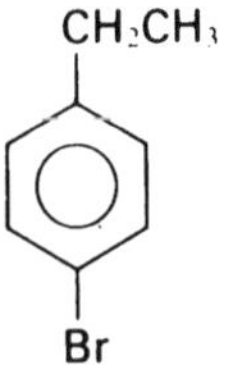

(d)

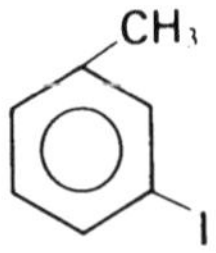

22.10 (a) *m*-ethyltoluene

(b) 1,3-dihydroxybenzene

(c) 1,3,5-trichlorobenzene

(d) *o*-*n*-propyltoluene

22.11 —C—C═C—C═C— is more stable because the π system is delocalized over four C atoms.

22.12 In a saturated hydrocarbon, each C atom bonds to four atoms; in an unsaturated hydrocarbon, C atoms form at least one double or triple bond.

22.13 (a) $CH_3—CH_3 + Cl_2 \xrightarrow{h\nu} CH_3—CH_2Cl + HCl$

(b) $CH_3—CH{=}CH_2 + H_2 \xrightarrow{Pt} CH_3—CH_2—CH_3$

(c)

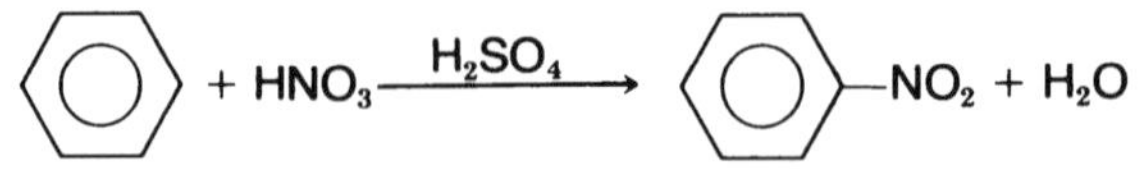

(d) $CH_3—CH{=}CH—CH_3 + HI \rightarrow CH_3—CHI—CH_2—CH_3$

(e) $CH_3—C{\equiv}C—CH_3 + HBr \rightarrow CH_3—CBr{=}CH—CH_3$

(f) $CH_3—C{\equiv}C—CH_3 + 2Br_2 \rightarrow CH_3—CBr_2—CBr_2—CH_3$

(g) $HC\equiv CH + 2HBr \rightarrow CHBr_2{-}CH_3$

(h) $CH_3{-}CH{=}CH{-}CH_3 + HOCl \rightarrow CH_3{-}\underset{\displaystyle OCl}{\underset{|}{CH}}{-}CH_2{-}CH_3$

(i) 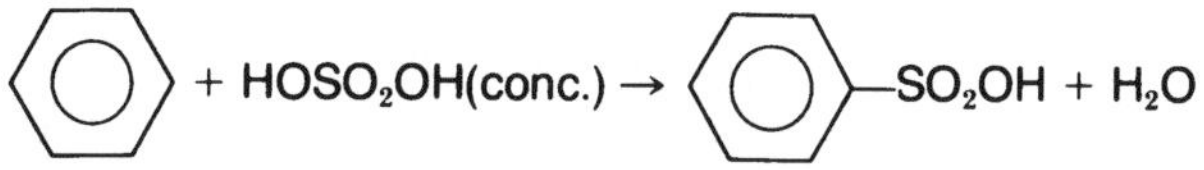

22.14 2-pentanol
2-methyl-3-pentanol
2-bromo-2-chlorobutane
2-chloro-1-propanol
$[CH_3{-}\underset{\displaystyle Cl}{\underset{|}{NH}}{-}CH_2CH_3]^+$

22.15 (a) $CH_3{-}\overset{\displaystyle O}{\overset{\|}{C}}{-}O{-}C_2H_5$

(b) $CH_3CH_2\underset{\displaystyle H}{\underset{|}{C}}{=}O$

(c) $C_3H_7\overset{\displaystyle O}{\overset{\|}{C}}{-}NH{-}C_2H_5$

(d) $CH_3CH{=}CH_2$

(e) $CH_3CH_2CH_2Cl$

(f) $CH_3CH_2C{=}O$ with $\backslash$ OH on C (i.e. $CH_3CH_2C(OH){=}O$)

(g) $C_4H_9\overset{\displaystyle O}{\overset{\|}{C}}{-}OH + C_2H_5OH$

(h) $CH_3(CH_2)_{16}\overset{\displaystyle O}{\overset{\|}{C}}{-}O^-Na^+$

(i) $[CH_3NH_3]^+Cl^-$

(j) $CH_3CH{=}CH{-}CH_3$ and $CH_3CH_2CH{=}CH_2$

(k) $CH_3CH_2CHNH_3{}^+Br^-$

(l) 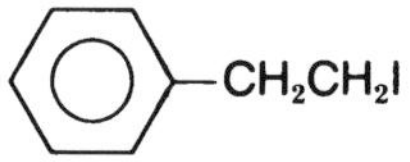

(m) $C_3H_7{-}O{-}C_3H_7$

(n) $CH_3CH_2{-}\overset{\displaystyle OH}{\overset{|}{CH}}{-}CH_3$

(o) 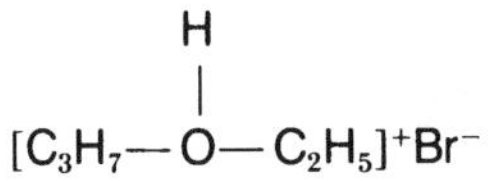

(p) 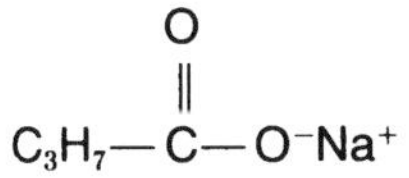

22.16 (a) 7 (f) 9

(b) 6 (g) 8

(c) 5 (h) 3

(d) 2 (i) 1

(e) 4

22.17 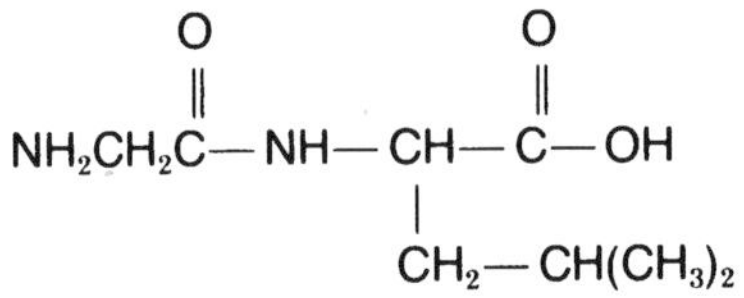

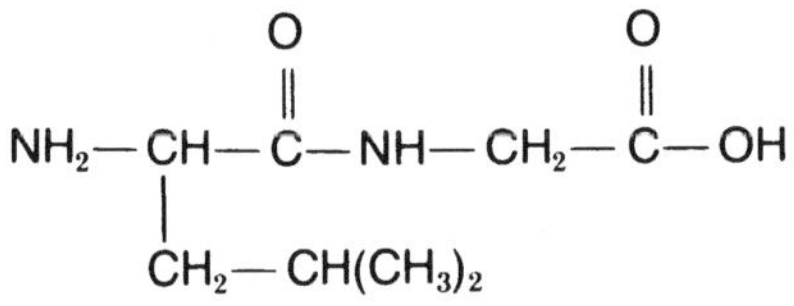

Chapter 23

POLYMERS AND BIOCHEMICALS

SECTION 23.1 SMALL MOLECULES AND LARGE MOLECULES

Objective

- To define a monomer, polymer, and copolymer.

Focus

Many high molecular weight molecules, called **macromolecules**, are composed of a number of low molecular weight molecules bonded together.

A molecule that enters into repetitive covalent bonding to form a macromolecule is a **monomer**. A dimer forms from the bonding of two monomers, a trimer from three monomers, and so on, in the formation of a **polymer**. If two or more *different* monomers bond repetitively, the resulting macromolecule is called a **copolymer**.

Polymers are macromolecules; molecular weights greater than 100,000 are quite common.

Example 23.1 Ethylene, $H_2C{=}CH_2$, is a monomer. When ethylene polymerizes, it forms polyethylene. Write the formula for the monomer of polypropylene and the monomer for polyvinylchloride (PVC).

Answer $H_2C{=}CH{-}CH_3$; $H_2C{=}CHCl$

Solution The monomer for polypropylene is propylene or propene, $H_2C{=}CH{-}CH_3$. The monomer for polyvinylchloride is vinyl chloride, $H_2C{=}CHCl$.

SECTION 23.2 THE DECOMPOSITION OF NATURAL HIGH POLYMERS: REPEATING UNITS

Objectives

- To identify the repeating monomer in a polymer.
- To label the characteristics of a monomer that affect the properties of the polymer.
- To explain the difference between addition and condensation polymers.

Focus

The polymer is made up of repeating units that resemble the monomer.

Example 23.2 Identify the repeating unit in the following polymer used in Acrilan and Orlon fibers.

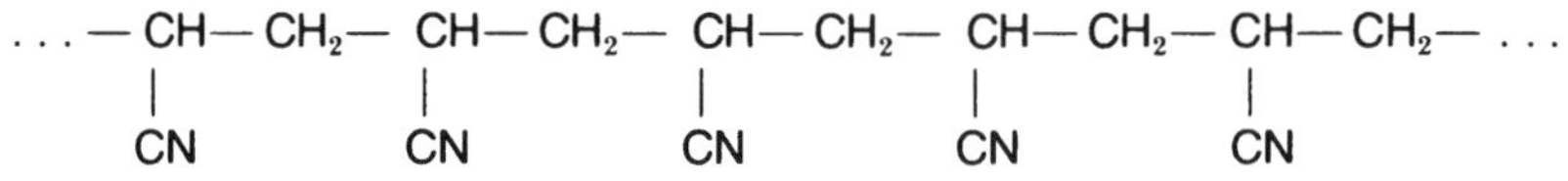

Answer

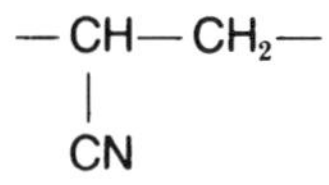

The manner in which monomers bond to form the polymer affects the properties of the polymer. Three factors are:

1. The mode of linkage.
2. The sequence and extent of polymerization.
3. The 3-D orientation of the polymer.

Polymers formed as a result of monomers bonding head to tail are **addition polymers**. Polymers formed from the bonding of two monomers with the elimination of a smaller molecule, such as H_2O, NH_3, or CH_3OH, are **condensation polymers.**

SECTION 23.3 THE SYNTHESIS OF POLYMERS BY CONDENSATION

Objective

- To describe the synthesis of a condensation polymer.

Focus

A condensation polymer is prepared from monomer molecules having two or more functional groups. These molecules form a bond, and in the reaction eliminate a low molecular weight molecule.

Example 23.3 Kodel polyester is a condensation copolymer formed by the reaction

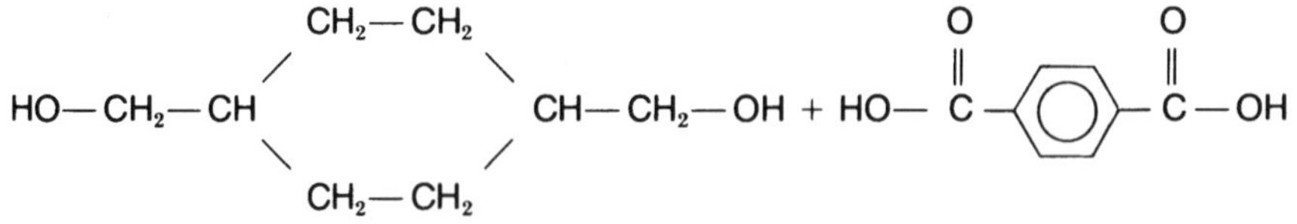

with the elimination of H_2O. Write the structural formula for the polymer and box the repeating unit.

Answer

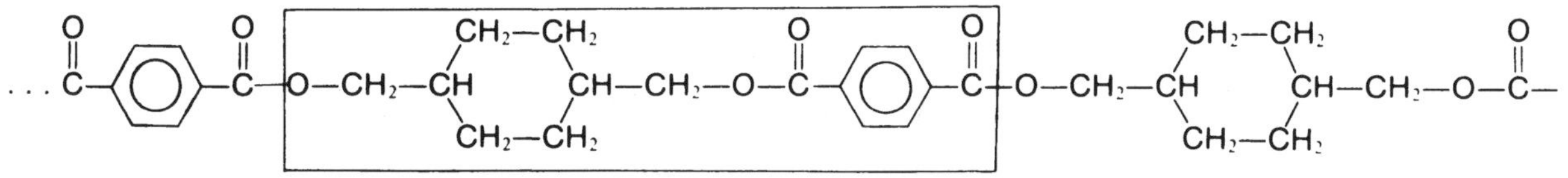

Solution The reaction of an alcohol, —OH, with an acid, —C(=O)—OH, produces an ester linkage, —C(=O)—O—, with the elimination of H_2O.

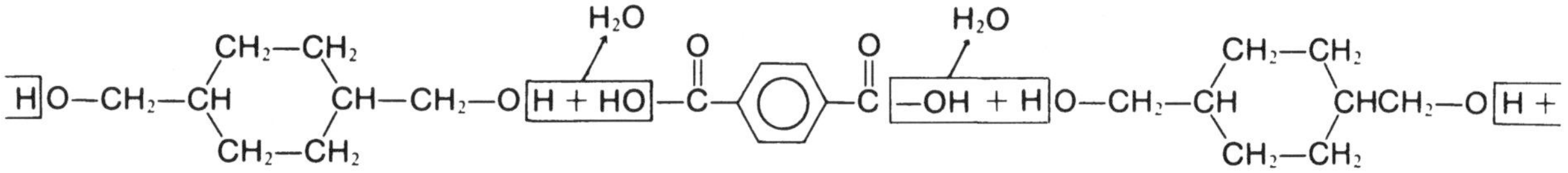

SECTION 23.4 ADDITION POLYMERIZATION

Objective

- To explain the mechanism for the preparation of addition polymers.

Focus

Most addition polymers are prepared from monomers having a carbon-to-carbon double bond, C=C. The π bond is easily broken with one p electron forming a new σ bond; the other p electron from the π bond then interacts with the π bond of another monomer. This highly reactive intermediate, with one unpaired electron, is called a free radical. Review the mechanism for the preparation of polyacrylonitrile in the text.

Example 23.4 The monomeric unit of polyethylene is $H_2C{=}CH_2$. Write the structural formula for polyethylene.

Answer $\ldots—CH_2—CH_2—CH_2—CH_2—CH_2—\ldots$

Solution A free radical starts the reaction; let's say ·OH.

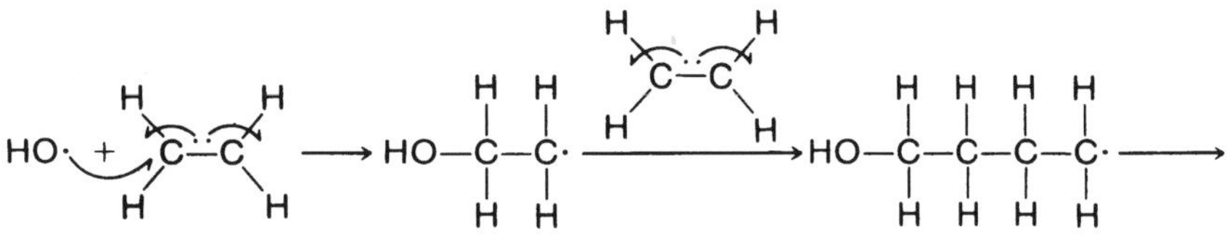

SECTION 23.5 LINKAGE OF MONOMERIC UNITS

Objectives

- To account for the formation of cross-linked polymers.
- To describe the general preparation of a cross-linked polymer.

Focus

Some monomers bond in several different ways, forming polymers with different chemical and physical properties.

The *cis*- or *trans*-polymer of isoprene, discussed in the text, illustrates how different geometric structures affect the properties of a polymer. Two polymeric chains are bonded together by hydrogen bonding or London forces. The better the fit of two polymeric chains, the greater is the magnitude of interpolymeric bonding (H bonding or London forces) and the more rigid the polymer is.

By using monomers with three or more active functional groups, instead of the required two, different properties for condensation polymers result. The bonds form not only at the head and tail of the molecule but also to the side. This 3-D polymeric network is called a **cross-linked polymer**.

SECTION 23.6 PROPERTIES OF POLYMERS

Objective

- To relate the properties of a polymer to its structure.

Focus

The properties of a polymer may be changed by:

1. The chemical attack of a functional group (such as —OH or C=C) on a polymeric chain.
2. The extent of cross-linkage between polymeric chains: the more cross-linkages, the more rigid and the more thermally stable is the polymer.
3. The extent of branching on the polymeric chain: the more branches, the poorer the fit between chains, and the more pliable is the polymer.
4. Changes in temperature: the higher the temperature, the less rigid is the structure.
5. The number of monomeric units that combine to form the polymer: the lower the molecular weight of the polymer, the more pliable it is.
6. Using mixtures of monomers in the polymerization reactions (copolymerization) for both the addition and condensation mechanisms.

SECTION 23.7 PROTEINS

Objective

- To define a protein in terms of structure and function.

Focus

A **protein** is a condensation polymer consisting of from approximately 50 to over 10,000 amino acid molecules. H_2O is eliminated when two monomeric units combine to form the *peptide linkage*,

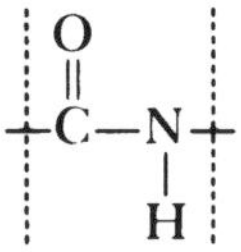

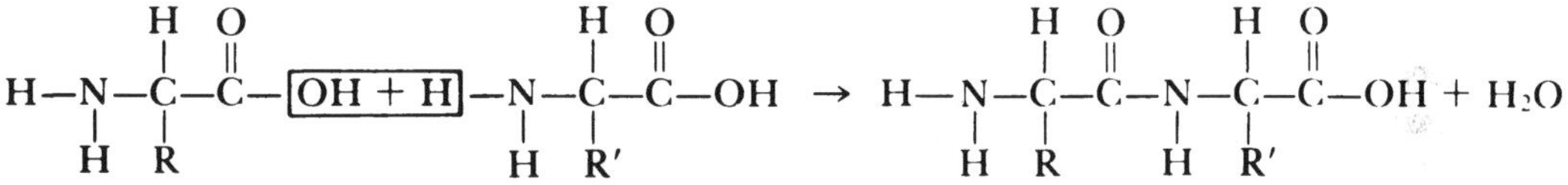

This dimer molecule called a *dipeptide*, which maintains the $-NH_2$ and the $-COOH$ groups at the ends of the molecule, undergoes additional condensation reactions to form a polypeptide (protein).

A protein molecule functioning as a catalyst in biochemical reactions is called an **enzyme**.

SECTION 23.8 NUCLEIC ACIDS

Objectives

- To define a nucleotide and its basic components.
- To summarize the structure of the DNA molecule.
- To explain the mechanism for the synthesis of a protein molecule.

Focus

A nucleic acid molecule, such as deoxyribonucleic acid (DNA), is made up of a large number of repeating units called **nucleotides**. Each nucleotide consists of

1. Phosphoric acid.
2. A five-carbon sugar, two of which condense with one H_3PO_4 to eliminate two molecules of H_2O.
3. An organic base which condenses with the sugar to eliminate H_2O.

Example 23.5 Draw the structural formula for the DNA nucleotide containing guanine and deoxyribose.

Answer

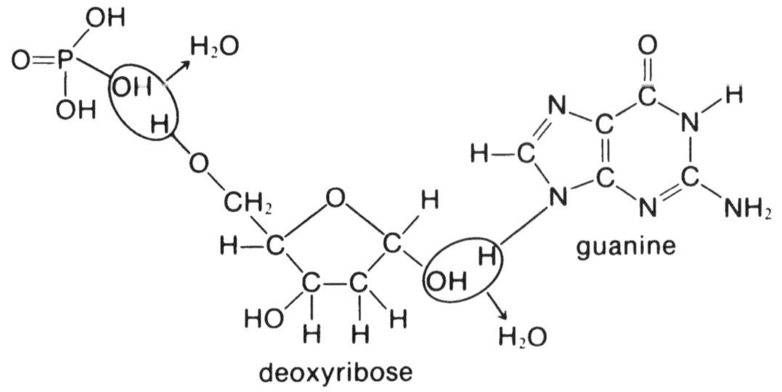

The DNA molecule consists of a double helix held together by H bonding between thymine-adenine (T-A) and cytosine-guanine (C-G) bases on the two chains of the double helix. This pairing of bases continues throughout the DNA molecule.

To begin preparation of a protein, the DNA molecule partially unravels. Read carefully through the text for the mechanism of protein synthesis. It is detailed, but interesting. Remember your life depends on it!

SECTION 23.9 EPILOGUE

Objective

- To discuss some advances in "recombinant DNA."

Focus

Recent research in the restructuring of the DNA molecule has caused some concern among scientists and nonscientists. Review the material in this section for your own information.

QUESTIONS

Section 23.2

23.1 What is the repeating unit in Dacron?

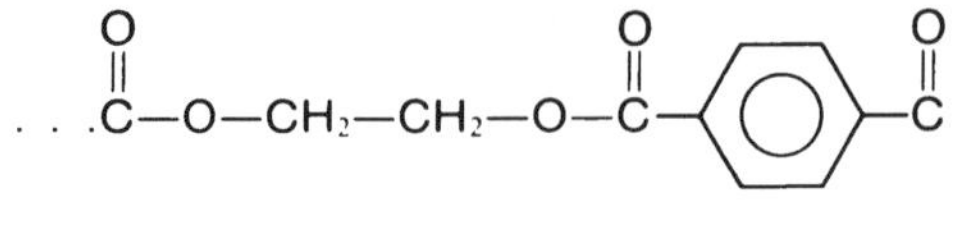

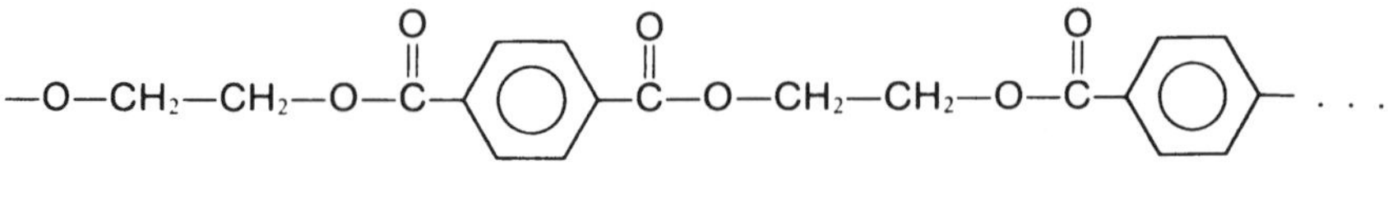

Section 23.3

23.2 Write the structural formula for the copolymer formed between the following molecules. H_2O is eliminated. Box the repeating unit.

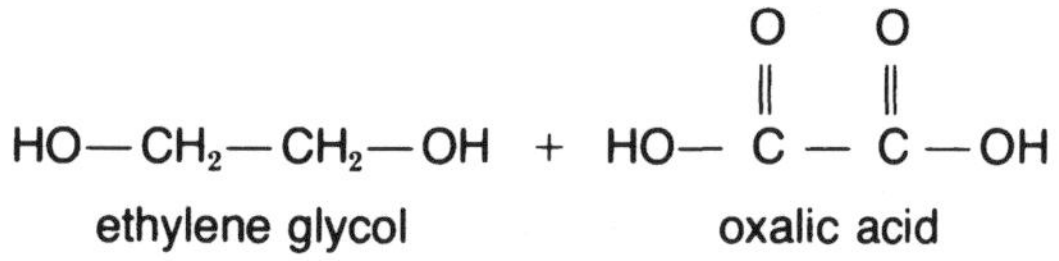

Section 23.4

23.3 Silicone rubber is a condensation polymer formed from the monomer $(CH_3)_2Si(OH)_2$ with the elimination of H_2O. Write the structural formula for silicone rubber.

23.4 Write the structural formulas for the polymer formed from styrene, 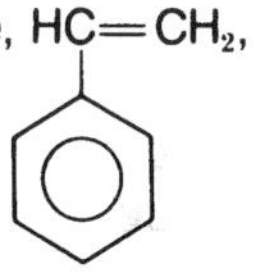

and the one formed from methyl methacrylate,

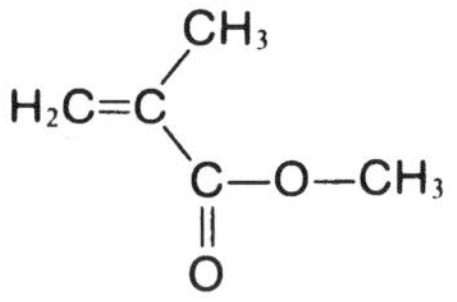

Section 23.5

23.5 A polymer formed between *p*-phenylenediamine, 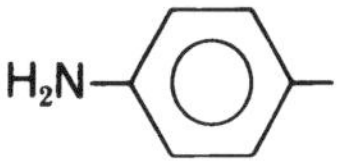 and terephthalic acid,

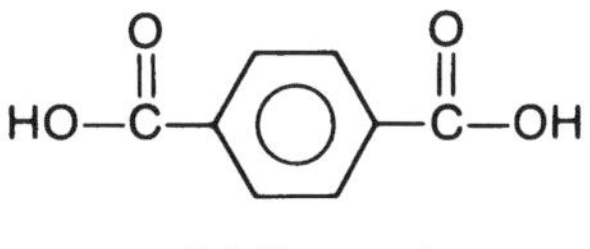

has a higher melting point, lower water solubility, and a greater tensile strength than when *o*- or *m*-phenylenediamine is used as a monomer. Explain.

Section 23.7

23.6 Write the formula for the repeating unit in the condensation of the amino acids alanine,

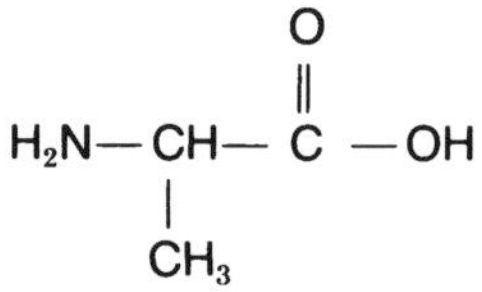 and phenylalanine 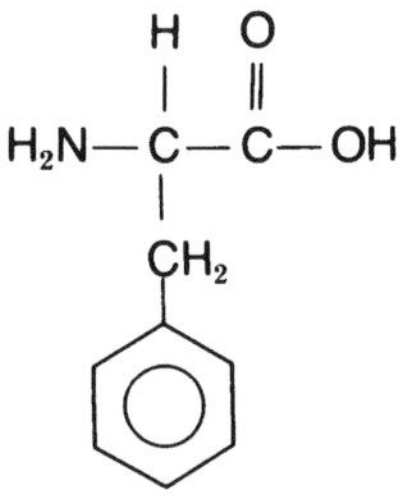

Section 23.8

23.7 To how many other molecules does a five-carbon sugar molecule bond in a nucleic acid?

23.8 Draw the structural formula of the nucleotide formed from H_3PO_4:

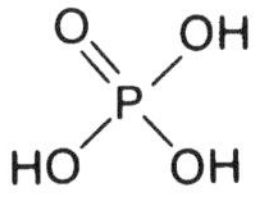

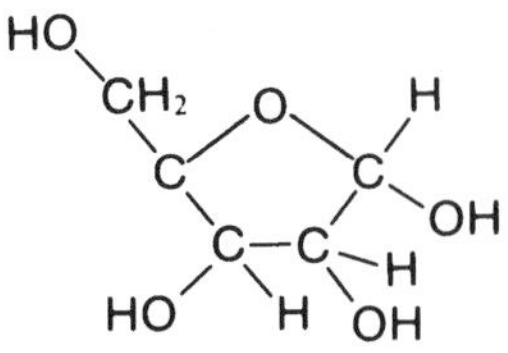

ribose

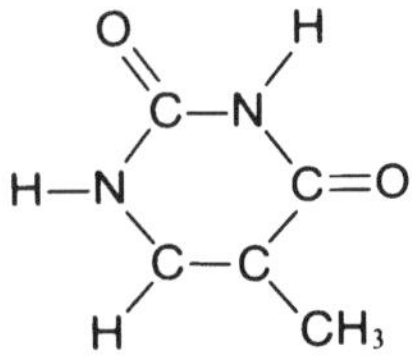

thymine

23.9 Using the concept of complementary base pairing, write the complementary strand for

(a) T A C G A T

(b) A G C T A C

23.10 How are the two polymeric chains in the DNA molecule held together?

23.11 Which organic bases are found in DNA? in RNA?

ANSWERS

23.1

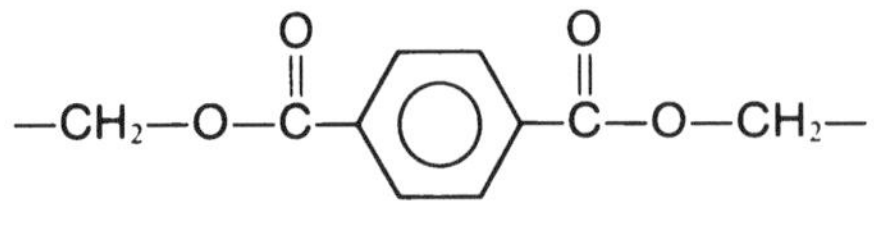

23.2

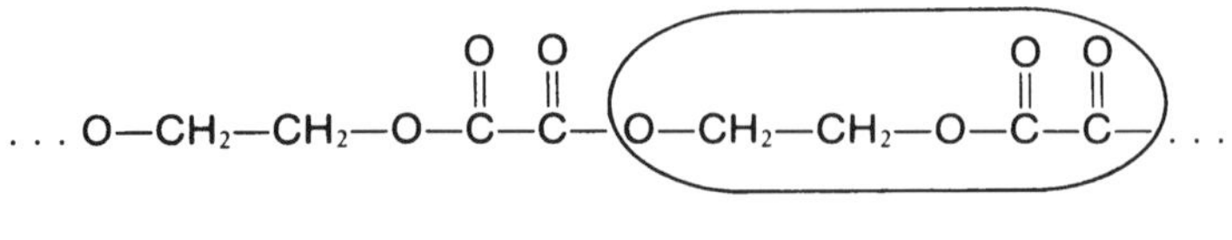

23.3

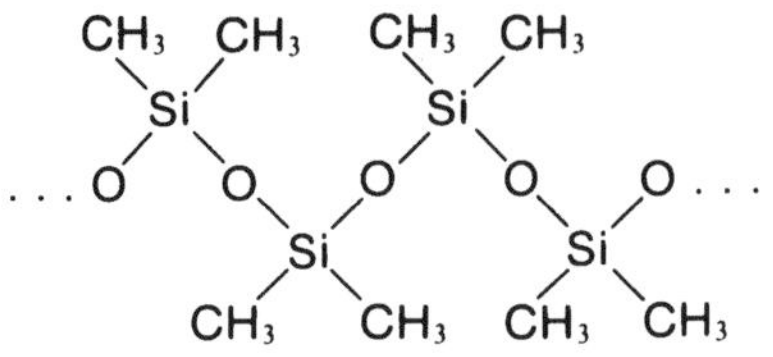

23.4

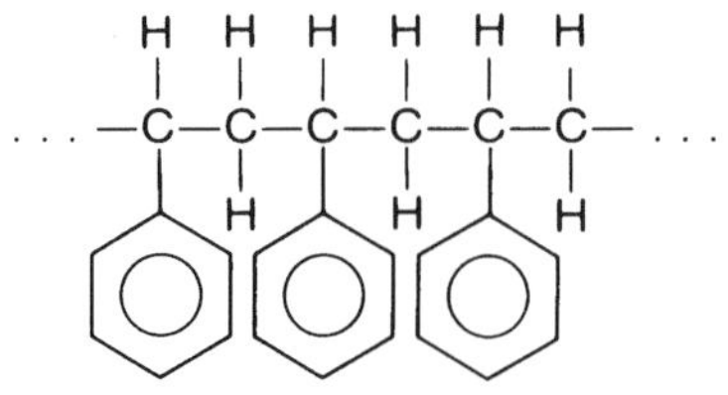

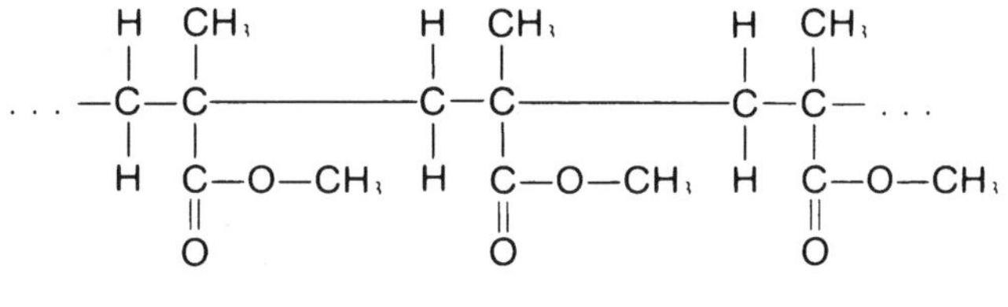

23.5 Because *p*-phenylenediamine is more symmetrical, the polymeric chains fit better, thus causing the observed properties.

23.6

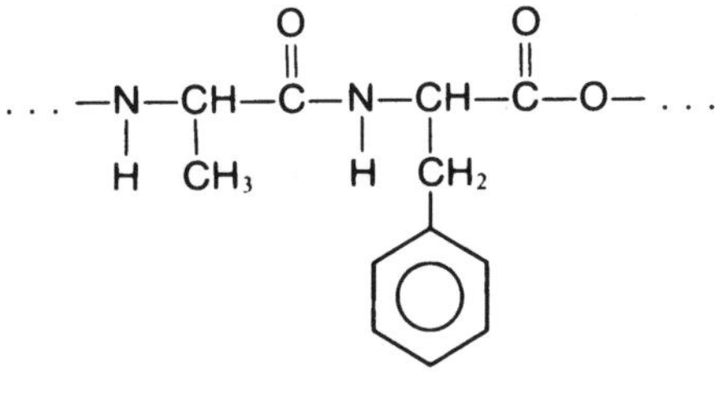

23.7 Three: two phosphoric acid molecules and an organic base.

23.8

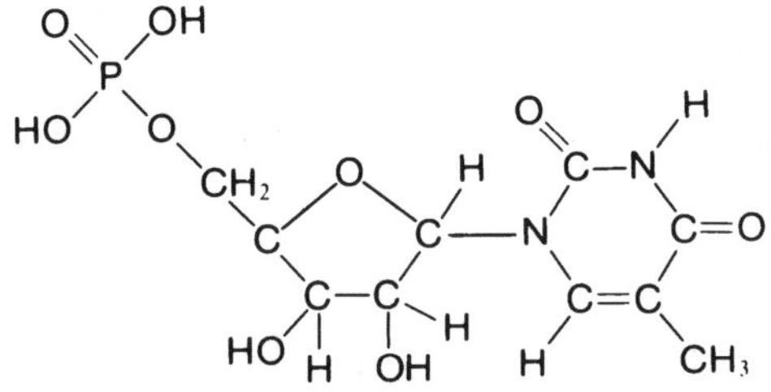

23.9 (a)

```
T A C G A T
| | | | | |
A T G C T A
```

(b)

```
A G C T A C
| | | | | |
T C G A T G
```

23.10 By H bonding between the pairs of bases along the two strands of the double helix.

23.11 DNA: cytosine(C), thymine(T), adenine(A), guanine(G)
RNA: uracil(U), cytosine(C), adenine(A), guanine(G)

Chapter 24

RADIATION AND MATTER

SECTION 24.1 MOLECULAR ENERGIES

Objectives

- To identify the various categories for energy absorption by a molecule.
- To make conversions between λ, ν, and $\bar{\nu}$ and to calculate the energy for an energy transition.

Focus

Over the past 20 years available laboratory equipment has improved and expanded a great deal. Most instruments take advantage of the various energy transitions for a molecule caused by photon absorption. The photons may have a very low energy, such as that needed to change the spin of a nucleus, or a very high energy, such as that required to excite an electron to a higher state. In both cases, the molecule remains intact but in an excited state.

$$X + \text{photon} \rightarrow X^*$$

See the text, Table 24.1, for the categories of energy associated with a molecule.

Chapter 24 discusses only electronic, vibrational, and nuclear spin energies. The ΔE (the difference between the excited-state and ground-state energies) for electronic excitation is greater than for vibrational, which in turn is greater than for nuclear spin excitation.

$$\Delta E_{\text{electronic}} > \Delta E_{\text{vibrational}} > \Delta E_{\text{nuclear spin}}$$

Only a photon of specific energy can be absorbed for a specific energy transition; in other words, energy transitions in a molecule or atom are *quantized*.

The energy of the absorbed photon is

$$\Delta E = h\nu = h\,\frac{c}{\lambda}\ (\text{J/photon})$$

$$\frac{c}{\lambda} = \nu = \text{frequency (sec}^{-1}\text{ or Hz)}$$

$$\frac{1}{\lambda} = \bar{\nu} = \text{wave numbers (cm}^{-1}\text{)}$$

Example 24.1 A photon with a wavelength of 3000Å is emitted in the deexcitation of a molecule.

(a) Calculate the energy of the photon.
(b) Calculate the energy when 1 mole of molecules deexcites by emitting 3000Å photons.
(c) What is the wave number for a 3000Å photon?

$$1\ \text{Å} = 10^{-10}\ \text{m}$$

Answer

(a) 6.62×10^{-19} J/photon
(b) 3.99×10^{5} J/mol
(c) 3.33×10^{4} cm^{-1}

Solution (a) The energy of a photon with a wavelength of 3000Å is calculated from the equation

$$\Delta E = h\frac{c}{\lambda}$$

λ must be expressed in meters: 1Å = 10^{-10} m.

$$\lambda(\text{m}) = 3000\cancel{\text{Å}} \times \frac{10^{-10}\ \text{m}}{1\cancel{\text{Å}}} = 3.00 \times 10^{-7}\ \text{m}$$

$$\Delta E_{3000\text{Å}} = 6.624 \times 10^{-34}\ \frac{\text{J}\cancel{\text{sec}}}{\text{photon}} \times \frac{3.00 \times 10^{8}\ \cancel{\text{m}}/\cancel{\text{sec}}}{3.00 \times 10^{-7}\ \cancel{\text{m}}}$$

$$= 6.62 \times 10^{-19}\ \text{J/photon}$$

(b) For 1 mole photons

$$6.62 \times \frac{10^{-19}\ \text{J}}{\cancel{\text{photon}}} \times 6.02 \times 10^{23}\ \frac{\cancel{\text{photons}}}{\text{mol}} = 3.99 \times 10^{5}\ \text{J/mol}$$

(c) A wave number is expressed in reciprocal centimeters.

$$3000\cancel{\text{Å}} \times \frac{10^{-10}\ \cancel{\text{m}}}{1\cancel{\text{Å}}} \times \frac{\text{cm}}{10^{-2}\ \cancel{\text{m}}} = 3.00 \times 10^{-5}\ \text{cm}$$

$$\bar{\nu} = \frac{1}{\lambda} = \frac{1}{3.00 \times 10^{-5}\ \text{cm}} = 3.33 \times 10^{4}\ \text{cm}^{-1}$$

SECTION 24.2 GENERAL FEATURES OF SPECTRA

Objective

- To explain the general shape of an energy absorption curve produced by a spectrophotometer.

Focus

A **spectrophotometer** is a laboratory instrument that records automatically the absorption of photons as a function of the wavelength, frequency, or wave numbers of incoming radiation.

The absorption curve shows that the molecules absorb photons at a specific wavelength. As a result, the molecule is excited to a higher energy state. The peak (or valley) in the absorption curve, λ_{max}, is the wavelength at which most photons are absorbed. Those photons with shorter or longer wavelengths pass unaffected through the sample and are recorded.

Example 24.2 What is the λ_{max} in the following spectrum? Calculate the energy of the photon that a molecule absorbs at λ_{max} (see Figure 24.1).

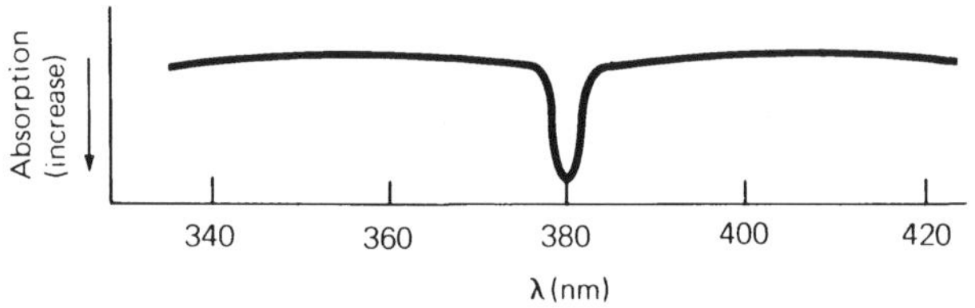

Figure 24.1 Absorption curve.

Answer $\lambda_{max} = 380$ nm

$$\Delta E = 5.2 \times 10^{-19} \text{ J/photon}$$

Solution λ_{max} from the curve is 380 nm. The energy of the absorbed photon is

$$\Delta E = h\frac{c}{\lambda} = 6.624 \times 10^{-34} \frac{\text{J sec}}{\text{photon}} \times \frac{3.0 \times 10^{8} \text{ m/sec}}{380 \text{ nm} \times \frac{10^{-9} \text{ m}}{\text{nm}}}$$

$$= 5.2 \times 10^{-19} \text{ J/photon}$$

SECTION 24.3 ELECTRONIC EXCITATION; ULTRAVIOLET (uv) SPECTROSCOPY

Objectives

- To account for electronic transitions in a molecule using the basic principles of MO theory.
- To predict the type of electronic transitions in the uv spectrum.
- To account for the effect of delocalized π electrons on electronic transitions.
- To distinguish between singlet and triplet states in an excited molecule and to describe the energy decay process for each.

Focus

When a molecule absorbs a photon in the ultraviolet (uv) or visible region of the electromagnetic spectrum, an electron is excited from the ground state to a higher energy state. The excited electron is a valence electron of which there are three types: σ bonding, π bonding, or nonbonding electrons.

The electronic excitation and the magnitude of ΔE depend on the type of MO's in a molecule. If the molecule has a σ bond (all molecules do), there are the σ bonding MO and the σ^* antibonding MO; if it has a π bond, there are the π bonding MO and π^* antibonding MO; and if it has nonbonding (n) pairs of electrons, there is a nonbonding MO. The n^* antibonding MO does not exist. Arranged in order of increasing energy, the MO's are

$$\sigma < \pi < n < \pi^* < \sigma^*$$

$$\xrightarrow[\text{above ground state}]{\text{increasing energy}}$$

See the text, Figure 24.7, for the relative energies of the five MO's and the relative ΔE's for the six electronic transitions between them. Remember that the $\sigma \rightarrow \sigma^*$ is the highest ΔE, the shortest λ electronic transition, and the $n \rightarrow \pi^*$ is the lowest ΔE, the longest λ transition.

Look closely at the text, Table 24.2. Notice that in CH_3OH, the lowest ΔE corresponds to the $n \rightarrow \sigma^*$ transition. There are no π bonds in the CH_3OH molecule. Therefore, no $n \rightarrow \pi^*$ transition can occur. In addition, only σ bonds and nonbonding pairs are in $C_6H_{13}\ddot{S}H$, $CH_3\ddot{C}l$:, $CH_3\ddot{B}r$:, $CH_3\ddot{I}$:, and $(CH_3)_3\ddot{N}$; thus the $n \rightarrow \sigma^*$ is the lowest ΔE electronic transition for these molecules. In the $H_2C{=}CH_2$ and $HC{\equiv}CH$ molecules, we find no nonbonding pairs and therefore no n MO's, only σ and π MO's; thus, the lowest ΔE is the $\pi \rightarrow \pi^*$ transition. But in $(CH_3)_2C{=}O$ and 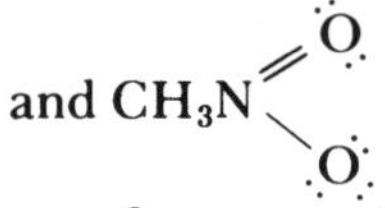, σ, π, and n electrons are present, causing the appearance of σ, π, n, π^*, and σ^* MO's; the lowest ΔE (longest λ) electronic transition is the $n \rightarrow \pi^*$, followed by $\pi \rightarrow \pi^*$.

Example 24.3 Identify the MO's in benzene, C_6H_6, and predict the lowest energy transition.

Answer σ, π, π^*, σ^*; the $\pi \rightarrow \pi^*$ is the lowest energy transition.

Solution The Lewis formula for benzene is

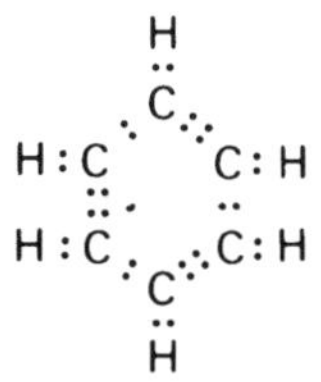

Only σ and π type electrons are in the molecule. Hence, the MO energy diagram is

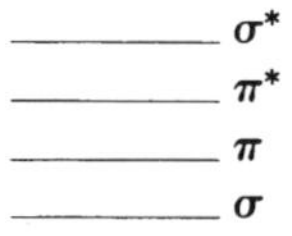

The possible electronic energy transitions are

increasing energy ↓

$\pi \rightarrow \pi^*$
$\pi \rightarrow \sigma^*$
$\sigma \rightarrow \pi^*$
$\sigma \rightarrow \sigma^*$

The photon energy a molecule absorbs depends upon the energy separation between the HOMO (*h*ighest *o*ccupied *MO*) and the LUMO (*l*owest *u*noccupied *MO*). Often if the π system is delocalized over a larger portion of the molecule, the energy difference for the $\pi \rightarrow \pi^*$ transition narrows. The wavelength of the absorbed photon is longer or toward the visible region of the spectrum.

Example 24.4 Which molecule absorbs at a longer wavelength?

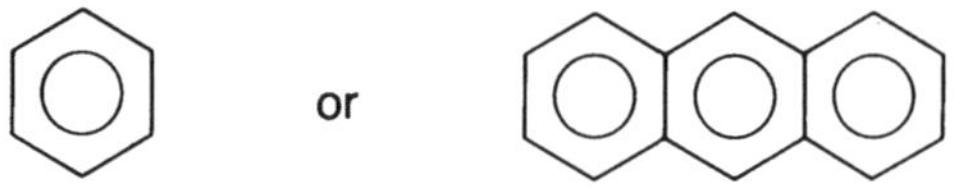

benzene or anthracene

Answer

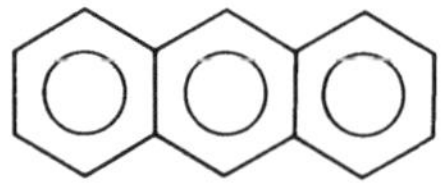

anthracene

Solution The π bonded system in anthracene is delocalized over 14 C atoms, but only over 6 C atoms in benzene. As a result, the $\pi \rightarrow \pi^*$ energy difference is less in anthracene. With a smaller ΔE, a longer wavelength photon is absorbed.

$$\lambda_{max}, \text{ benzene} = 260 \text{ nm}$$

$$\lambda_{max}, \text{ anthracene} = 375 \text{ nm}$$

When an electron is excited to a higher energy state, it maintains the same spin. This excited electronic state is called a **singlet state**. Two principal modes of decay occur for an electron in a singlet state:

1. The lifetime of a singlet state is relatively short; the electron may lose either all or part of its energy as a heat loss through intermolecular collisions and the remainder through photon emission. Thus the emitted photon has less energy than the one which excited the electron and often is in the visible region. This type of photon emission is called **fluorescence**. As a result of this energy decay process, "black lights" cause some colors to fluoresce.
2. The electron can *flip* while in the singlet state, resulting in two electrons in different MO's having the same spin. This **triplet state** for the excited atom is of lower energy than the singlet state. Some energy is released when the excited electron *flips*. In a triplet state the excited molecule may also deexcite by emitting heat through intermolecular collisions or by delayed photon emission. Because the triplet state has a longer

lifetime than a singlet state, photons in the visible region continue to be emitted even after the light source is turned off. This is called **phosphorescence**.

SECTION 24.4 VIBRATIONAL EXCITATION; INFRARED SPECTROSCOPY

Objectives

- To explain the origin of infrared spectra.
- To use characteristic stretching and bending vibration frequencies for the identification of a molecule.

Focus

The energy a photon needs to excite a molecule to a higher vibrational energy state is in the infrared (ir) region. The photon may excite the molecule to higher bending vibrational or higher stretching vibrational energy states.

Various groups of atoms have nearly the same ΔE for a vibrationally excited state in any molecule where the group is present. For example, to vibrationally excite the $-C{\equiv}N$ group of atoms in any molecule, a photon must have a wave number of 2000 to $2300 cm^{-1}$ (see text, Table 24.3). The energy of the photon is in the range of

$$\Delta E = h\frac{c}{\lambda} = hc\left(\frac{1}{\lambda}\right) = hc\bar{\nu}$$

$$= \frac{6.624 \times 10^{-34}\, J\, \cancel{sec}}{photon} \times 3.0 \times 10^{10}\, \frac{\cancel{cm}}{\cancel{sec}} \times 2000\, \cancel{cm^{-1}}$$

$$= 4.0 \times 10^{-20}\, J/photon$$

and

$$\Delta E = 6.624 \times 10^{-34}\, \frac{J\, \cancel{sec}}{photon} \times 3.0 \times 10^{10}\, \frac{\cancel{cm}}{\cancel{sec}} \times 2300\, \cancel{cm^{-1}}$$

$$= 4.6 \times 10^{-20}\, J/photon$$

For each bending and stretching vibrational energy transition in a molecule, a photon is absorbed. Therefore, we see many more absorption peaks in the ir spectrum than in a uv spectrum for the same molecule.

Study Table 24.3 so that you can identify some of the *fingerprint* regions of the ir spectrum characteristic of groups of atoms.

Example 24.5 Where in the ir spectrum are you able to distinguish between each of the following molecules? Use text Table 24.3.

(a) $CH_3CH_2-\overset{\overset{\large O}{\|}}{C}-OH$ and $CH_3-\overset{\overset{\large O}{\|}}{C}-O-CH_3$

(b) CH_3OH and CH_3CH_2OH

Answer (a) 3590 to 3650 cm^{-1}
(b) 600 to 1500 cm^{-1}

Solution (a) In $CH_3CH_2—\overset{\overset{O}{\|}}{C}—OH$, the 3590 to 3650 cm^{-1} region shows a peak because of an —OH vibrational excitation which does not appear in $CH_3—\overset{\overset{O}{\|}}{C}—O—CH_3$.

(b) Both CH_3OH and CH_3CH_2OH have the —OH group of atoms which gives rise to an absorption band in the 3590 to 3650 cm^{-1} region. However, CH_3CH_2OH has a C—C single bond that is not present in CH_3OH. Therefore, an absorption peak in the 600 to 1500 cm^{-1} region is predicted for CH_3CH_2OH and not for CH_3OH.

SECTION 24.5 NUCLEAR SPIN EXCITATION: NMR SPECTROSCOPY

Objectives

- To describe the principle of nmr spectroscopy.
- To identify the number of protons in different chemical environments in a molecule from the ^1H nmr spectrum.

Focus

Like electrons, nucleons have a spin. An unpaired proton and/or neutron, like an unpaired electron, causes the nucleus to behave as a tiny magnet that interacts with an externally applied magnetic field. If the mini-magnetic field from the nucleus is in the same direction as the external magnetic field, the spin energy state of the nucleus is aligned with the field. The nucleus is of lower energy than when the mini-magnetic field is in opposition to the external magnetic field (see text, Figure 24.11). The stronger the external magnetic field, the larger is the energy differences between these two spin states.

The energy, ΔE, necessary to *flip the spin* of the nucleus from the lower nuclear spin energy state to a higher one is the basis of nuclear magnetic resonance (nmr) spectroscopy. The ΔE is proportional to the strength of the external magnetic field and to the frequency (or wave numbers, cm^{-1}) of the electromagnetic radiation. In the nmr technique, the external magnetic field strength changes while constant energy photons irradiate the sample. The energy of the photons is in the radio wave region of the electromagnetic spectrum.

Proton or ^1H nmr is a very powerful technique in determining the molecular structure of organic compounds. The H atom has 1 proton which is obviously unpaired.

The magnetic field strength at which the photon *flips* the proton depends on the concentration of electrons near the proton. The electrons that bond the proton set up a magnetic field opposite to the external magnetic field; therefore, the more electrons in the area of the proton, the greater must be the strength of the external magnetic field. A large number of electrons *shield* the proton, causing the absorption signal to appear *upfield* in the spectrum.

A lower electron density in the bonding environment of the proton requires a weaker external magnetic field to *flip* the proton. Photon absorption occurs *downfield* because the nucleus is shielded less.

The position or **chemical shift** of the signal in the nmr spectrum therefore depends entirely on the proton's bonding environment. Some facts about chemical shifts are:

1. All protons in the same bonding environment have the same chemical shift or absorb photons at the same external magnetic field strength.
2. Highly electronegative atoms bonded to or close to a H atom, for example, $-\overset{|}{\underset{|}{C}}-O-H$ or $-O-\overset{|}{\underset{|}{C}}-H$, reduce the electron density in the proton's bonding environment; the proton is deshielded, and the signal appears more downfield.
3. When a H atom is the more electronegative atom in the bond, the proton is heavily shielded and a stronger magnetic field is needed to absorb a photon. The signal appears more upfield.
4. A H atom bonded to a double bonded C atom is also deshielded, and the absorption peak or signal appears downfield.

Example 24.6 List the order of the appearance of the absorption peaks in the nmr spectrum starting with the one that occurs farthest downfield.

HBr H_2O HCl HF H_2S

Answer HF, H_2O, HCl, HBr, H_2S

Solution In order of decreasing electronegativity,

$$F > O > Cl > Br > S$$

Therefore the proton in HF is most *de*shielded by electrons and requires the lowest external magnetic field strength for absorption. The HF proton appears farthest downfield and the H_2S proton farthest upfield in the nmr spectrum.

What does an nmr spectrum tell us?

1. The number of peaks identifies the number of protons in different chemical environments of a molecule.
2. The position (upfield or downfield) of the signal indicates the bonding environment of each proton.
3. The area under the peaks represents the relative number of protons in the same bonding environment.

Example 24.7 How many different bonding environments for protons are there, and what is the ratio for the area under each peak for the following compounds?

(a) $CH_3-CH_2-CH_2-CH_3$

(b) $ClCH_2-CH_2-CH\begin{matrix}\diagup CH_3\\ \diagdown CH_3\end{matrix}$

(c) $CH_3-O-CH_2-CH_2Cl$

Answer (a) The 4 H atoms in the two CH_2 groups are equivalent and the 6 H atoms in the two CH_3 groups are equivalent. Thus there are two different bonding environments

	$-CH_2-CH_2-$	and	CH_3, CH_3
Ratio of peak areas	4	:	6
or		2 : 3	

(b) Four different bonding environments

	$ClCH_2-$	and	$-CH_2-$	and	$-CH$	and	CH_3, CH_3
Ratio of peak areas	2	:	2	:	1	:	6

(c) Three different bonding environments

	CH_3-	and	$-CH_2-$	and	$-CH_2Cl$
Ratio of peak areas	3	:	2	:	2

Example 24.8 How do you distinguish between $CH_3-\overset{O}{\overset{\|}{C}}-CH_3$ and $CH_3-\overset{O}{\overset{\|}{C}}-O-CH_3$ in the nmr spectrum?

Answer In $CH_3-\overset{O}{\overset{\|}{C}}-CH_3$, all protóns are equivalent, and only one signal appears in the spectrum. But in $CH_3-\overset{O}{\overset{\|}{C}}-O-CH_3$ the highly electronegative O atom reduces the electron density in the bonding environment of the protons in the $-O-CH_3$ group. These protons are therefore shielded less and absorb downfield from the $CH_3-\overset{O}{\overset{\|}{C}}-$ protons.

SECTION 24.6 RADIATION CHEMISTRY AND PHOTOCHEMISTRY

Objective

- To explain five ways in which high energy radiation interacts with matter.

Focus

The absorption of high energy photons by a molecule can cause several reactions. The molecule either ionizes, fragments into smaller groups of atoms, rearranges its atoms, transfers energy to another molecule and simultaneously causes ionization, or weakens the bonds in another molecule and then adds to it.

1. *Ionization*. A high energy photon can cause ejection of an electron from the HOMO of a molecule.

$$\text{R:H} \xrightarrow{h\nu} \text{R}\cdot\text{H}^+ + e^-$$

2. *Fragmentation*. A molecule absorbing a high energy photon is excited to a very high energy state. The molecule can deexcite by splitting into two fragments:

$$\text{R:H} \xrightarrow{h\nu} [\text{R:H}]^* \rightarrow \text{R}\cdot + \text{H}\cdot$$

   The two free radicals carry away the excess energy in the form of kinetic energy. The weakest bond in the molecule is the one that is usually broken.

   Ionization sometimes precedes fragmentation:

$$\text{R:H} \xrightarrow{h\nu} [\text{R:H}]^* \rightarrow [\text{R}\cdot\text{H}^+]^* + e^-$$

$$[\text{R}\cdot\text{H}^+]^* \begin{cases} \rightarrow \text{R}^\cdot + \text{H}^+ \\ \rightarrow \text{R}^+ + \text{H}\cdot \end{cases}$$

   The free radical is associated with either fragment.

3. *Rearrangement*. A *cis-trans* rearrangement occurs in a π bonded molecule in which a $\pi \rightarrow \pi^*$ electron transition occurs. When this happens, the π bond is broken, allowing free rotation around the remaining σ bond, Deexcitation then occurs and the π bond is reformed. However, the molecule may now have a different configuration; the *cis*-isomer may have rearranged to the *trans*-isomer or vice versa. See text, Figure 24.17, for the sequence of steps.

4. *Electron Transfer*. An excited molecule may deexcite by colliding with a second molecule and transferring an electron.

$$\text{R:H} \xrightarrow{h\nu} [\text{R:H}]^*$$

$$[\text{R:H}]^* + \text{Y} \rightarrow \text{R}\cdot\text{H}^+ + \text{Y}^-$$

   or

$$[\text{R:H}]^* + \text{Y}^+ \rightarrow \text{R}\cdot\text{H}^+ + \text{Y}$$

5. *Addition*. An excited molecule uses its excess energy to weaken a bond in a second molecule and then adds to it forming a larger single molecule.

$$[\text{R:H}]^* + \text{X:Y} \rightarrow \text{X:}\underset{\text{H}}{\underset{\cdot\cdot}{\text{R}}}\text{:Y}$$

Example 24.9 Complete the following reactions, and describe the type of deexcitation reaction that occurs:

$$\text{(a) } CH_3-\overset{\overset{\displaystyle O}{\|}}{C}-H \xrightarrow{h\nu} \cdot CH_3 + ?$$

(b) $O_2 \xrightarrow{\lambda = 135 \text{ to } 175 \text{ nm}} O\cdot + ?$
(upper atmosphere reaction)

(c) $N_2 \xrightarrow{h\nu} N_2^+ + ?$
(upper atmosphere reaction)

(d) $O_2 \xrightarrow{h\nu} O_2^*$
$O_2^* + O_3 \rightarrow 2O_2 + ?$

(e) $O \xrightarrow{h\nu} O^+ + e^-$
$O^+ + N_2 \rightarrow NO^+ + ?$

Answer

(a) $\cdot\overset{\overset{\large O}{\|}}{C}—H$, a fragmentation reaction
(b) ·O, a fragmentation reaction
(c) e^-, an ionization reaction
(d) ·O, a transfer + fragmentation (similar to type 4)
(e) ·N, ionization and charge transfer reaction

SECTION 24.7 MASS SPECTRAL ANALYSIS

Objectives

- To explain the use of mass spectrometry in identifying the structure of a molecule.
- To predict the structure of some simple compounds from mass spectra.

Focus

In a mass spectrometer medium energy electrons collide with molecules, stripping each molecule of an electron to form a positive ion. If the resulting ion is in a low energy state, it is detectable as the **parent ion** in the mass spectrum. The parent ion peak is usually the one with the largest mass. If the ion is in a vibrationally excited state, it can fragment further, forming lower mass ions that are also detectable in the mass spectrum.

$$R{:}R' + e^- \rightarrow R{\cdot}R'^+ + 2e^-$$

or

$$R{:}R' + e^- \rightarrow [R{\cdot}R'^+]^* + 2e^-$$

$$\swarrow \qquad \searrow$$

$$R\cdot + R'^+ \qquad R^+ + \cdot R'$$

Neutral particles are not detected in a mass spectrum. The mass of each fragment ion is recorded on the mass spectrum. The more stable structures will have fragmented the least and will therefore be more abundant. The height of each peak is a measure of the relative abundance of each ion.

What does a mass spectrum tell us?

1. The mass of the parent ion which is the molecular weight of the molecule.
2. The masses of the various stable fragments in the molecule.
3. When the fragments are placed together, the molecular structure can possibly be established.

Example 24.10 The prominent peaks in the mass spectrum of CF_3—CF_3 occur at masses 138, 69, 119, 31, 50, 19, and 12. Identify the mass fragments in the spectrum.

Answer

138	$CF_3:CF_2^+$ (with :F: above)
69	CF_3^+
119	$CF_3:CF_2^+$
31	$:CF^+$
50	$\cdot CF_2^+$
19	$:\ddot{F}^+$
12	$:C\cdot^+$

Solution Rewrite the structure for clarity as $F_3C:CF_2$ (with :F: above)

$$F_3C:CF_2 + e^- \rightarrow F_3C:CF_2^+ \ (m/e = 138) + 2e^-$$

$$[F_3C:\dot{C}F_2^+]^* \rightarrow F_3C:CF_2^+ \ (m/e = 119) + :\ddot{F}\cdot$$

$$\rightarrow :\dot{F}:^+ \ (m/e = 19) + F_3C:\dot{C}F_2$$

$$\rightarrow \cdot CF_2^+ \ (m/e = 69) + \cdot CF_3$$

$$[\cdot \dot{C}F_2^+]^* \rightarrow \cdot CF_2^+ \ (m/e = 50) + :\dot{F}:$$

$$\rightarrow :CF^+ \ (m/e = 31) + 2:\dot{F}:$$

$$\rightarrow :C\cdot^+ \ (m/e = 12) + 3:\dot{F}:$$

Example 24.11 Identify the fragment ion with the indicated mass from the mass spectra of the following molecules:

	Molecule	Mass of fragment ion	Fragment ion
(a)	$H_3C—CH_2—CHCl$	48	
(b)	$H_3C—CH_2—C(=O)—CH_3$	15	
(c)	$H_3C—CH_2—C(=O)—H$	43	
(d)	$(CH_3)_3N$	29	

Answer

(a) $CHCl^+$ (c) $H_2C—C(=O)—H^+$

(b) CH_3^+ (d) $CH_3\ddot{N}^+$

QUESTIONS

Section 24.1

24.1 A molecule absorbs a photon with an energy of 3.20×10^{-19} J. What is the wavelength (nm) and the frequency (Hz) of the absorbed photon? Using the text, Table 24.1, in what region of the spectrum is this energy for the photon?

Section 24.3

24.2 What types of electronic transitions occur in the uv spectrum for the following?

(a) $CH_3—\ddot{\underset{..}{O}}—CH_3$

(b) 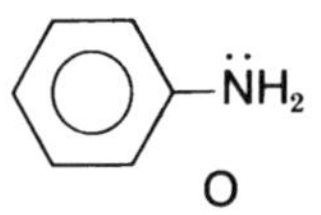

(c) $CH_3CH_2—C(=O)—\ddot{N}H_2$

(d) $CH_2{=}CH—CH{=}CH_2$

(e) $CH_3—C(=O)—CH_3$

24.3 List the following compounds in each set in order of increasing λ_{max} in the uv spectrum. List the shortest λ_{max} (largest ΔE) first.

Set A

$CH_3—CH_3$; 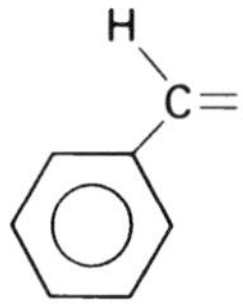 ; $H_2C{=}CH_2$; 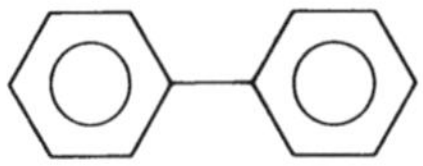 ; 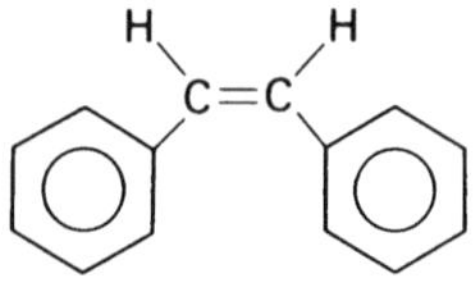

Set B

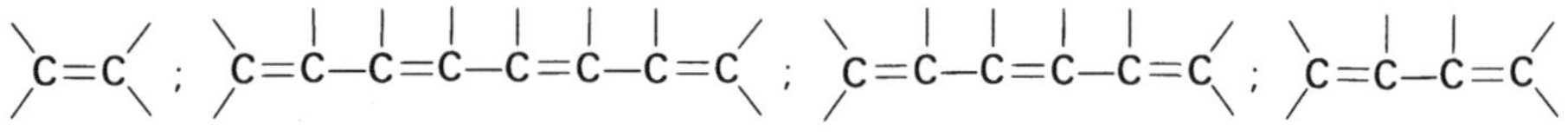

Section 24.4

24.4 Calculate the λ, ν, and ΔE for the —N—H stretching frequency. From Table 24.3, ν = 3300 to 3500 cm^{-1}.

24.5 From what region of the ir spectrum do you distinguish between the following compounds that have the same molecular formula?

Molecular Formula	
$C_4H_{11}N$	(a) $CH_3CH_2—N(CH_3)—CH_3$ and $CH_3CH_2CH_2—N(H)—CH_3$
C_3H_8O	(b) $CH_3CH_2—O—CH_3$ and $CH_3CH_2CH_2OH$
C_3H_6O	(c) $CH_3—C(=O)—CH_3$ and $CH_3CH_2C(=O)—H$
C_3H_5N	(d) $HC{\equiv}CCH_2NH_2$ and $CH_3CH_2—C{\equiv}N$
C_5H_8	(e) $H_2C{=}CH—CH_2CH{=}CH_2$ and $CH_3C{\equiv}C—CH_2CH_3$

Section 24.5

24.6 Describe the difference in the ^1H nmr spectra of the following pairs:

(a) $CH_3—CH_2—CH_3$ and $CH_3—CH_2—CH_2—CH_3$

(b) $CH_3—C(=O)—CH_3$ and $CH_3—CH_2—C(=O)—H$

(c) $CH_3—CH_2—OH$ and $CH_3—O—CH_3$

(d) $CH_3—CH_2—OH$ and $CH_3—OH$

(e) $CH_3—CH_2—CH_2Cl$ and $CH_3—CHCl—CH_3$

Section 24.6

24.7 Complete the following reactions and describe the type of deexcitation reaction that occurs.

(a) $N_2 \xrightarrow{h\nu} N_2^+ + e^-$

$N_2^+ + O_2 \rightarrow N_2 +$ ________

(b) $H_2O_2 \xrightarrow{h\nu} \cdot OH +$ ________

(c) $H_2O_2 \xrightarrow{h\nu} \cdot OH + \cdot OH$

$HO\cdot +$ ________ $\rightarrow HNO_2$

(d) $O_3 \xrightarrow{h\nu} O^* +$ ________

(e) 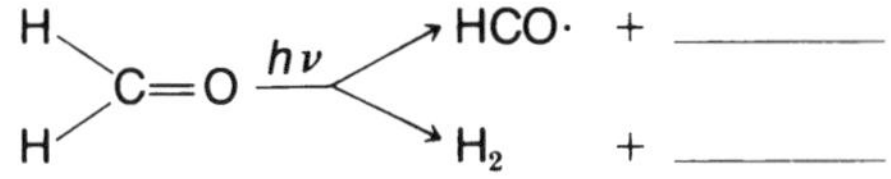

(f) $C_3H_6 + Hg^* \rightarrow C_3H_6^* + Hg$

$C_3H_6^* \rightarrow C_3H_5\cdot +$ ________

Section 24.7

24.8 The mass spectrum of an inorganic compound shows major peaks at 71, 52, 33, 19, and 14. What is the compound?

24.9 The mass spectrum of a compound has prominent peaks at 60, 32, 28, and 12. Identify the compound.

24.10 The mass spectrum of $C_2H_6N_2$ has prominent peaks at 15, 43, 28, 42, 58, 29, and 14. Identify the compound.

ANSWERS

24.1 622 nm, 4.82×10^{14} Hz; visible

24.2 (a) $\sigma \rightarrow \sigma^*$

$n \rightarrow \sigma^*$

(b), (c) $\sigma \rightarrow \sigma^*$; $\sigma \rightarrow \pi^*$; $\pi \rightarrow \sigma^*$;

$\pi \rightarrow \pi^*$; $n \rightarrow \sigma^*$; $n \rightarrow \pi^*$

(d), (e) $\sigma \rightarrow \sigma^*$; $\sigma \rightarrow \pi^*$; $\pi \rightarrow \sigma^*$;

$\pi \rightarrow \pi^*$;

24.3 Set A. CH_3CH_3; $CH_2{=}CH_2$; 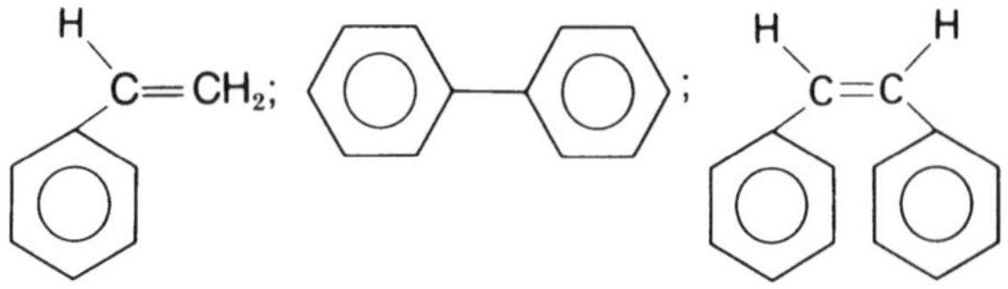

Set B. 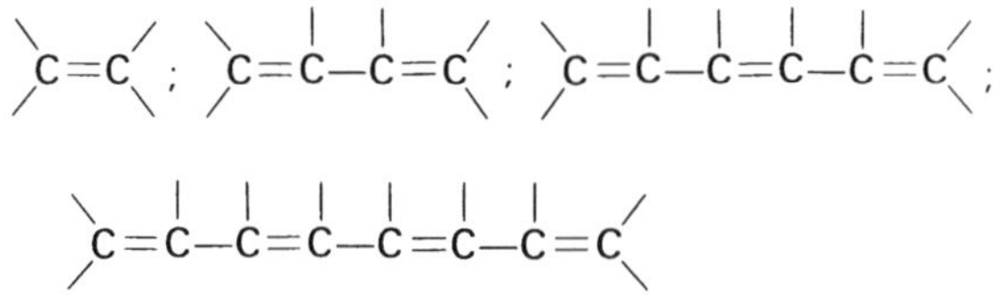

24.4 $\lambda = 3.0 \times 10^{-4}$ cm to 2.9×10^{-4} cm

$\nu = 9.9 \times 10^{13}$ Hz to 1.0×10^{14} Hz

$\Delta E = 6.6 \times 10^{-20}$ J/photon to 7.0×10^{-20} J/photon

24.5 (a) 3300 to 3500 cm^{-1}; N—H stretching frequency in $CH_3CH_2CH_2{-}N(H){-}CH_3$

(b) 3590 to 3650 cm^{-1}; O—H stretching frequency in $CH_3CH_2CH_2OH$

(c) 600 to 1500 cm^{-1}; C—C stretching frequency in $CH_3CH_2C({=}O){-}H$

(d) 3300 to 3500 cm^{-1}; N—H stretching frequency in $HC{\equiv}CH_2NH_2$ and
2000 to 2300 cm^{-1}; —C≡N stretching frequency in $CH_3CH_2—C{\equiv}N$ and
2100 to 2260 cm^{-1}; —C≡C— stretching frequency in $HC{\equiv}CH_2NH_2$

(e) 1620 to 1680 cm^{-1}; C=C stretching frequency in $H_2C{=}CH—CH_2CH{=}CH_2$

24.6 (a) Two peaks in each. A 6:2 ratio in $CH_3—CH_2—CH_3$ and a 6:4 ratio in $CH_3—CH_2—CH_2—CH_3$.

(b) One peak in $CH_3—\overset{\overset{O}{\|}}{C}—CH_3$; three peaks in $CH_3—CH_2—\overset{\overset{O}{\|}}{C}—H$ with a ratio of 3:2:1.

(c) Three peaks in $CH_3—CH_2—OH$ with a ratio of 3:2:1; one peak in $CH_3—O—CH_3$.

(d) Three peaks in $CH_3—CH_2—OH$ with a ratio of 3:2:1; two peaks in $CH_3—OH$ having a ratio of 3:1.

(e) Three peaks in $CH_3—CH_2—CH_2Cl$ with a 3:2:2 ratio; two peaks in $CH_3—CHCl—CH_3$ with a 6:1 ratio.

24.7

(a)	O_2^+	electron transfer
(b)	·OH	fragmentation
(c)	NO	addition
(d)	O_2	fragmentation
(e)	H·	fragmentation
	CO	fragmentation
(f)	H·	fragmentation

24.8 NF_3

24.9 COS

24.10 $CH_3—N{\equiv}N—CH_3$

Chapter 25

ENVIRONMENTAL CHEMISTRY

SECTION 25.1 GEOCHEMISTRY

Objectives

- To state the major components of air.
- To classify the materials formed in natural waters.

Focus

Dry air is composed primarily of N_2, 78.09% by volume, and O_2, 20.94% by volume. The text, Table 25.1, lists the gaseous components of dry air. Notice that the percent by volume of gases other than N_2 and O_2 is very small. A major environmental concern is the low concentrations of substances that are not natural components of dry air.

Materials formed in natural waters are classified by source (the atmosphere, minerals, soils, rocks, live organisms), by their decomposition products, and by their chemical state in water (suspended, collodial, or dissolved). Review the text, Table 25.2, for the classification of various ions and molecules in natural waters.

SECTION 25.2 THE OXYGEN-CARBON CYCLE

Objective

- To discuss the oxygen-carbon cycle.

Focus

Study the text, Figure 25.1. The decay of organic material and the respiration of plants remove O_2 from the atmosphere and convert it to CO_2:

$$C_6H_{12}O_6(aq) + 6O_2\,(g) \rightarrow 6CO_2(g) + 6H_2O(\ell)$$

CO_2 is also a product of volcanic eruptions. But the photosynthesis process and some marine organisms return O_2 to the atmosphere:

$$6CO_2\,(g) + 6H_2O\,(\ell) \rightarrow C_6H_{12}O_6\,(aq) + 6O_2\,(g)$$

Photosynthesis consumes CO_2, the ocean absorbs it to produce H_2CO_3 and insoluble carbonates, and some marine organisms absorb CO_2. This carbon-oxygen cycle maintains a delicate natural balance of CO_2 and O_2 in the atmosphere.

SECTION 25.3 THE NITROGEN CYCLE

Objective

- To discuss the nitrogen cycle.

Focus

Become familiar with the text, Figure 25.2.

The N_2 cycle involves the nitrogen-containing compounds NH_3, N_2O, NO, and NO_2. Microbial action converts much of the N_2 in the atmosphere to NH_3. Modern technology, however, has converted large quantities of N_2 to NO and NO_2 by direct reaction of N_2 with O_2 at the high temperatures within combustion chambers.

SECTION 25.4 ENVIRONMENTAL DEGRADATION

Objective

- To define the three areas of disruption of natural environments and processes.

Focus

Pollution, depletion of resources, and disturbances of natural conditions tend to disrupt natural environmental processes. SO_2 pollution, for example, causes acid rain, which in turn affects the pH of delicate aquatic environments.

The mining of iron ore results in the dispersal of iron from a concentrated area to one less defined; the result is a large increase in entropy. This dispersal of natural resources affects the environment esthetically, physically, and chemically. Any attempt to decrease the entropy of a natural resource requires work or energy. Because the source of energy or work also comes from a natural resource, the entropy sum of all natural resources increases.

The introduction of synthetic organic compounds into the environment is of great concern. The controversies over the Freons (chlorofluoromethanes) and their effect on the ozone layer, the effects of numerous insecticides, industrial synthesis of pharmaceuticals, and so forth, are a concern to many scientists, environmentalists, and the public.

We shall look at each of these areas of disruption in more detail later in this chapter.

SECTION 25.5 ENVIRONMENTAL POLLUTION

Objectives

- To describe the pollution resulting from SO_2 emissions and the internal combustion engine.
- To explain the formation of photochemical smog.

Focus

Two major sources of SO_2 emissions into the atmosphere are fossil fuel combustion (about 78%) and Cu, Pb, and Zn roasting (about 13%).

Fossil fuels were formed from living organisms containing S atoms. In high temperature combustion chambers these S atoms oxidize to produce SO_2 molecules that frequently adsorb on the surface of particulate matter, also emitted during fossil fuel combustion. The smaller particulates, 1 to 4 μm, can become lodged deep within the lungs and cause permanent damage to lung tissue.

SO_2 emissions from smelters arise from the roasting (converting a sulfide to an oxide) of a metallic sulfide. For Cu_2S,

$$2Cu_2S + 3O_2 \rightarrow 2Cu_2O + 2SO_2$$

The subsequent fate of SO_2 in the atmosphere is its oxidation to SO_3. Generally, SO_3 is absorbed by H_2O to form sulfuric acid which can react with metals (corrosion) to form metal sulfates.

Currently, in industrial combustion processes, basic (pH > 7) scrubbers reduce some SO_2 emissions.

The internal combustion engine emits unburned hydrocarbons, CO, NO, and other pollutants into the atmosphere. The short residence time of the air/fuel mixture, the high ignition temperature, and the cool cylinder walls are the primary causes of these emissions.

An oxidizing Pt or Pd catalyst, located in the exhaust system of the automobile, converts the hydrocarbons and CO to harmless CO_2 and H_2O. However, the addition of tetraethyl lead (TEL) to gasoline to provide a smoother burn in the combustion chamber poisons the catalyst by covering the Pt- or Pd-coated catalytic converter. Therefore, only unleaded gasoline should be used if hydrocarbons and CO emissions are to be decreased. TEL provides for a smoother burn (or a more even wall of fire from the spark plugs) by releasing $C_2H_5\cdot$ and $\cdot\dot{Pb}$: free radicals which react with hydrocarbon free radicals generated during the combustion of the gasoline.

Photodissociation of NO_2 initiates **photochemical smog** production, a free radical reaction.

$$NO_2 \xrightarrow{h\nu} NO + O\cdot$$

The $O\cdot$ free radical and O_3 (formed from $O\cdot + O_2 \rightarrow O_3$) react readily with hydrocarbons (from automobile exhaust) to produce oxyhydrocarbon free radicals. These further react to produce a vast array of photochemical pollutants. Of these the peroxyacylnitrates (PAN's) have the most noticeable and immediate effects. These severe eye irritants are toxic to plants.

To control smog formation, either the nitrogen oxide emissions and/or the hydrocarbon emissions must be reduced.

SECTION 25.6 DEPLETION OF RESOURCES—THERMODYNAMIC CONSIDERATIONS

Objective

- To predict the thermodynamic stability of oxides at various temperatures.

Focus

From the text, Figure 25.5, the thermodynamic stability of various oxides is predictible. This figure is a plot of ΔG^0 versus T, producing a straight line because of the equation

$$\Delta G^0 = \Delta H^0 + (-\Delta S^0)T$$

$$y = b + mx$$

Example 25.1 Using Figure 25.5, estimate the values of ΔS^0 and ΔH^0 for the reaction $Ti + O_2 \rightarrow TiO_2$ at 1800 K

Answer $\Delta S^0 = -0.16$ kJ/K; $\Delta H^0 = -890$ kJ

Solution $-\Delta S^0$ is the slope of the line m equal to

$$\frac{y_2 - y_1}{x_2 - x_1} \quad \text{or} \quad \frac{\Delta G_2{}^0 - \Delta G_1{}^0}{T_2 - T_1}$$

	$\Delta G^0(y)$	$T(x)$
point 2	−600 kJ	1800 K
point 1	−830 kJ	400 K

or

$$-\Delta S^0 = \frac{-600\text{ kJ} - (-830\text{ kJ})}{1800\text{ K} - 400\text{ K}} = \frac{230\text{ kJ}}{1400\text{ K}} = 0.16\text{ kJ/K}$$

$$\Delta S^0 = -0.16\text{ kJ/K}$$

$$\Delta H^0 = \Delta G^0 + T\Delta S^0 = -600\text{ kJ} + (1800\cancel{\text{K}})\frac{(-0.16\text{ kJ})}{\cancel{\text{K}}}$$

$$= -600\text{ kJ} + (-288\text{ kJ})$$

$$= -888\text{ kJ} = -890\text{ kJ}$$

Example 25.2 ΔG_f^0 at 25°C for Cu_2S is −86.2 kJ/mol, and for Cu_2O is −146 kJ/mol. Two pathways for the production of copper are:

(a) $Cu_2S(c) + O_2(g) \rightarrow 2Cu(c) + SO_2(g)$

(b) $2Cu_2S(c) + 3O_2(g) \rightarrow 2Cu_2O(g) + 2SO_2(g)$
$2Cu_2O(c) + C(c) \rightarrow 4Cu(c) + CO_2(g)$

Calculate ΔG^0 at 25°C for each process. Use the text, Appendix D, for additional ΔG^0 values.

Answer (a) −107 kJ/mol Cu; (b) −205 kJ/mol Cu

Solution For reaction (a),

$$\Delta G^0 = \Delta G^0_{f,\ SO_2(g)} - \Delta G^0_{f,\ Cu_2S(c)}$$

$$= [1\ \text{mol} \times (-300\ \text{kJ/mol})] - [1\ \text{mol} \times (-86.2\ \text{kJ/mol})]$$

$$= -214\ \text{kJ} \quad \text{or} \quad \left(\frac{-214\ \text{kJ}}{2\ \text{mol Cu}} = \right) -107\ \text{kJ/mol Cu produced.}$$

For reaction (b), the sum of the two equations gives

$$2Cu_2S(c) + 3O_2(g) + C(c) \rightarrow 4Cu(c) + CO_2(g) + 2SO_2(g)$$

$$\Delta G^0 = [\Delta G^0_{f,\ CO_2(g)} + 2\Delta G^0_{f,\ SO_2(g)}] - [2\Delta G^0_{f,\ Cu_2S(c)}]$$

$$= \left[1\ \text{mol} \times \left(\frac{-394.4\ \text{kJ}}{\text{mol}}\right) + 2\ \text{mol} \times \left(\frac{-300\ \text{kJ}}{\text{mol}}\right)\right] - \left[2\ \text{mol} \times \left(\frac{-86.2\ \text{kJ}}{\text{mol}}\right)\right]$$

$$= -821\ \text{kJ} \quad \text{or} \quad \left(\frac{821\ \text{kJ}}{4\ \text{mol Cu}} = \right) -205\ \text{kJ/mol Cu produced}$$

Pathway 2 has a more negative ΔG^0 value and is thus thermodynamically more advantageous.

Example 25.3 Using Figure 25.5, calculate ΔG^0 for the reduction of aluminum at 1600 K.

$$Al_2O_3 + 3C \rightarrow 2Al + 3CO$$

Answer $\Delta G^0 = +450$ kJ

Solution

$$\Delta G^0 = 3\Delta G^0_{f,\ CO} - \Delta G^0_{f,\ Al_2O_3}$$

$$\Delta G^0_{f,CO} \text{ at } 1600\ \text{K} \simeq \frac{-500\ \text{kJ}}{2\ \text{mol CO}} = -250\ \text{kJ/mol CO}$$

$$\Delta G^0_{f,Al_2O_3} \text{ at } 1600\ \text{K} \simeq \frac{-800\ \text{kJ}}{2/3\ \text{mol } Al_2O_3} = -1200\ \text{kJ/mol } Al_2O_3$$

$$\Delta G^0 = \left[3\ \text{mol CO} \times \left(\frac{-250\ \text{kJ}}{\text{mol CO}}\right)\right] - \left[1\ \text{mol } Al_2O_3 \times \left(\frac{-1200\ \text{kJ}}{\text{mol } Al_2O_3}\right)\right]$$

$$= +450\ \text{kJ}$$

The positive ΔG^0 value indicates that the reduction of Al_2O_3 with C is not thermodynamically feasible at 1600 K.

SECTION 25.7 IRON

Objective

- To explain how iron is produced in a blast furnace.

Focus

CO, produced from the reaction of O_2 with C (coke), reduces concentrated iron ore to metallic iron in the blast furnace. The ore is generally in some form of the oxide, either Fe_2O_3 or Fe_3O_4 ($FeO \cdot Fe_2O_3$). Limestone, $CaCO_3$, converted to CaO in the heated blast furnace, removes acidic impurities (nonmetallic oxides) by the reactions

$$CaCO_3 \xrightarrow{\Delta} CaO + CO_2$$

$$CaO + SO_2 \rightarrow CaSO_3$$

$$CaO + SiO_2 \rightarrow CaSiO_3$$

$$3CaO + P_2O_5 \rightarrow Ca_3(PO_4)_2$$

These by-products are collectively called "slag." The product of the blast furnace, **pig iron**, contains 4 to 5% C and other metals commonly found with iron ore.

Different grades of steel have varying amounts of carbon mixed with the iron, but usually a 2% or less C composition is in the mixture: the more C, the more brittle the steel.

Pollution from the blast furnace comes from CO and particulate emissions at the top of the furnace. Some CO is recycled back through it as a valuable reducing agent for the iron ore. A large quantity of the particulates are removed by pollution-control devices. Water, for cooling the hot iron, also becomes contaminated with iron and other heavy metals. Stockpiles of slag are also an environmental nuisance.

SECTION 25.8 ALUMINUM

Objective

- To explain the purification of bauxite and the electrolytic reduction of aluminum.

Focus

Bauxite, an ore rich in Al_2O_3, is treated with strong hot alkali to dissolve the amphoteric Al_2O_3,

$$Al_2O_3 + 2OH^-(aq) + 7H_2O \rightarrow 2[Al(OH)_4(H_2O)_2]^-(aq)$$

Other metal impurities, such as Fe, precipitate as insoluble hydroxides,

$$Fe^{2+}(aq) + 2OH^-(aq) \rightarrow Fe(OH)_2(c)$$

The aluminate solution, $[Al(OH)_4(H_2O)_2]^-$, is then acidified with CO_2 to precipitate $Al(OH)_3$,

$$[Al(OH)_4(H_2O)_2]^- + H_2O + CO_2 \rightarrow Al(OH)_3(H_2O)_3 + HCO_3^-$$

which is heated to dryness, producing alumina, Al_2O_3.

$$2Al(OH)_3(H_2O)_3 \xrightarrow{\Delta} Al_2O_3 + 9H_2O$$

The alumina is melted in cryolite, Na_3AlF_6, and electrolyzed; Al^{3+} is reduced to aluminum metal at the cathode.

Example 25.4 Calculate the mass in kg of Al produced in an electrolytic cell operating at 100,000 amperes for 24 hours.

Answer 806 kg Al/day

Solution Remember a few things about Faraday's laws of electrolysis:

$$1 \text{ A (amp)} = 1 \text{ C/sec}$$

$$1\,\mathcal{F} = 96{,}500 \text{ C} = 1 \text{ mol } e^-$$

$$Al^{3+} + 3e^- \rightarrow Al$$

Therefore, $3\mathcal{F}$ produce 1 mol Al.

Use the dimensional analysis for problem solving.

Given quantity: 100,000 A = 100,000 C/sec

Desired quantity: $\dfrac{\text{mass Al (kg)}}{\text{day}}$

$$\frac{\text{mass Al (kg)}}{\text{day}} = \frac{100{,}000 \text{ C}}{\cancel{\text{sec}}} \times \frac{3600 \cancel{\text{sec}}}{\cancel{\text{hr}}} \times \frac{24 \cancel{\text{hr}}}{\text{day}} \times \frac{\cancel{\mathcal{F}}}{96{,}500 \text{ C}} \times \frac{\cancel{\text{mol Al}}}{3\cancel{\mathcal{F}}} \times \frac{27.0 \cancel{\text{g Al}}}{\cancel{\text{mol Al}}} \times \frac{\text{kg Al}}{10^3 \cancel{\text{g Al}}}$$

$$= 806 \text{ kg Al/day}$$

SECTION 25.9 CORROSION

Objectives

- To explain the corrosion of iron.
- To cite methods that reduce corrosion.

Focus

The natural corrosion of iron or steel is due to the atmospheric oxidation of Fe. The oxidation-reduction reaction is similar to that occurring in a electrochemical cell in which O_2 is at two different concentrations.

Consider the H_2O droplet on the surface of the iron (Figure 25.1) and the corresponding anode and cathode reactions:

O_2 oxidizes Fe^{2+} at the surface of the droplet

$$Fe^{2+} \rightarrow Fe^{3+} + e^-$$

precipitating Fe^{3+} as $Fe_2O_3(H_2O)_x$ or *rust*.

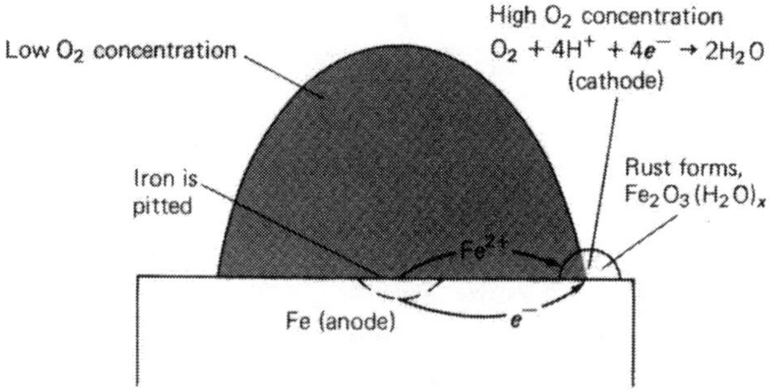

Figure 25.1

This reaction also happens when wet dirt or sand is left on a shovel for a period before being washed off. A "rust ring" defines the position of the wet dirt.

The rate of corrosion, that is, the oxidation of iron, is decreased by **cathodic protection** in which a metal with a lower reduction potential (greater oxidizing potential or more easily oxidized), such as Zn or Mg, is placed in contact with iron. Cathodic protection means that the Fe is more cathodic (a higher reduction potential) than the other metal. For example, the Zn coating on galvanized iron is oxidized more readily (less cathodic) than Fe.

Example 25.5 Explain why a "tin can" (iron coated with tin) rusts.

Answer Tin metal is relatively passive to oxidation. When the Sn metal coating over the Fe can is broken, Fe, having a lower reduction potential than Sn, serves as an anode and is oxidized quickly. The Sn metal plating, however, is preserved.

Some metals oxidize quickly, but produce an impervious oxide coating which inhibits the oxidation of the lower layers of metal atoms (how we wish iron would do this!) Stainless steel (18% Cr, 8% Ni, and 74% Fe) resists corrosion because Cr and Ni form this impervious surface coating.

SECTION 25.10 DISTURBANCE OF NATURAL CONDITIONS: O_2, O_3, AND CO_2

Objective

- To cite the causes and effects for potential changes in the atmospheric concentrations of O_2, O_3, and CO_2.

Focus

There appears to be little danger of any critical changes in the level of the atmospheric concentration of O_2 because of the natural photosynthesis, respiration, and decay cycles (see Section 25.2). The combustion of large quantities of fossil fuels has had little effect on the O_2 concentration in the atmosphere.

However, the level of O_3 concentration in the stratosphere could be in danger if certain photochemical processes become significant. Any one of three mechanisms can reduce O_3 levels (see text):

1. Reaction with hydroxyl radicals (from H_2O).
2. Reaction with NO (from combustion).
3. Reaction with chlorine atoms (from photodissociation of C—Cl bonds in Freons or chlorofluoromethanes).

Each reaction is a "free radical" reaction consuming O atoms and O_3 molecules. Remember that free radicals are very reactive intermediates that cause chain and/or catalyzed reactions. Review the mechanism of each process whereby the end product is a result of O_3 and $O\cdot$ consumption.

The concentration of CO_2 in the Earth's atmosphere has slowly increased over the past 50 years. CO_2 molecules absorb radiation reflected from the Earth's surface; a portion of this is emitted by the CO_2 molecules back toward the Earth's surface. This helps maintain a warmer, more stable temperature at its surface: the higher the CO_2 concentration, the greater this warming effect.

The increased amount of particulate matter partially offsets this apparent warming effect. The particulate matter reflects a portion of the sun's incident radiation back into space, and therefore never reaches the Earth's surface.

SECTION 25.11 AIR POLLUTION CONTROL

Objective

- To define methods used for the removal of particulates and exhaust gases from stationary sources.

Focus

Particulates are removed from stack gases by:

1. Gravity-settling chambers—good for large particulates.
2. Bag houses—like a vacuum cleaner except that a gas pressure is used rather than a vacuum.
3. Cyclone separators—the dense, heavier particles are forced along the circumference of the separator and settle out.
4. Electrostatic precipitators—a large electric potential discharge ionizes the particulates. Review this method in more detail.

Stack gases are cleaned by wet scrubbers in which the gaseous pollutants

1. Merely dissolve in water, such as NH_3.
2. React with an acid or a base, such as the scrubbing of $SO_2(g)$ with a lime slurry, $Ca(OH)_2$:

$$Ca(OH)_2(aq) + SO_2(g) \rightarrow CaSO_3(c) + H_2O$$

3. React with an oxidizing or reducing agent; these scrubbers tend to be slow and costly.

The gaseous pollutant is also

4. Adsorbed by a dry scrubber, such as activated charcoal (carbon).
5. Oxidized by air or O_2-enriched air.

Some industries concentrate their organic waste gases and burn them to increase fuel efficiency.

SECTION 25.12 WATER PURIFICATION

Objective

- To describe the processes used in purifying domestic sewage.

Focus

Three stages in the treatment of domestic sewage are:

1. The primary stage removes the boards, sticks, rubber ducks, and floating objects with a bar screen. Then it passes through grit and sedimentation chambers where the heavier particles settle out.
2. The secondary stage removes the biodegradable organic material. Here bacteria *eat* the organic material and some N- and P- nutrients, producing CO_2 and H_2O. This biodegradation is done by the **trickling filter** process (see text, Figure 25.14) in which bacteria adhere to the rocks of the filter, or by the **activated sludge** process in which air is blown through the waste in the presence of the bacteria-laden sludge. In most sewage treatment plants, the water is now chlorinated to kill any pathogenic (disease-causing) bacteria before it is discharged into the environment.
3. A few sewage treatment plants in the United States have tertiary treatment capabilities. This stage removes specific substances, such as synthetic organics, phosphates, nitrates, soluble inorganics, and so on. Some of the more common types of tertiary treatments are described in the text.
   (a) *Coagulation and Sedimentation.* A flocculating agent, such as $Al_2(SO_4)_3$ or $Fe_2(SO_4)_3$, neutralizes the charge on suspended inorganic collodial particles and causes them to precipitate.
   (b) *Adsorption.* Many dissolved nonbiodegradable organics can be adsorbed to activated charcoal.
   (c) *Chemical Oxidizing Agents.* Strong oxidizing agents, such as O_3, OCl^-, or MnO_4^-, can oxidize nonbiodegradable organics.
   (d) *Reverse Osmosis.* The use of pressure reverses natural osmosis, forcing the dirty water to the pure water side of a semipermeable membrane.
   (e) *Deionization.* Offensive cations and/or anions are removed by passing the water through a cation and/or anion exchange column. This replaces "bad" ions for "good" ones.

QUESTIONS

Section 25.2

25.1 The solubility of CO_2 in water at 25°C and 1 atm is 3.38×10^{-2} mol/L. Assuming an excess of Ca^{2+} and no change in pH, calculate the maximum mass in grams of $CaCO_3$ that precipitates in 1 liter water. Refer to the text, Equation (2).

Section 25.3

25.2 Alfalfa root nodules contain N-fixing bacteria. What does this mean?

25.3 What N-compound is a by-product of combustion? Write an appropriate equation.

25.4 What role do the bacteria have in plant roots if the plant assimilates NO_3^-, but only NH_3 fertilizer is used?

Section 25.6

25.5 Is TiO_2 more or less stable at higher temperatures? Explain. Use text, Figure 25.5 in the explanation.

25.6 Explain how CaO reduces the viscosity of a silica melt.

25.7 What is pig iron?

Section 25.9

25.8 Name the metals that commonly form impervious oxide coatings.

Section 25.10

25.9 What effects result from the reduction of the O_3 concentration in the upper atmosphere?

25.10 NO and Cl· serve as catalysts to the photochemical decomposition of O_3. Show this by means of a chemical equation(s).

Section 25.12

25.11 What type of pollutant is removed in the secondary stage of sewage treatment?

25.12 In which stage of sewage treatment are the nitrates from runoff fertilizers and the phosphates from detergents removed?

25.13 Explain how $Al_2(SO_4)_3$ is used in tertiary water treatment.

25.14 Carbohydrate waste, represented by the formula $C_6H_{10}O_5$, is consumed by bacteria to produce CO_2 and H_2O. How much O_2 is required to degrade one metric ton (1000 kg) of waste?

ANSWERS

25.1 3.38 g $CaCO_3$

25.2 The bacteria of the alfalfa catalyze the reaction of N_2 from the air into ammonia.

25.3 NO; $N_2 + O_2 \rightarrow 2NO$

25.4 Bacteria oxidize NH_3 to NO_3^-

25.5 Less stable. A positive slope for the $Ti + O_2 \rightarrow TiO_2$ reaction in Figure 25.5 shows that at higher temperatures, ΔG becomes more positive, thus indicating greater *in*stability.

25.6 SiO_2 is a major impurity in the iron ore. CaO breaks down the three-dimensional SiO_2 network into smaller $CaSiO_3$ units. Since the SiO_2 polymeric chain is broken down, the melt is less viscous.

25.7 Pig iron is the form of the reduced iron obtained from the blast furnace.

25.8 Al, Cr, Ni, Co

25.9 More high energy ultraviolet radiation reaches the Earth's surface. This increases skin cancers and warms the Earth's surface.

25.10 $NO + O_3 \rightarrow NO_2 + O_2$

$NO_2 + O\cdot \rightarrow NO + O_2$

$Cl\cdot + O_3 \rightarrow ClO\cdot + O_2$

$ClO\cdot + O\cdot \rightarrow Cl\cdot + O_2$

25.11 Biodegradable organic material

25.12 Tertiary stage

25.13 $Al_2(SO_4)_3$ is used as a flocculating agent, which sweeps together collodial particles to clarify the water.

25.14 1184 kg O_2

FORMULA WEIGHTS OF COMMON INORGANIC COMPOUNDS

$AgCl$	143.32	$HC_2H_3O_2$	60.052
$AgNO_3$	169.87	HNO_3	63.013
$AlCl_3$	133.34	H_3PO_4	97.995
Al_2O_3	101.96	H_2SO_4	98.07
$Al_2(SO_4)_3$	342.16	$Hg(NO_3)_2$	324.60
$BaCl_2$	208.24	$KClO_3$	122.55
$BaSO_4$	233.39	K_2CrO_4	194.19
$Bi(NO_3)_3$	395.10	$K_2Cr_2O_7$	294.18
Bi_2S_3	514.20	$KMnO_4$	158.03
$CaCO_3$	100.09	NH_3	17.03
$Ca(NO_3)_2$	164.09	NH_4Cl	53.49
Ca_3N_2	148.26	$(NH_4)_2SO_4$	132.15
$Ca_3(PO_4)_2$	310.19	$NaC_2H_3O_2$	82.04
Cr_2O_3	151.99	Na_2CO_3	105.99
$CuSO_4 \cdot 5H_2O$	249.69	$NaOH$	40.00
$FeCl_2$	126.75	Na_3PO_4	163.95
$FeCl_3$	162.22	Na_2SO_4	142.04
Fe_2O_3	159.70	$NiCl_2$	129.62
$Fe_2(SO_4)_3$	399.90	$PbSO_4$	303.3
H_2O	18.02		

NOTES

NOTES

NOTES

NOTES

NOTES

NOTES

NOTES

NOTES

Printed in Dunstable, United Kingdom

91343232R10243